Electromagnetism

The Manchester Physics Series

General Editors
F. MANDL : R. J. ELLISON : D. J. SANDIFORD
Physics Department, Faculty of Science,
University of Manchester

Published

Properties of Matter: B. H. Flowers and E. Mendoza

Optics: F. G. Smith and J. H. Thomson

Statistical Physics: F. Mandl

Solid State Physics: H. E. Hall

Electromagnetism: I. S. Grant and W. R. Phillips

In preparation

Atomic Physics: J. C. Willmott

Electronics: J. M. Calvert and M. A. H. McCausland

ELECTROMAGNETISM

I. S. Grant
W. R. Phillips

Department of Physics,
University of Manchester

John Wiley & Sons Ltd.

CHICHESTER   NEW YORK   BRISBANE   TORONTO

Copyright © 1975, by John Wiley & Sons, Ltd.

Reprinted April 1976.
Reprinted June 1978.
Reprinted March 1979.
Reprinted with corrections October 1980

Library of Congress Cataloging in Publication Data:

Grant, I. S.
Electromagnetism.

(The Manchester physics series)
Bibliography: p.
1. Electromagnetism. I. Phillips, William Robert, joint author. II. Title.

QC760.G76   537   73-17668
ISBN 0 471 32245 8 (Cloth bound)
ISBN 0 471 32246 6 (Paper bound)

Printed in Great Britain by J. W. Arrowsmith Ltd.,
Winterstoke Road, Bristol.

Editors' Preface to the Manchester Physics Series

In devising physics syllabuses for undergraduate courses, the staff of Manchester University Physics Department have experienced great difficulty in finding suitable textbooks to recommend to students; many teachers at other universities apparently share this experience. Most books contain much more material than a student has time to assimilate and are so arranged that it is only rarely possible to select sections or chapters to define a self-contained, balanced syllabus. From this situation grew the idea of the Manchester Physics Series.

The books of the Manchester Physics Series correspond to our lecture courses with about fifty per cent additional material. To achieve this we have been very selective in the choice of topics to be included. The emphasis is on the basic physics together with some instructive, stimulating and useful applications. Since the treatment of particular topics varies greatly between different universities, we have tried to organize the material so that it is possible to select courses of different length and difficulty and to emphasize different applications. For this purpose we have encouraged authors to use flow diagrams showing the logical connection of different chapters and to put some topics into starred sections or subsections. These cover more advanced and alternative material, and are not required for the understanding of later parts of each volume.

Since the books of the Manchester Physics Series were planned as an integrated course, the series gives a balanced account of those parts of physics which it treats. The level of sophistication varies: '*Properties of Matter*' is for the first year, '*Solid State Physics*' for the third. The other volumes are inter-mediate, allowing considerable flexibility in use. '*Electromagnetism*', '*Optics*', '*Electronics*' and '*Atomic Physics*' start from first year level and progress to material suitable for second or even third year courses. '*Statistical Physics*' is suitable for second or third year. The books have been written in such a way that each volume is self-contained and can be used independently of the others.

Although the series has been written for undergraduates at an English university, it is equally suitable for American university courses beyond the Freshman year. Each author's preface gives detailed information about the pre-requisite material for his volume.

In producing a series such as this, a policy decision must be made about units. After the widest possible consultations we decided, jointly with the authors and the publishers, to adopt SI units interpreted liberally, largely following the recommendations of the International Union of Pure and Applied Physics. Electric and magnetic quantities are expressed in SI units. (Other systems are explained in the volume on electricity and magnetism.) We did not outlaw physical units such as the electron-volt. Nor were we pedantic about factors of 10 (is 0.012 kg preferable to 12 g?), about abbreviations (while s or sec may not be equally acceptable to a computer, they should be to a scientist), and about similarly trivial matters.

Preliminary editions of these books have been tried out at Manchester University and circulated widely to teachers at other universities, so that much feedback has been provided. We are extremely grateful to the many students and colleagues, at Manchester and elsewhere, who through criticisms, suggestions and stimulating discussions helped to improve the presentation and approach of the final versions of these books. Our particular thanks go to the authors, for all the work they have done, for the many new ideas they have contributed, and for discussing patiently, and frequently accepting, our many suggestions and requests. We would also like to thank the publishers, John Wiley and Sons, who have been most helpful in every way, including the finan-cing of the preliminary editions.

Physics Department   F. MANDL
Faculty of Science   R. J. ELLISON
Manchester University   D. J. SANDIFORD

Authors' Preface

This book is based on lectures on classical electromagnetism given at Manchester University. The level of difficulty is suitable for honours physics students at a British University or physics majors at an American University. A-level or high school physics and calculus are assumed, and the reader is expected to have some elementary knowledge of vectors. Electromagnetism is often one of the first branches of physics in which students find that they really need to make use of vector calculus. Until one is used to them, vectors are difficult, and we have accordingly treated them rather cautiously to begin with. Brief descriptions of the properties of the differential vector operators are given at their first appearance. These descriptions are not intended to be a substitute for a proper mathematical text, but to remind the reader what div, grad and curl are all about, and to set them in the context of electromagnetism. The distinction between macroscopic and microscopic electric and magnetic fields is fully discussed at an early stage in the book. It is our experience that students do get confused about the field $\mathbf{E}$ and $\mathbf{D}$, or $\mathbf{B}$ and $\mathbf{H}$. We think that the best way to help them overcome their difficulties is to give a proper explanation of the origin of these fields in terms of microscopic charge distributions or circulating currents.

The logical arrangement of the chapters is summarized in a flow diagram on the inside of the front cover. Provided that one is prepared to accept Kirchhoff's rules and the expressions for the e.m.f.'s across components before discussing the laws on which they are based, the A.C. theory in Chapters 7 and 8 does not require any prior knowledge of the earlier chapters. Chapters 7 and 8 can therefore be used at the beginning of a course on electromagnetism. Sections of the book which are starred may be omitted at a first reading, since they do not contain material needed in order to understand later chapters.

We should like to thank the many colleagues and students who have helped with suggestions and criticisms during the preparation of this book; any errors which remain are our own responsibility. It is also a pleasure to thank Mrs. Margaret King and Miss Elizabeth Rich for their rapid and accurate typing of the manuscript.

May, 1974.   I. S. GRANT
Manchester, England.   W. R. PHILLIPS

Contents

★ Starred sections may be omitted as they are not required later in the book.

CHAPTER 1

Force and energy in electrostatics

The only laws of force which are known with great precision are the two laws describing the gravitational forces between different masses and the electrical forces between different charges. When two masses or two charges are stationary, then in either case the force between them is inversely proportional to the square of their separation. These inverse square laws were discovered long ago: Newton's law of gravitation was proposed in 1665, and Coulomb's law of electrostatics in 1785. This chapter is concerned with the application of Coulomb's law to systems containing any number of stationary charges. Before studying this topic in detail, it is worth pausing for a moment to consider the consequences of the law in the whole of physics.

In order to make full use of our knowledge of a law of force, we must have a theory of mechanics, that is to say, a theory which describes the behaviour of an object under the action of a known force. Large objects which are moving at speeds small compared to the speed of light obey very closely the laws of classical Newtonian mechanics. For example, these laws and the gravitational force law together lead to accurate predictions of planetary motion. But classical mechanics does not apply at all to observations made on particles of atomic scale or on very fast-moving objects. Their behaviour can only be understood in terms of the ideas of quantum theory and of the special theory of relativity. These two theories have changed the framework of discussion in physics, and have made possible the spectacular advances of the twentieth century.

It is remarkable that while mechanics has undergone drastic amendment, Coulomb's law has stood unchanged. Although the behaviour of atoms does not fit into the framework of the old mechanics, when the Coulomb force is used with the theories of relativity and quantum mechanics, atomic interactions are explained with great precision in every instance when an accurate comparison has been made between experiment and theory. In principle, atomic physics and solid state physics, and for that matter the whole of chemistry, can be derived from Coulomb's law. It is not feasible to derive everything in this way, but it should be borne in mind that atoms make up the world around us, and that its rich variety and complexity are governed by electrical forces.

1.1 ELECTRIC CHARGE

Most of this book applies electromagnetism to large-scale objects, where the atomic origin of the electrical forces is not immediately apparent. However, to emphasize this origin, we shall begin by consideration of atomic systems. The simplest atom of all is the hydrogen atom, which consists of a single proton with a single electron moving around it. The hydrogen atom is stable because the proton and the electron attract one another. In contrast, two electrons repel one another, and tend to fly apart, and similarly the force between two protons is repulsive*. These phenomena are described by saying that there are two different kinds of *electric charge*, and that like charges repel one another, whereas unlike charges are attracted together. The charge carried by the proton is called *positive*, and the charge carried by the electron *negative*.

The magnitude and direction of the force between two stationary particles, each carrying electric charge, is given by Coulomb's law. The law summarizes four facts:

  (i) Like charges repel, unlike charges attract.
 (ii) The force acts along the line joining the two particles.
(iii) The force is proportional to the magnitude of each charge.
(iv) The force is inversely proportional to the square of the distance between the particles.

The mathematical statement of Coulomb's law is:

$$\mathbf{F}_{21} \propto \frac{q_2 q_1}{r_{21}^3} \mathbf{r}_{21}. \tag{1.1}$$

The vector $\mathbf{F}_{21}$ in Figure 1.1 represents the force on particle 1 (carrying a charge q_1) exerted by the particle 2 (carrying charge q_2). The line from q_2 to q_1 is represented by the vector $\mathbf{r}_{21}$, of length r_{21}: since the unit vector along the direction $\mathbf{r}_{21}$ can be written $\mathbf{r}_{21}/r_{21}$, Equation '(1.1) is an *inverse square law* of force, although r_{21}^3 appears in the denominator. Notice that the equation automatically

---

* If the protons are separated by a distance less than 10^{-14} m, they are affected by the very short range nuclear forces. Unlike gravitational and electrical forces, nuclear forces are not known precisely. However, in the study of atoms one does not need to know anything about nuclear forces beyond the fact that they are strong enough to bind together the constituent parts of the atomic nucleus.

accounts for the attractive or repulsive character of the force if q_1 and q_2 include the sign of the charge. When the charges q_1 and q_2 are both positive or both negative, the force on q_1 is along $\mathbf{r}_{21}$, i.e. it is repulsive. On the other hand, when one charge is positive and the other negative, the force is in the direction opposite to $\mathbf{r}_{21}$, i.e. it is attractive.

To complete the statement of the force law, we must decide what units to use, and hence determine the constant of proportionality in Equation (1.1). We shall use SI (Système Internationale) units, which are favoured by most physicists

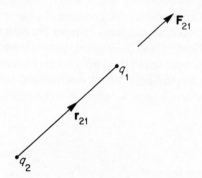

Figure 1.1. The force between two
charges.

and engineers applying electromagnetism to problems involving large-scale objects. A different system of units, called the Gaussian system, is almost universally used in atomic physics and solid state physics, and it is an unfortunate necessity for students to become reasonably familiar with both systems. (The two systems of units are discussed in Appendix A.) In SI units, Coulomb's law is written as

$$\mathbf{F}_{21} = \frac{1}{4\pi\varepsilon_0} \frac{q_2 q_1}{r_{21}^3} \mathbf{r}_{21} \qquad (1.2)$$

where

q_1 and q_2 are measured in coulombs,

$\mathbf{r}_{21}$ is measured in metres

and

$\mathbf{F}_{21}$ is measured in newtons.

The magnitude of the unit of charge, which is called the *coulomb*, is actually defined in terms of magnetic forces, and we shall leave discussion of the definition until Chapter 4. The factor 4π in the constant of proportionality in Coulomb's law is introduced in order to simplify the form of some important equations. The constant ε_0, which is called the *permittivity of free space*, has the value

$$\varepsilon_0 = 8.85 \times 10^{-12} \text{ coulomb}^2 \text{ newton}^{-1} \text{ m}^{-2}.$$

This value is determined experimentally, though the best result is found indirectly, and not by measurement of the force between stationary charges.

Electrostatic forces are two-body forces, which means that the force between any pair of charges is unaltered by the presence of other charges in their neighbourhood*. In a system containing many charges, the electrostatic force between each pair is given by Coulomb's law. To find the total force on any one particle, one simply makes a vector sum of the forces it experiences due to all the others separately. This rule for the addition of electrostatic forces is known as the Principle of Superposition. Figure 1.2 illustrates the application of the principle

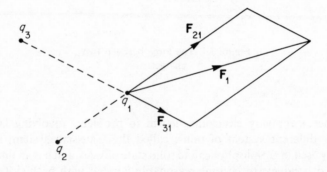

Figure 1.2. Superposition of electrostatic forces.

of superposition to a system of three charges. The forces on q_1 due to the presence of the charges q_2 and q_3 are $\mathbf{F}_{21}$ and $\mathbf{F}_{31}$ and the total force on q_1 is

$$\mathbf{F}_1 = \mathbf{F}_{21} + \mathbf{F}_{31}$$

$$= \frac{q_2 q_1}{4\pi\varepsilon_0 r_{21}^3}\mathbf{r}_{21} + \frac{q_3 q_1}{4\pi\varepsilon_0 r_{31}^3}\mathbf{r}_{31}.$$

* The forces between complete atoms are many-body forces, since the force between two atoms does depend on where other atoms are situated. The many-body nature of the force arises because the distribution of the constituent charged particles within each atom is changed by the presence of other atoms. But if the distribution of all the charge within each atom were specified, then the total force could be found by application of the principle of superposition.

In general, the force $\mathbf{F}_j$ on a charge q_j due to a number of other charges q_i is

$$\mathbf{F}_j = \sum_{i \neq j} \mathbf{F}_{ij} = \frac{1}{4\pi\varepsilon_0} \sum_{i \neq j} \frac{q_i q_j}{r_{ij}^3} \mathbf{r}_{ij}.$$

This equation can be written in another way in terms of the position vectors of the charges with respect to a fixed origin O. If the position vectors of the charges $q_1, q_2 \ldots q_i \ldots$ are $\mathbf{r}_1, \mathbf{r}_2 \ldots \mathbf{r}_i$, then the vector joining charges i and j is $\mathbf{r}_{ij} = \mathbf{r}_j - \mathbf{r}_i$. The total force on q_j is thus

$$\mathbf{F}_j = \frac{1}{4\pi\varepsilon_0} \sum_{i \neq j} \frac{q_i q_j}{|\mathbf{r}_j - \mathbf{r}_i|^3} (\mathbf{r}_j - \mathbf{r}_i). \tag{1.3}$$

A trivial example of the application of the principle of superposition is in working out the electrostatic forces exerted by atomic nuclei containing many protons on the electrons surrounding them. Nuclei are much smaller than atoms, and for this purpose can be regarded as point charges. The principle of superposition then tells us that the attractive force between an electron and a nucleus containing Z protons is Z times as great as that between an electron and a single proton.

It turns out that apart from the sign, the charge carried by electrons and protons is the same, and has the magnitude $e = 1.602 \times 10^{-19}$ coulombs: the charge on the proton is $+e$, that on the electron is $-e$. The strength of atomic interactions is governed by the size of the electronic charge e. Although e is a very small number when expressed in coulombs, this does not imply that electrostatic forces are feeble. On the contrary, they are immensely strong. For example, electrostatic forces are responsible for the great strength of solids under compression. When neighbouring atoms are close together, their electron clouds begin to overlap, and the mutual repulsion of these clouds opposes any compressing force.

Another example of the strength of the electrostatic force acting on the atomic scale is given by the experiment which led Rutherford to propose the nuclear model of the atom. He found that when swiftly-moving α-particles are allowed to collide with gold atoms, they are sometimes deflected through $180°$, implying that a strong force is at work. The force is just the electrostatic repulsion experienced by an α-particle when it chances to approach close to the nucleus of a gold atom. Let us calculate the magnitude of the force. An α-particle is a helium nucleus, containing two protons and carrying a charge $+2e$, and the gold nucleus carries a charge $+79e$. In Rutherford's experiment, the α-particles were energetic enough to approach within 2×10^{-14} m of the nucleus (still well outside the range of the nuclear forces). Substituting in Equation (1.3), the repulsive force at this distance is

$$F = \frac{2 \times 79 \times e^2}{4\pi\varepsilon_0 (2 \times 10^{-14})^2} = 110 \text{ newtons}.$$

This force, acting within a single atom, is more than the weight of a mass of 10 kilogrammes!

Normally we do not notice that electrostatic forces are so powerful, because matter is usually electrically neutral, carrying equal amounts of positive and negative charge. Not only are large lumps of matter electrically neutral, but the positive charge on the nucleus of a single isolated atom is precisely cancelled out by the negative charge of the surrounding electrons. So far as we know the cancellation is exact, and it has been shown experimentally that the magnitude of the net residual charge on a neutral atom is less than $10^{-20}e$. This is very remarkable, since apart from their electrical behaviour, protons and electrons are totally dissimilar particles. Many elementary particles besides electrons and protons have been discovered by nuclear physicists, and all share the property of carrying charges $\pm e$ or zero. It follows that the total charge carried by any piece of matter must be an integral multiple of the electronic charge e. A situation like this one, in which a physical quantity is not allowed to have a continuous range of values, but is restricted to a set of definite discrete values, is referred to as a quantum phenomenon. No one knows why electric charge should obey this quantum rule; it is an experimental fact. Nevertheless, because the rule is universal in its application, we can be sure that the electronic charge is a physical quantity of fundamental importance.

1.2   THE ELECTRIC FIELD

According to the superposition principle, the total force on a charged particle is the vector sum of the forces exerted on it by all other charges. Usually there is an enormous number of charged particles present in real matter. When considering the forces acting on any one of them, it is helpful to distract attention from the multitude of sources contributing to the net force by introducing the concept of the *electric field*. If a charge q experiences a force $\mathbf{F}$, then the ratio $\mathbf{F}/q$ is called the electric field at the point where q is located. The dimensions of electric field are [force] [charge]$^{-1}$, and in SI units the electric field is measured in newton coulomb^{-1}. (An equivalent unit, which will be explained when we come to deal with electrostatic energy, is the volt m^{-1}; 1 volt m$^{-1} \equiv$ 1 newton coulomb^{-1}.)

The electric field acting on q can be expressed in terms of the magnitudes of the other charges in the neighbourhood of q and their relative positions with respect to q. Let us assume that q is a test charge which can be put anywhere, and that its magnitude is very small, so that it exerts negligible forces on the other charges and can be moved about without altering their positions. Now we can evaluate the electric field at a point P, caused by an assembly of charges q_i, by placing the test charge at P. In Figure 1.3 P has a position vector $\mathbf{r}$, and the charges q_i have position vectors $\mathbf{r}_i$ with respect to an origin at O. The force on the test charge is given by Equation (1.3) as

$$\mathbf{F} = \frac{1}{4\pi\varepsilon_0} \sum_i \frac{qq_i}{|\mathbf{r} - \mathbf{r}_i|^3}(\mathbf{r} - \mathbf{r}_i),$$

where $(\mathbf{r} - \mathbf{r}_i)$ represents the vector joining q_i to the point P. The factor q is common to all terms in the sum, and it follows that

$$\frac{\mathbf{F}}{q} = \frac{1}{4\pi\varepsilon_0} \sum_i \frac{q_i}{|\mathbf{r} - \mathbf{r}_i|^3}(\mathbf{r} - \mathbf{r}_i).$$

The magnitude q of the test charge does not appear on the right-hand side of this equation. We can therefore allow q to become vanishingly small—then we are

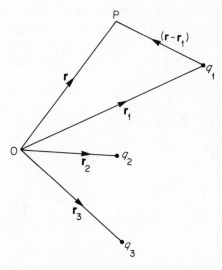

Figure 1.3. Vectors used in the definition
of the electric field.

quite sure that the presence of the test charge does not modify the position of the other charges. The electric field $\mathbf{E}(\mathbf{r})$ at the point P with position vector $\mathbf{r}$ is now

$$\mathbf{E}(\mathbf{r}) = \operatorname*{Lim}_{q \to 0} \frac{\mathbf{F}}{q},$$

or

$$\mathbf{E}(\mathbf{r}) = \frac{1}{4\pi\varepsilon_0} \sum_i \frac{q_i}{|\mathbf{r} - \mathbf{r}_i|^3}(\mathbf{r} - \mathbf{r}_i). \tag{1.4}$$

Equation (1.4) is the definition of the electric field $\mathbf{E}(\mathbf{r})$, and it contains no reference to a test charge q. The test charge was introduced because it illustrates that the electric field is a force per unit charge, and because it helps one to visualize the electric field if one imagines a test charge which can be moved around to sample the strength of the field at any position. The electric field $\mathbf{E}(\mathbf{r})$ is a function of position; just as the value of a function $f(x)$ is determined by the argument x, so the value of $\mathbf{E(r)}$ is given by Equation (1.4) in terms of its argument, which is the position vector $\mathbf{r}$. The function $\mathbf{E(r)}$ is itself a vector, specifying the direction as well as the magnitude of the force per unit charge on a point charge at $\mathbf{r}$. The electric field is only the first of a number of functions of position which are useful in electromagnetism. Those functions of position which are themselves vectors are called *vector fields*. When discussing vector fields we shall frequently omit any reference to the argument, writing the electric field, for example, simply as '$\mathbf{E}$', leaving it to be understood that the field is a function of position. Whenever there is some uncertainty about the position at which the function is to be evaluated, we shall always write out the vector field in full, including the argument.

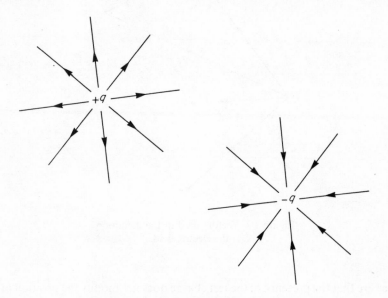

Figure 1.4. Field lines around point charges.

Now let us investigate the properties of the electric field in the neighbourhood of isolated point charges. From Equation (1.4), the magnitude of the field at a distance r from a positive point charge $+q$ is $q/4\pi\varepsilon_0 r^2$, and the field points away from the charge. The field around a negative point charge $-q$ has the same magnitude, but it points towards the charge. In Figure 1.4 the direction of the field around positive and negative charges is indicated by the arrowed lines. These continuous lines, everywhere following the direction of the field, are called *lines of force* or *field lines*. Lines of force may begin on positive charges and end on negative charges, but they may also go to infinity without terminating, as in Figure 1.4. Notice that the lines of force are close together near the point charges where the field is strong, and far apart at large distances where the field is weak.

We can also draw diagrams of lines of force to illustrate the electric field when there are many charges present. Lines of force are continuous, except where they terminate on positive or negative charges, and they never cross one another, since the direction of the field is unique at every point. One can often get a rough idea of the field around a distribution of charges simply by sketching lines of force, and without doing any mathematics. For example, Figure 1.5 shows the lines of force near a pair of point charges of equal magnitude, one positive and one negative. Such a pair of equal and opposite charges is called

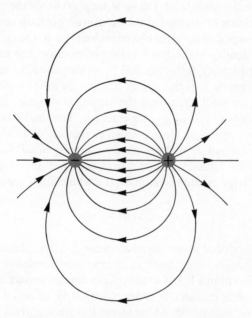

Figure 1.5. Field lines around an electric dipole.

an electric dipole. Very close to each charge, the field is almost the same as for isolated point charges, but the field lines starting off at the positive charge curve round to finish at the negative charge. The diagram gives an indication of the strength of the field as well as its direction, because lines of force are always densely packed in regions where the field is strong. The total number of lines on the diagram is not significant; if twice as many lines were drawn terminating on each charge, regions of relatively high or low density of lines of force would still correspond to regions of high or low field strength. The example of the electric dipole is an important one and later on we shall obtain a mathematical expression for the field at some distance from a dipole.

1.3 ELECTRIC FIELDS IN MATTER

1.3.1 The atomic charge density

So far electric fields have been dealt with in terms of idealized particles which are stationary and which carry point charges. This is not satisfactory if we want to discuss the electric field within an actual atom, since electrons are certainly not stationary point charges. Indeed, according to quantum mechanics the position of an electron cannot even be sharply defined. Instead of imagining the electron as a point, it is more realistic to regard its charge as being smeared out in a cloud around the nucleus. The electric field in and around a hydrogen atom, for example, behaves as though only part of the electronic charge is contained by any section of the electron cloud. Let us write $\rho_{el}(\mathbf{r})\,\delta\tau$ for the charge contained within a small volume $\delta\tau$ situated at the point with position vector $\mathbf{r}$. The quantity $\rho_{el}(\mathbf{r})$ is the *charge density* of the electron cloud, i.e. its charge per unit volume at $\mathbf{r}$. The charge density is another function of position, but unlike the electric field it is a scalar quantity, fully specified by its magnitude at each point. Scalar functions of position such as the charge density are called *scalar fields*. As with the vector fields, we shall often omit the argument of scalar fields, writing the electron charge density just as ρ_{el}, for example. The total charge $(-e)$ carried by the electron in a hydrogen atom is equal to the sum of the charges contained in all the small volumes $\delta\tau$ making up the electron cloud. In the limit as the volumes $\delta\tau$ become infinitesimal, the sum becomes an integral; integrating over a volume V large enough to contain the whole atom, we write

$$\int_V \rho_{el}(\mathbf{r})\,d\tau = -e.$$

Here we have introduced a shorthand notation for the volume integral, which is really a triple integral over the three components needed to specify position vectors $\mathbf{r}$ within the volume V. In Cartesian coordinates $d\tau$ is $dx\,dy\,dz$, the volume of the rectangular box enclosed by sides of length dx, dy and dz. The volume of integration V must include the whole atom, but since $\rho_{el}(\mathbf{r})$ is zero outside the atom, it can be made as large as we please without affecting the result. If we extend V to cover the whole of space, in Cartesian coordinates the integral becomes

$$\int_{\text{all space}} \rho_{el}(\mathbf{r})\,d\tau = \int_{-\infty}^{\infty}\int_{-\infty}^{\infty}\int_{-\infty}^{\infty} \rho_{el}(\mathbf{r})\,dx\,dy\,dz = -e.$$

In atoms more complex than hydrogen, the electron clouds of different electrons overlap, and a volume element may contain parts of the charge of more than one electron. Even the charge carried by the nucleus should be represented by a charge density since although the nucleus is small compared to the whole atom, it does have a definite size. Now we define the atomic charge density $\rho_{atomic}(\mathbf{r})$ in a piece of matter as the net charge density at $\mathbf{r}$, including positive contributions from nuclei. Where the charge densities associated with different particles overlap, they are added together to form the *atomic charge density*. The atomic charge density has large positive values inside the nucleus of each atom and is negative in the electron cloud. The volume integral $\int_V \rho_{\text{atomic}} \, d\tau$ represents the net charge within V: thus for a volume V containing an electrically neutral piece of matter, the value of the integral must be zero.

1.3.2   The atomic electric field

The electric field due to an assembly of point charges q_i was found from the principle of superposition by making a vector sum of the fields due to each charge separately:

$$\mathbf{E}(\mathbf{r}) = \frac{1}{4\pi\varepsilon_0} \sum_i \frac{(\mathbf{r} - \mathbf{r}_i) \, q_i}{|\mathbf{r} - \mathbf{r}_i|^3}.$$

Applying the same procedure to a continuous charge density $\rho_{\text{atomic}}(\mathbf{r})$, the sum becomes an integral, and the *atomic electric field* is

$$\mathbf{E}_{\text{atomic}}(\mathbf{r}) = \frac{1}{4\pi\varepsilon_0} \int_{\text{all space}} \frac{(\mathbf{r} - \mathbf{r}') \, \rho_{\text{atomic}}(\mathbf{r}')}{|\mathbf{r} - \mathbf{r}'|^3} \, d\tau'. \tag{1.5}$$

Here $\rho_{\text{atomic}}(\mathbf{r}') \, d\tau'$ is the charge within a volume element $d\tau'$ located at the point with position vector $\mathbf{r}'$, as illustrated in Figure 1.6. Both sides of Equation (1.5) are vectors, and to work out the volume integral in practice it is necessary to integrate each component separately.

Although it is easy to write down this equation for the atomic field*, it is not very useful, because in practice one does not know the charge density ρ_{atomic}.

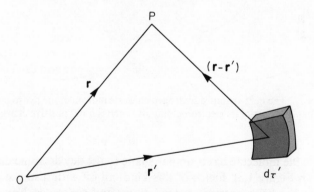

Figure 1.6. The charges within each volume element $d\tau'$ contributes to the electric field at P.

---

* Strictly speaking Equation (1.5) only applies to assemblies of stationary atoms. The exact expression for the electric field, due to moving particles, which is discussed in Chapter 13, differs appreciably only for speeds close to the speed of light. In ordinary matter Equation (1.5) is a good approximation to the atomic electric field at a particular moment.

Not only does the charge density within each atom have a complicated form, but even in a solid, atoms are always in thermal motion. The charge density ρ_{atomic} and the field $\mathbf{E}_{atomic}$ at a fixed point are therefore continually changing. However, although the exact atomic field is rarely known, it is possible to describe its main qualitative features. Each atom has an internal field generated by its own nucleus and electron cloud. This field is very strong near the concentration of positive charge in the atomic nucleus. Superimposed on the internal field there may be fields caused by other atoms and molecules. Not only do charged atoms and molecules generate external electric fields, but even outside neutral atoms and molecules there may be fields which are significant over distances several times as large as a typical atomic diameter. A hydrogen chloride molecule, for example, is electrically neutral, but in the chemical bond which holds it together part of the electron cloud of the hydrogen atom is transferred to the chlorine atom. The hydrogen atom is thus left with an excess of positive charge, while the chlorine acquires a negative charge. Lines of force leave the positively charged hydrogen and end on the negatively charged chlorine as sketched in Figure 1.7 (notice that the lines of force indicate that the electric

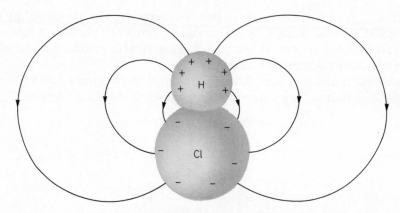

Figure 1.7. The field around a hydrogen chloride molecule: the hydrogen atom has lost part of its electron cloud and carries a net positive charge.

field outside the molecule has a similar shape to the field at some distance from the dipole in Figure 1.5). Fields of this kind fall off with distance much more quickly than the fields outside charged atoms and molecules. Nevertheless the local average field acting on a particular atom is often dominated by the contributions from neutral neighbours.

1.3.3 The macroscopic electric field

Fortunately, when one is considering large systems, the details of the atomic field are not important, and all that is needed is an average field. The value of the field averaged over a region much bigger than a single atom, but still small compared to the size of the whole system, is called the *macroscopic electric field* (the word macroscopic means large-scale, and is used to make a contrast with the microscopic scale of the atomic field). In the macroscopic field all the sharp variations of the atomic field within each atom are smoothed out: the macroscopic field is the one which would exist if matter were continuous and had no atomic structure.

By comparison with the strong atomic fields near nuclei, macroscopic fields are of modest magnitude. In a good electrical insulator like porcelain, electrons are bound tightly to their atoms, and any local excesses of positive or negative charge can remain immobile for a long time, maintaining a macroscopic field. But if this field becomes too large catastrophic breakdown occurs. Typically a good insulator breaks down in fields of about 10^9 volts/metre. Static fields as great as this do not occur over extensive regions, and even over distances of a few atomic diameters, they are found only in some special situations*.

Net charges may be distributed throughout the volume of an insulator, or may be concentrated in a thin layer at a surface. Only surface charges occur in a material like copper which is a good electrical conductor. In a conductor, some electrons, which are referred to as *conduction electrons*, are free to move in the presence of a macroscopic electric field. It follows that *in static equilibrium the macroscopic electric field within a conductor is everywhere zero.* What happens when a conductor is placed in an electric field generated by outside charges? Consider a slab of conductor in a uniform field directed perpendicularly to the surface of the slab. When the field is first applied, a macroscopic field extends throughout the conductor. The conduction electrons move under the influence of this field, leaving net charges on the surface as shown in Figure 1.8. The surface

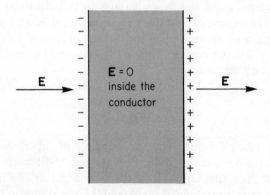

Figure 1.8. Induced surface charges appear on a
slab of conductor when it is placed in an electric
field.

* Very high fields are maintained in the field emission microscope, which is described in an example at the end of this chapter.

charges generate a field inside the conductor which is *opposed* to the external field, and the surface charges build up until the macroscopic field within the bulk of the conductor is exactly zero. Charges appearing on the surface of a material in this way, due to the presence of an external field, are called *induced charges*.

We now relate the macroscopic field to the charges which generate it, taking into account both distributed volume charges and surface charges. First consider the volume charges. Imagine a volume δV, small compared with a macroscopic system, but still large enough to contain many atoms. The net charge within δV is $\int_{\delta V} \rho_{\text{atomic}} \, d\tau$. Neighbouring regions of the same volume δV may not contain precisely the same net amount of charge. However, the relative differences between regions are small, provided that each region contains a large enough number of atoms. We can define the macroscopic charge density $\rho(\mathbf{r})$ at the point with position vector $\mathbf{r}$ as the net charge per unit volume contained in a volume δV in the vicinity of $\mathbf{r}$, i.e.

$$\rho(\mathbf{r}) = \frac{1}{\delta V} \int_{\delta V} \rho_{\text{atomic}}(\mathbf{r}) \, d\tau. \tag{1.6}$$

The charge density $\rho(\mathbf{r})$ is a function of position which varies smoothly inside matter, except at boundaries between media, where ρ may change discontinuously. Near a boundary there may also be a concentration of surface charges. Surface charges always occupy a very thin layer, no more than a few atoms thick, and for the purposes of discussing the macroscopic field they may be represented by an infinitesimally thin sheet of charge lying exactly on the boundary surface. A surface charge density $\sigma(\mathbf{r})$ is defined, analogously to the volume density $\rho(\mathbf{r})$, as the net charge per unit area on an area δS which is small on the macroscopic scale, but large enough to include many surface atoms. Consider a slab of volume δV just thick enough to enclose all the surface charge $\sigma(\mathbf{r}) \, \delta S$ associated with an area δS in the vicinity of a point on the surface with position vector $\mathbf{r}$. The net charge within δV is a volume integral over the atomic charge density, and we have*

$$\sigma(\mathbf{r}) = \frac{1}{\delta S} \int_{\delta V} \rho_{\text{atomic}} \, d\tau. \tag{1.7}$$

Macroscopic electric fields obey the principle of superposition just as do atomic fields. To find the macroscopic field at a point with position vector $\mathbf{r}$, it is necessary to add the contributions from the net charges $\rho(\mathbf{r}') \, d\tau'$ in all the volume elements $d\tau'$ covering the whole of space, and the contributions from the surface charges $\sigma(\mathbf{r}') \, dS'$ on the surface elements dS' over all boundary surfaces. The field generated by the volume charge density ρ is given by a volume integral of

---

* A small part of the charge within V should be ascribed to the macroscopic density ρ, which is assumed to continue smoothly right up to the boundary.

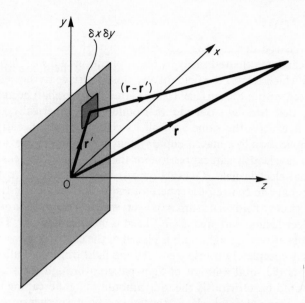

Figure 1.9. The vector addition of the fields due to all
surface elements $\delta x\,\delta y$ leads to a *surface integral*.

the same form as Equation (1.5) for the atomic field, namely

$$\frac{1}{4\pi\varepsilon_0}\int_{\text{all space}}\frac{(\mathbf{r}-\mathbf{r}')\,\rho(\mathbf{r}')}{|\mathbf{r}-\mathbf{r}'|^3}\,\mathrm{d}\tau'.$$

Similarly the surface charge density σ generates a field given in shorthand notation as

$$\frac{1}{4\pi\varepsilon_0}\int_{\text{all surfaces}}\frac{(\mathbf{r}-\mathbf{r}')\,\sigma(\mathbf{r}')}{|\mathbf{r}-\mathbf{r}'|^3}\,\mathrm{d}S'.$$

The surface integral is a double integral, and for example if the rectangular
surface in the x–y plane shown in Figure 1.9 carries a uniform charge density
σ, its contribution to the field at $\mathbf{r}$ is

$$\frac{1}{4\pi\varepsilon_0}\int_{x=-a}^{a}\int_{y=-b}^{b}\frac{(\mathbf{r}-\mathbf{r}'\,\sigma\,\mathrm{d}x\,\mathrm{d}y}{|\mathbf{r}-\mathbf{r}'|^3}.$$

Finally, adding the terms from volume and surface charge densities, the macroscopic electric field is defined to be

$$\mathbf{E}(\mathbf{r})=\frac{1}{4\pi\varepsilon_0}\int_{\text{all space}}\frac{(\mathbf{r}-\mathbf{r}')\,\rho(\mathbf{r}')\,\mathrm{d}\tau'}{|\mathbf{r}-\mathbf{r}'|^3}+\frac{1}{4\pi\varepsilon_0}\int_{\text{all surfaces}}\frac{(\mathbf{r}-\mathbf{r}')\,\sigma(\mathbf{r}')\,\mathrm{d}S'}{|\mathbf{r}-\mathbf{r}'|^3}. \qquad (1.8)$$

1.4 GAUSS' LAW

When the charge density has a distribution with a simple symmetry, the electric field can be evaluated easily by the application of *Gauss' law*, an important theorem which relates the field on any closed surface to the net amount of charge enclosed within the surface. Using Gauss' law we shall be able to prove, for example, that the field outside an isolated and spherically symmetrical charged ion is exactly the same as if all its charge were concentrated at the centre; an application to a macroscopic system is at the surface of a conductor, where Gauss' law leads to an expression for the surface charge density in terms of the external electric field. Gauss' law can be illustrated by considering a point charge q at the centre of a sphere of radius r. The area of the sphere is $4\pi r^2$, and the electric field on its surface is everywhere of magnitude $q/4\pi\varepsilon_0 r^2$. The product of electric field and area is q/ε_0, i.e. it is independent of r. This is rather like what happens when a light bulb is placed at the centre of a spherical lampshade which is completely transparent. All the light from the bulb escapes, or in other words the total amount of light passing through the shade is independent of its radius. Obviously the total amount of light escaping is also independent of the shape of the shade; all the light gets out whether or not the shade is spherical. Gauss' law is the analogous result in electrostatics applied to a quantity called the *flux* of the electric field; the total flux out of a surface enclosing a charge q is q/ε_0, whatever the shape of the surface. In the rest of this section we shall explain what is meant by flux, and then go on to prove Gauss' law rigorously.

1.4.1 The flux of a vector field

Flux may be defined for any vector function of position, but it is most easily visualized in the flow of fluids. Imagine that a fluid is flowing at a speed v through a small flat surface δS bounded by a wire. In Figure 1.10a the shaded area δS is perpendicular to the direction of flow of the fluid. The flux of fluid through the area δS is the *rate of flow* of fluid through the wire, which in this case is the product $v \, \delta S$ of speed and area. Now suppose that the wire is rotated to the position shown in Figure 1.10b, where the normal to the surface δS is at an angle ψ to the direction of flow. Looking along the direction of flow, the projected area within the wire is now only $\delta S_{\mathrm{p}} = \delta S \cos \psi$, and the flux, i.e. the rate of flow of fluid through the wire, is reduced to

$$v \, \delta S_{\mathrm{p}} = v \, \delta S \cos \psi.$$

This flux can be written in vector notation if we represent the surface by a vector $\delta \mathbf{S}$, of magnitude δS and directed along the normal to the surface. The velocity vector $\mathbf{v}$ is then at an angle ψ to $\delta \mathbf{S}$, and the flux through $\delta \mathbf{S}$ is given by the scalar product

$$\mathbf{v} \cdot \delta \mathbf{S} = v \, \delta S \cos \psi.$$

The flux has a sense as well as a magnitude, and in Figure 1.10b the sense is

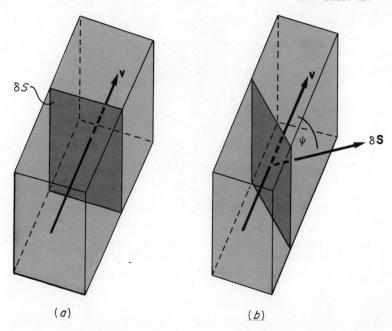

Figure 1.10. A column of fluid of length v passes through the area δS every second. The volume of fluid passing is larger when **v** is perpendicular to δS, as in (a), than when it is at an angle as in (b).

from the left hand side of the wire to the right hand side. If we choose to represent the surface by the vector $(-\delta S)$, still normal to the surface but pointing from right to left in the opposite direction to δS, the scalar product $\mathbf{v} \cdot (-\delta S)$ becomes negative. A negative flux from right to left is equivalent to a positive flux from left to right, but to avoid ambiguity the sense of surface vectors must always be carefully specified.

In an electrostatic field there is of course nothing actually flowing, but the flux of an electric field $\mathbf{E}$ through δS is defined in the same way as $\mathbf{E} \cdot \delta S$, the value of $\mathbf{E}$ being taken at the position of the surface δS. In terms of components the flux of $\mathbf{E}$ through δS can be expressed as

$\mathbf{E} \cdot \delta S$ = (magnitude of $\mathbf{E}$) × (projected area δS_p)

= (magnitude of $\mathbf{E}$) × (component of δS along the direction of $\mathbf{E}$),

or equivalently

$\mathbf{E} \cdot \delta S$ = (component of $\mathbf{E}$ normal to the surface) × (area δS).

So far we have only defined the flux through a small flat surface. To find the flux of the electric field through a surface S of arbitrary shape, we first approximate the shape of S by dividing it up into a lot of small flat surfaces. The flat

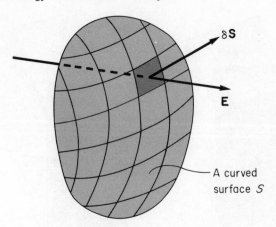

Figure 1.11. Splitting up a surface S into a lot of little
flat surfaces in order to work out the flux of $\mathbf{E}$
through S.

surfaces are represented by vectors such as $\delta\mathbf{S}$ in Figure 1.11, all pointing outwards from the same side of the whole surface. The flux of the field $\mathbf{E}$ through $\delta\mathbf{S}$
is $\mathbf{E} \cdot \delta\mathbf{S}$, and the total flux through the surface made up of the mosaic of flat
surfaces is the sum of the fluxes through each separately, i.e.

$$\sum_{\text{all surfaces } \delta\mathbf{S}} \mathbf{E} \cdot \delta\mathbf{S}.$$

If the size of each of the flat surfaces is made smaller and smaller, the mosaic
approaches more and more closely to the smooth surface S, and the sum
approaches a limiting value which is equal to the flux through S. The limit is a
two-dimensional surface integral written as

$$\int_S \mathbf{E} \cdot d\mathbf{S} = \text{flux through } S = \lim_{\delta\mathbf{S}\to 0} \sum_{\text{all } \delta\mathbf{S}} \mathbf{E} \cdot \delta\mathbf{S}.$$

The subscript S under the integral sign indicates that the area of integration is
the whole of the surface S.

1.4.2 The flux of the electric field out of a closed surface

Let us now evaluate the flux through a closed surface S for the electric field
$\mathbf{E}$ generated by a point charge q enclosed by S. We shall use spherical polar
coordinates referred to an origin at q, and concentrate attention for the moment
on a surface element of area δS at the position with coordinates (r, θ, ϕ). The
surface element is represented by the vector $\delta\mathbf{S}$ directed along the *outward
normal* to S. The element is of such a size and shape that its projection along the
radius vector from q lies between the polar angles θ and $(\theta + \delta\theta)$ and the azi-

muthal angles ϕ and $(\phi + \delta\phi)$. As illustrated in Figure 1.12, the projected area is

$$\delta S_{\mathrm{p}} = r^2 \sin\theta \, \delta\theta \, \delta\phi.$$

The field at δS is directed outwards along the radius vector and has a magnitude

$$E = \frac{q}{4\pi\varepsilon_0 r^2}.$$

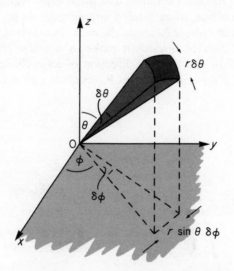

Figure 1.12. The cone subtends angles $\delta\theta$ and $\delta\phi$ at the origin, and its cross-sectional area at a distance r from the apex is $r^2 \sin\theta \, \delta\theta \, \delta\phi$.

Hence the outward flux through $\delta \mathbf{S}$ is

$$\mathbf{E} \cdot \delta\mathbf{S} = E \, \delta S_{\mathrm{p}} = \frac{q}{4\pi\varepsilon_0} \sin\theta \, \delta\theta \, \delta\phi$$

$$= \frac{q}{4\pi\varepsilon_0} \delta\Omega, \tag{1.9}$$

where the element $\delta\Omega$ is termed the *solid angle* of the cone with apex at the origin and base δS_{p}. Using Equation (1.9), we can now transform the surface integral expression for the flux through the complete surface S into an integral over angular variables only. The range of variation required to cover the whole of the closed surface is $\theta = 0$ to π and $\phi = 0$ to 2π. Taking the limit of infinitesimal area δS, the total flux through S is

$$\int_S \mathbf{E} \cdot d\mathbf{S} = \frac{q}{4\pi\varepsilon_0} \int_0^\pi \sin\theta \, d\theta \int_0^{2\pi} d\phi$$

$$= \frac{q}{\varepsilon_0}. \tag{1.10}$$

In this derivation it has been assumed that the flux through $d\mathbf{S}$ is always positive. Sometimes, however, when S has a complicated shape, the radius vector from q may pass from outside to inside the enclosing surface, as in Figure 1.13. Where the field is directed inwards it makes a negative contribution to the net outward flux. Provided that the direction of $d\mathbf{S}$ is chosen to be along the *outward* normal to S, the scalar product $\mathbf{E} \cdot d\mathbf{S}$ always gives the correct sign for

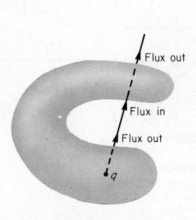

Figure 1.13. The radius vector from a charge inside a surface always makes one more outward crossing than the number of inward crossings.

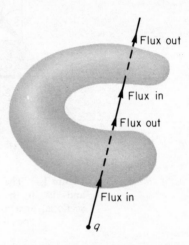

Figure 1.14. For a charge outside the surface the number of outward crossing is the same as the number of inward crossings.

the flux *out of* $d\mathbf{S}$. But Equation (1.9) should be modified to

$$\mathbf{E} \cdot d\mathbf{S} = \pm \frac{q}{4\pi\varepsilon_0} d\Omega, \tag{1.11}$$

with the negative sign applying for those parts of the surface where the field is directed inwards. Notice that however many times a radius vector may cross the surface S, if the origin lies within S the radius vector must always make one more outward crossing than the number of inward crossings. It follows that no matter how many times the cone of solid angle $d\Omega$ is cut by the surface S, the *net* flux leaving S within $d\Omega$ is $(q/4\pi\varepsilon_0)\, d\Omega$, and Equation (1.10) is correct for any surface whatever which encloses q.

For a charge q which is *outside* a closed surface S, on the other hand, the radius vector must cross S an even number of times, with equal contributions to the inward and outward fluxes, as illustrated in Figure 1.14. The net outward flux through a closed surface due to any charge not enclosed by the surface is therefore exactly zero. Now we can use the principle of superposition to deduce the outward flux through a closed surface S of the electric field $\mathbf{E}$ due to an arbitrary distribution of charges q_i. Only those charges enclosed by S make a contribution to the flux, and

$$\int_S \mathbf{E} \cdot d\mathbf{S} = \frac{1}{\varepsilon_0} \sum_i q_i, \tag{1.12a}$$

where the summation is restricted to the charges within S. If the charge has a continuous distribution with a charge density ρ, then the summation is replaced by an integral, and

$$\int_S \mathbf{E} \cdot d\mathbf{S} = \frac{1}{\varepsilon_0} \int_V \rho \, d\tau, \tag{1.12b}$$

where the integral on the right-hand side extends over the whole of the volume V enclosed by S. Here we have related a two-dimensional surface integral to a three-dimensional volume integral. To illustrate how to apply the relation to a particular coordinate system, let us write it out in full for a sphere of radius R. In spherical polar coordinates a position vector $\mathbf{r}$ is specified by the three components r, θ and ϕ. The charge density and electric field are both functions of position, and are written as $\rho(r, \theta, \phi)$ and $\mathbf{E}(r, \theta, \phi)$. The electric field at the surface of the sphere may not be normal to the surface, but only the radial component E_r contributes to the flux of $\mathbf{E}$ through the sphere. The element of surface subtending angles $d\theta$ and $d\phi$ at the origin is $R^2 \sin\theta \, d\theta \, d\phi$ (see Figure 1.12), and the volume element within dr, $d\theta$ and $d\phi$ is $r^2 \sin\theta \, dr \, d\theta \, d\phi$. With limits which extend the integrations over the whole of the surface and volume of the sphere Equation (1.12b) becomes

$$\int_{\theta=0}^{\pi} \int_{\phi=0}^{2\pi} E_r(R, \theta, \phi) \, R^2 \sin\theta \, d\theta \, d\phi$$

$$= \frac{1}{\varepsilon_0} \int_{r=0}^{R} \int_{\partial=0}^{\pi} \int_{\phi=0}^{2\pi} \rho(r, \theta, \phi) \, r^2 \sin\theta \, dr \, d\theta \, d\phi. \tag{1.13}$$

Whether it is applied to point charges or a continuous charge distribution, Equation (1.12) can be expressed in words as

$$\text{(Flux of } \mathbf{E} \text{ out of } S) = \frac{1}{\varepsilon_0} \text{(total charge within } S).$$

This equation, which is *Gauss' law*, is a direct result of the inverse square dependence of the electric field. Only with an inverse square law does a point charge generate a flux per unit solid angle which is independent of distance from the source. The analogy between Gauss' law and the flux of light through a closed surface surrounding a lamp also rests on the inverse square law. Light intensity diminishes as the inverse square of the distance from the source of light, and in this case the constancy of flux confined within a fixed solid angle, as for example in the beam of a flashlight, is a consequence of the conservation of energy.

As has already been mentioned, Gauss' law can be used to calculate the field when the charge distribution has a simple symmetry. An example of spherical symmetry is provided by an isolated ion of singly ionized sodium, that is, a sodium atom which has lost one of its electrons and carries a net charge $+e$. The field outside the ion can be calculated at once using Gauss' law, without any knowledge of the distribution of charge inside. Choose a spherical surface of radius R centred on the ion, and large enough to contain the whole of the ion. Gauss' law then takes the form of Equation (1.13). Because of the spherical symmetry, the electric field on the surface does not depend on θ and ϕ, but has a constant magnitude $E(R)$, say, and is directed radially away from the origin. Taking $E(R)$ outside the surface integral, the integral over θ and ϕ is simply the area $4\pi R^2$ of the sphere, and the total flux is $E(R) \times 4\pi R^2$. The total charge within R must be $+e$, whatever may be the radial distribution of the charge carried by the electrons. Hence

$$E(R) \times 4\pi R^2 = e/\varepsilon_0,$$

or

$$E(R) = \frac{e}{4\pi\varepsilon_0 R^2}.$$

the same field as for a point charge $+e$ located at the origin.

An isolated neutral atom of an inert gas provides another example of spherical symmetry. Since the atom carries no net charge, it follows that the field outside it is exactly zero. Even when they are in the neighbourhood of other atoms, inert gas atoms depart only slightly from spherical symmetry, and the electric fields near them remain small. There are other interactions between atoms besides those due to their electrostatic fields, but it is nevertheless true to say that the feeble chemical activity of inert gases is related to the fact that they are always very nearly spherically symmetrical.

Gauss' law also applies to macroscopic fields and charge densities, since they too are connected by the inverse square law. Figure 1.15 represents a slab of conductor in an external field perpendicular to its surface. Induced surface charge densities $\pm \sigma$ generate an equal and opposite field inside the conductor, where the macroscopic electric field must be zero. Now imagine a disc enclosing an area δS of the surface, as shown in the figure. The flat surfaces of the disc are parallel to the conducting surface while the curved surface is parallel to the

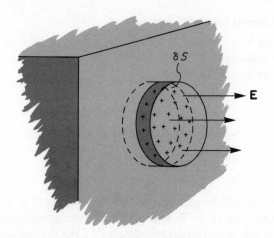

Figure 1.15. A disc enclosing some of the induced charge on a conductor.

external field. The flux of $\mathbf{E}$ out of the flat surface of the disc which is outside the conductor is $E \delta S$. Inside the conductor the field is zero, and there is no flux out of the other flat surface of the disc; nor is there any flux out of the curved surface, since the field lies along this surface. The total charge enclosed by the disc is $\sigma \delta S$, and application of Gauss' law leads to

$$E \delta S = \frac{1}{\varepsilon_0} \sigma \delta S, \quad \text{or} \quad \sigma = \varepsilon_0 E. \tag{1.14}$$

The example of a slab placed perpendicular to an external field is a special case, but it is easy to show that the induced charges are disposed in such a way that the electric field is always perpendicular to the surface. For if there were a component of the field parallel to the surface, conduction electrons would move until they set up a field which exactly cancelled this component. Over a small part of a conducting surface of any shape, the argument given above is valid, and Equation (1.14) holds at each point of the surface, whether or not it is plane.

Since macroscopic electric fields are limited to a maximum value of about 10^9 newtons/coulomb, it follows that practically realizable surface charge densities are less than $10^9 \varepsilon_0 \simeq 10^{-2}$ coulomb m^{-2}. Atomic diameters are about 2×10^{-10} m, and the area associated with each atom on the surface of a conductor is about 4×10^{-20} m^2. The maximum surface charge thus corresponds to 4×10^{-22} coulombs, or less than a hundredth of the electronic charge, for each surface atom. The appearance of induced charges represents only a slight perturbation of the outermost layers of atoms in a conductor, and from the macroscopic point of view it is well justified to regard surface charges as infinitesimally thin sheets.

1.4.3   The differential form of Gauss' law

Gauss' law can be expressed in a differential form as well as in the integral form given in Equation (1.12). If δS is the surface enclosing a volume element $\delta \tau$, then Equation (1.12b) becomes

$$\int_{\delta S} \mathbf{E} \cdot \mathbf{dS} = \frac{1}{\varepsilon_0} \rho \, \delta \tau.$$

In the limit when $\delta \tau$ becomes infinitesimally small, we have

$$\lim_{\delta \tau \to 0} \frac{\int_{\delta S} \mathbf{E} \cdot \mathbf{dS}}{\delta \tau} = \frac{\rho}{\varepsilon_0}. \tag{1.15}$$

The quantity on the left-hand side is called the *divergence* of $\mathbf{E}$, and is usually written div $\mathbf{E}$. The form of div $\mathbf{E}$ depends on the coordinate system used, but it is easy to work it out in Cartesian coordinates*. Take the volume element $\delta \tau$ to be a rectangular box with sides parallel to the axes and of lengths δx, δy and δz as shown in Figure 1.16. The flux of $\mathbf{E}$ out of the sides of the box normal to the

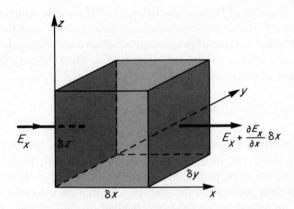

Figure 1.16. The flux out of a rectangular box in a non-uniform field.

---

* The forms of div $\mathbf{E}$ in other coordinate systems are given in Appendix B.

x-axis is (remembering to take δS always to be the *outward* normal to the surface)

$$\left(E_x + \frac{\partial E_x}{\partial x}\delta x\right)\delta y\,\delta z - E_x\,\delta y\,\delta z = \frac{\partial E_x}{\partial x}\delta x\,\delta y\,\delta z.$$

There are similar contributions from the sides normal to the y and z axes, and the total flux out of the box is

$$\int_{\delta S}\mathbf{E}\cdot d\mathbf{S} = \left(\frac{\partial E_x}{\partial x} + \frac{\partial E_y}{\partial y} + \frac{\partial E_z}{\partial z}\right)\delta x\,\delta y\,\delta z.$$

Thus

$$\lim_{\delta x\,\delta y\,\delta z\to 0}\frac{\int_{\delta S}\mathbf{E}\cdot d\mathbf{S}}{\delta x\,\delta y\,\delta z} = \left(\frac{\partial E_x}{\partial x} + \frac{\partial E_y}{\partial y} + \frac{\partial E_z}{\partial z}\right) = \text{div }\mathbf{E}.$$

Comparing with Equation (1.15), we find that in Cartesian coordinates

$$\boxed{\frac{\partial E_x}{\partial x} + \frac{\partial E_y}{\partial y} + \frac{\partial E_z}{\partial z} = \text{div }\mathbf{E} = \rho/\varepsilon_0.}\qquad(1.16)$$

This is the differential form of Gauss' law, which is equivalent to the integral form given in Equation (1.12). The differential form is the more useful one except when the electric field has a very simple symmetry, and we shall use it to solve some realistic problems in Chapter 3.

From Equation (1.16) it follows that the divergence of the electric field $\mathbf{E}$ is zero in all regions where there is no charge. In the neighbourhood of a point charge, for example, the field is non-uniform, but although the individual terms on the left-hand side of Equation (1.16) may be different from zero, they always sum to zero. To show this, let us place a point charge q at the origin. The position vector $\mathbf{r}$ has Cartesian coordinates x, y and z, and the x-component of the electric field at $\mathbf{r}$ is

$$E_x = \frac{q}{4\pi\varepsilon_0}\left(\frac{(\mathbf{r})_x}{r^3}\right) = \frac{q}{4\pi\varepsilon_0}\left(\frac{x}{r^3}\right)$$

$$= \frac{qx}{4\pi\varepsilon_0(x^2 + y^2 + z^2)^{3/2}}.$$

Differentiating with respect to x (which is possible everywhere except right at the origin, where the field is singular),

$$\frac{\partial E_x}{\partial x} = \frac{q}{4\pi\varepsilon_0}\left\{\frac{1}{(x^2 + y^2 + z^2)^{3/2}} - \frac{3x^2}{(x^2 + y^2 + z^2)^{5/2}}\right\}$$

$$= \frac{q}{4\pi\varepsilon_0}\left(\frac{r^2 - 3x^2}{r^5}\right).$$

The terms $\partial E_y/\partial y$ and $\partial E_z/\partial z$ have the same form, with x replaced by y and z in turn. Hence

$$\text{div } \mathbf{E} = \frac{\partial E_x}{\partial x} + \frac{\partial E_y}{\partial y} + \frac{\partial E_z}{\partial z}$$

$$= \frac{q}{4\pi\varepsilon_0} \left\{ \frac{3r^2 - 3x^2 - 3y^2 - 3z^2}{r^5} \right\} = 0.$$

The divergence of any vector field $\mathbf{A(r)}$ is defined in the same way as for the electric field, and we can write

$$\text{div } \mathbf{A} = \underset{\delta\tau\to 0}{\text{Lim}} \frac{\int_{\delta S} \mathbf{A} \cdot d\mathbf{S}}{\delta\tau}.$$

By integrating this equation over a finite volume, the function div $\mathbf{A}$ can be related to the flux of $\mathbf{A}$ out of the volume, just as the integral form of Gauss' theorem relates ρ to the flux of $\mathbf{E}$:

$$\int_V \text{div } \mathbf{A} \, d\tau \equiv \int_S \mathbf{A} \cdot d\mathbf{S}. \tag{1.17}$$

As before, S is the surface enclosing the volume V. This result, an *identity* which holds whatever the form of $\mathbf{A}$, is often called Gauss' theorem. To avoid confusion with Gauss' law we shall refer to the identity as the *divergence theorem*.

Because Coulomb's law has a straightforward physical meaning, we have chosen to adopt it as the fundamental law of electrostatics. Then we have derived the integral and differential forms of Gauss' law as a consequence of Coulomb's law. But it is also possible to start at the other end of the logical chain, postulating at the very beginning that the electric field is the vector field which is the solution of Equation (1.16). The integral form of Gauss' law can then be deduced very simply. Applying the divergence theorem to the electric field, we have

$$\int_S \mathbf{E} \cdot d\mathbf{S} \equiv \int_V \text{div } \mathbf{E} \, d\tau.$$

Replacing div $\mathbf{E}$ by ρ/ε_0, it follows at once that

$$\int_S \mathbf{E} \cdot d\mathbf{S} = \frac{1}{\varepsilon_0} \int \rho \, d\tau,$$

which is the required result. Now the integral form of Gauss' law implies that Coulomb's law is valid, since if a point charge is placed at the centre of a sphere, only for an inverse square law of force is the flux of the electric field out of the sphere independent of its radius. Coulomb's law, the integral form of Gauss' law and the differential form of Gauss' law are thus equivalent statements, and any one of them may be taken as the fundamental law of electrostatics.

Many books on electromagnetism choose Equation (1.16) as the fundamental law; although Equation (1.16) is more difficult to understand than Coulomb's law, it has the advantage that it is the equation most often used to solve practical problems.

1.5   ELECTROSTATIC ENERGY

Energy changes are associated with alterations in either atomic or macroscopic charge density. The energy liberated in chemical reactions, for example, is almost entirely electrostatic in origin, and is made available by the rearrangement of charges within the atoms undergoing reaction. A flash of lightning, on the other hand, derives its energy from the net charge stored by a thundercloud. Later in section 1.5 we shall meet practical examples of energy changes in both microscopic and macroscopic physics, but we begin by discussing work and energy in terms of idealized point charges, in order to introduce the concept of electrostatic potential.

1.5.1   The electrostatic potential

External work must be done to move a small test charge in opposition to the force it experiences in an electrostatic field. A charge moved in this way acquires potential energy, which may be recovered by allowing the charge to retrace its path under the action of the electrostatic force. We shall show that the net work done when a charge is moved from one place to another does not depend on the path taken by the charge but only on its initial and final positions. For example, if a test charge in an electric field is moved from A to B (Figure 1.1(a)), the work

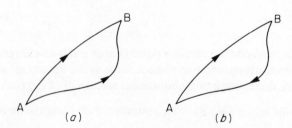

Figure 1.17. Possible paths for a test charge moving between A and B.

done is the same along both the paths shown. It follows that no work at all is done when a test charge moves along a path which returns it to its original position: in Figure 1.1(b) the work done from A to B is exactly cancelled by the return path from B to A. A force which has this property of doing no work around a closed path (and thus dissipating no energy) is called *conservative*. In electrostatics the field as well as the force is described as being conservative.

To show that the electrostatic field is conservative, it is sufficient to prove that the field around an isolated charge is conservative. The proof can then immediately be extended to all possible fields, since it follows from the principle of superposition that when a test charge is taken round a closed loop in any field whatever, no net amount of work is done on it by any of the other charges in the field. In Figure 1.18 the line joining A and B represents the path along which a

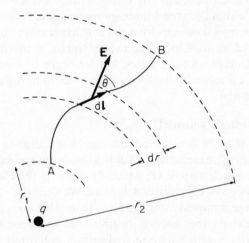

Figure 1.18. Working out the line integral along the path AB.

test charge q_t is moved in the field of a point charge q. If both charges are positive, work is done by the repulsive electrostatic force as the test charge moves from A to B. The work done along the infinitesimal section d**l** of the path is

$q_t \times$ (magnitude of the field **E**) $\times$ (component of d**l** along the direction of **E**)

$$= q_t E \, dl \cos \theta = q_t E \, dr = \frac{q_t q \, dr}{4\pi\varepsilon_0 r^2}.$$

The total work done when the test charge is moved from A to B is thus

$$\int_A^B q_t E \, dl \cos \theta = \frac{q_t q}{4\pi\varepsilon_0} \int_{r_1}^{r_2} \frac{dr}{r^2}$$

$$= \frac{q_t q}{4\pi\varepsilon_0} \left\{ \frac{1}{r_1} - \frac{1}{r_2} \right\} \qquad (1.18)$$

whatever path is used to join A and B. In vector notation the element $E \, dl \cos \theta$ occurring in Equation (1.18) is written as $\mathbf{E} \cdot \mathbf{dl}$, and the total work done as $q_t \int_A^B \mathbf{E} \cdot \mathbf{dl}$. The integral $\int_A^B \mathbf{E} \cdot \mathbf{dl}$ is called the *line integral* of the field $\mathbf{E}$ along the path AB. The line integral is independent of the path between A and B for the field in the neighbourhood of a point charge, and we deduce from the principle of superposition that it is independent of the path for all electrostatic fields. We have therefore proved that the electrostatic field is conservative, or in full: *In any electrostatic field* $\mathbf{E}$, *the line integral* $\int_A^B \mathbf{E} \cdot \mathbf{dl}$ *has a constant value independent of the path joining the points* A *and* B.

Work done by the electrostatic force represents a loss of potential energy of the test charge, for which

$$\text{(Potential energy at B)} - \text{(Potential energy at A)} = -q_t \int_A^B \mathbf{E} \cdot \mathbf{dl}.$$

Because the electrostatic field is conservative, the potential energy of the test charge depends only on its position, and not on the route taken to get there. In other words, the potential energy is a function of position, and at a point with position vector $\mathbf{r}$ it may be written as $q_t \phi(\mathbf{r})$. The function $\phi(\mathbf{r})$ is called the *electrostatic potential*. Putting $\mathbf{r}_A$ and $\mathbf{r}_B$ for the position vectors of the points A and B, the potential energy difference between B and A becomes

$$q_t \phi(\mathbf{r}_B) - q_t \phi(\mathbf{r}_A) = -q_t \int_A^B \mathbf{E} \cdot \mathbf{dl},$$

or, dividing by q_t,

$$\phi(\mathbf{r}_B) - \phi(\mathbf{r}_A) = -\int_A^B \mathbf{E} \cdot \mathbf{dl}. \tag{1.19}$$

Since the electric field $\mathbf{E}$ in this equation occurs only in a scalar product, the potential is a scalar function of position, which has a magnitude at any point but no direction. Only *differences* of potential are defined by Equation (1.19), which is still satisfied if an arbitrary constant is added to $\phi(\mathbf{r})$. The choice of the zero of potential is simply a matter of convenience. For an isolated system of charges, it is usual for the potential at infinity to be chosen as zero. Thus for an isolated point charge situated at the origin, the potential $\phi(\mathbf{r})$ is found from the work done in bringing up a test charge q_t from infinity to the point with position vector $\mathbf{r}$. The work done depends only on the magnitude r of the vector $\mathbf{r}$, and is

$$q_t \phi(\mathbf{r}) - q_t \phi(\infty) = -\int_\infty^r \frac{q_t q \, dr}{4\pi\varepsilon_0 r^2}$$

$$= \frac{q_t q}{4\pi\varepsilon_0 r}.$$

Taking $\phi(\infty)$ to be zero, we have

$$\phi(\mathbf{r}) = \frac{q}{4\pi\varepsilon_0 r}. \tag{1.20}$$

The dimensions of potential are energy per unit charge, and the unit of potential is the *volt*. One joule of work is done when a charge of one coulomb is moved through a potential difference of one volt.

1.5.2 The electric field as the gradient of the potential

Just as Gauss' theorem could be expressed in integral or differential form, so there is a differential relationship between field and potential which is equivalent to Equation (1.19). The differential relation is found by evaluating the potential difference $\delta\phi$ between the two points a small distance apart, labelled by position vectors $\mathbf{r}$ and $(\mathbf{r} + \delta\mathbf{r})$:

$$\delta\phi = \phi(\mathbf{r} + \delta\mathbf{r}) - \phi(\mathbf{r}) = -\int_{\mathbf{r}}^{\mathbf{r}+\delta\mathbf{r}} \mathbf{E} \cdot d\mathbf{l}$$

$$= -\mathbf{E} \cdot \delta\mathbf{r}.$$

Working in Cartesian coordinates, the vector $\delta\mathbf{r}$ has components $(\delta x, \delta y, \delta z)$, and $\delta\phi$ becomes

$$\delta\phi = -E_x \,\delta x - E_y \,\delta y - E_z \,\delta z.$$

By going to the limit as δx, δy and δz tend to zero we find that the components of the field are partial derivatives of the potentials:

$$E_x = -\frac{\partial\phi}{\partial x}, \qquad E_y = -\frac{\partial\phi}{\partial y}, \qquad E_z = -\frac{\partial\phi}{\partial z}.$$

The vector with components $\partial\phi/\partial x$, $\partial\phi/\partial y$ and $\partial\phi/\partial z$ is called the *gradient* of ϕ, and is written as grad ϕ or $\nabla\phi$. The differential form of Equation (1.19) is thus

$$\boxed{\mathbf{E} = -\text{grad } \phi.} \tag{1.21}$$

As a check, let us work out the field near a point charge from the potential $\phi(\mathbf{r}) = q/4\pi\varepsilon_0 r$ given in Equation (1.20). In Cartesian coordinates, the position vector $\mathbf{r}$ has components x, y and z, and

$$\phi(\mathbf{r}) = \frac{q}{4\pi\varepsilon_0(x^2 + y^2 + z^2)^{1/2}}.$$

Hence

$$E_x = -\frac{\partial\phi}{\partial x} = \frac{qx}{4\pi\varepsilon_0(x^2 + y^2 + z^2)^{3/2}} = \frac{qx}{4\pi\varepsilon_0 r^3}.$$

Similarly

$$E_y = \frac{qy}{4\pi\varepsilon_0 r^3} \quad \text{and} \quad E_z = \frac{qz}{4\pi\varepsilon_0 r^3}.$$

These components are proportional to the components of the position vector $\mathbf{r}$, and returning to the vector notation, we recover the correct expression for the field due to a point charge at the origin:

$$\mathbf{E}(\mathbf{r}) = \frac{q\mathbf{r}}{4\pi\varepsilon_0 r^3}.$$

Although we have used Cartesian coordinates for definiteness, the gradient, like the divergence, is a general differential operator. It can be applied to any scalar function of position, and the components of grad ϕ may be expressed in any coordinate system. Commonly occurring forms of grad ϕ are given in Appendix B.

What happens to the potential when two electric fields, generated by different charges, are superimposed? If the fields are $\mathbf{E}_1(\mathbf{r})$ and $\mathbf{E}_2(\mathbf{r})$, and the corresponding potentials are $\phi_1(\mathbf{r})$ and $\phi_2(\mathbf{r})$, then the x-component of the field $\mathbf{E}_1 + \mathbf{E}_2$ is

$$E_{1x} + E_{2x} = -\frac{\partial \phi_1}{\partial x} - \frac{\partial \phi_2}{\partial x} = -\frac{\partial}{\partial x}(\phi_1 + \phi_2).$$

Similar equations hold for the y and z components of the field. The potential for the combined field $(\mathbf{E}_1 + \mathbf{E}_2)$ is simply $(\phi_1 + \phi_2)$, or in other words, the principle of superposition holds for the potential as well as for the field. The potential at a point P, with position vector $\mathbf{r}$, due to an assembly of charges q_i at positions $\mathbf{r}_i$, is the sum of the potentials due to each charge separately:

$$\phi(\mathbf{r}) = \frac{1}{4\pi\varepsilon_0} \sum_i \frac{q_i}{|\mathbf{r} - \mathbf{r}_i|}. \tag{1.22}$$

To find the potential due to a continuous charge distribution, the summation in Equation (1.22) is replaced by a volume integral. As with the field, we distinguish between atomic and macroscopic potentials. By comparison with Equations (1.5) and (1.8) for $\mathbf{E}_{\text{atomic}}$ and $\mathbf{E}$, the atomic potential is

$$\phi_{\text{atomic}}(\mathbf{r}) = \frac{1}{4\pi\varepsilon_0} \int_{\text{all space}} \frac{\rho_{\text{atomic}}(\mathbf{r}')\, d\tau'}{|\mathbf{r} - \mathbf{r}'|}, \tag{1.23}$$

and the macroscopic potential is

$$\phi(\mathbf{r}) = \frac{1}{4\pi\varepsilon_0} \int_{\text{all space}} \frac{\rho(\mathbf{r}')\, d\tau'}{|\mathbf{r} - \mathbf{r}'|} + \frac{1}{4\pi\varepsilon_0} \int_{\text{all surfaces}} \frac{\sigma(\mathbf{r}')\, dS'}{|\mathbf{r} - \mathbf{r}'|}. \tag{1.24}$$

Around a point charge $+q$ located at the origin, the potential is $\phi = q/4\pi\varepsilon_0 r$. Spheres centred on the origin are surfaces of constant potential, or *equipotential surfaces.* The lines of force of the electric field are directed outwards from the point charge, and pass perpendicularly through the equipotential surfaces. Lines of force and equipotentials are always perpendicular to one another. To see that this is so, imagine a set of coordinates with origin at the point O on the equipotential surface shown shaded in Figure 1.19. The x and y axes are tangential to the surface at O, and near O, the potential ϕ is then constant in the x

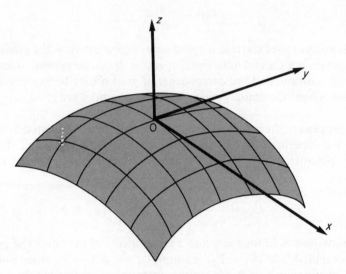

Figure 1.19. The electric field is always normal to equipotential surfaces.

and y directions. The field is the vector with components $-(\partial\phi/\partial x)$, $-(\partial\phi/\partial y)$ and $-(\partial\phi/\partial z)$, and at O the x and y components are zero. The field is therefore in the z-direction normal to the equipotential plane.

A section through a series of equipotentials constitutes a contour map of the potential in the plane of the section. Lines of force follow the components of the electric field in this plane, and they run in the direction in which the potential decreases most rapidly, down the steepest slope of the contour map. Lines of force and equipotentials on a section through an isolated point charge are shown in Figure 1.20 with a sketch illustrating the variation of potential in the plane of the section.

1.5.3 The dipole potential

The spherical equipotential surfaces and radial field lines around a point charge are very simple, but when more than one charge is present the field pattern becomes much more complicated. As an example, let us calculate the

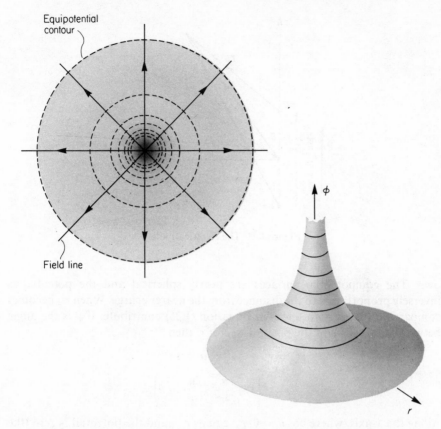

Equipotential
contour

Field line

Figure 1.20. The equipotentials and field lines around a point charge.

potential and electric field of an electric dipole consisting of two equal and
opposite point charges. The dipole in Figure 1.21 has two charges $\pm q$ displaced
from each other in the z-direction, and each at a distance $a/2$ from the origin.
By symmetry the field must have the same pattern in any plane containing the
z-axis, but to be definite we shall work in the x–z plane. The potential at the
point P lying in this plane with position vector **r** is the sum of the potentials of
each charge separately. Writing $r_\pm$ for the distances from $\pm q$ to P, the potential
is

$$\phi = \frac{q}{4\pi\varepsilon_0}\left\{\frac{1}{r_+} - \frac{1}{r_-}\right\}. \tag{1.25}$$

Equipotential surfaces cutting the x–z plane are shown in Figure 1.22, together
with the field lines, everywhere perpendicular to the equipotentials. What is the
magnitude of the potential in different parts of this diagram? Very close to
either of the charges, the potential is almost the same as if the charge were on its

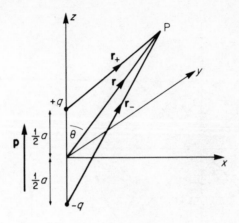

Figure 1.21. The electric dipole.

own. The equipotential surfaces are nearly spherical and the potential is inversely proportional to the distance from the nearer charge. When r_+ becomes comparable to r_-, both terms in Equation (1.25) contribute. If θ is the angle between the z-axis and the position vector $\mathbf{r}$, then

$$r_\pm^2 = r^2 + \tfrac{1}{4}a^2 \mp ar \cos \theta$$

$$= r^2\left(1 + \frac{a^2}{4r^2} \mp \frac{a}{r} \cos \theta\right).$$

Along the x-axis, where $\cos \theta = 0$, r_+ equals r_-, and the potential is zero (the potential is in fact zero everywhere on the x–y plane at $z = 0$).

At some distance from the dipole, where $r \gg a$, the potential can be expanded in powers of a/r.

$$\frac{1}{r_\pm} = (r_\pm^2)^{-1/2} = \frac{1}{r}\left(1 + \frac{a^2}{4r^2} \mp \frac{a}{r} \cos \theta\right)^{-1/2}$$

$$= \frac{1}{r} \pm \frac{a}{2r^2} \cos \theta + \text{higher order terms},$$

writing out only the term to first order in a/r. Substituting in Equation (1.25) we find

$$\phi = \frac{qa \cos \theta}{4\pi\varepsilon_0 r^2} + \text{higher order terms}.$$

The first term by itself is often referred to as the 'dipole potential'. It can be regarded as the potential due to a dipole in which the charges $\pm q$ have approached within an infinitesimal distance of one another while the product qa of charge and separation remains finite. The 'dipole potential' in this sense is

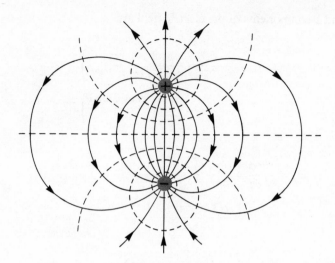

Figure 1.22. Equipotentials and field lines near
an electric dipole.

exactly proportional to $1/r^2$, but it should be remembered that the potential near a dipole consisting of two separated point charges also includes the higher order terms.

The vector **p** drawn from the negative to the positive point charge and of magnitude qa is called the *dipole moment* of the electric dipole. In Figure 1.21 the dipole moment **p** is directed along the z-axis at an angle θ to the position vector **r**. Hence $\mathbf{p} \cdot \mathbf{r} = qar \cos \theta$, and, dropping the higher order terms, the dipole potential can be written as

$$\phi(\mathbf{r}) = \frac{p \cos \theta}{4\pi\varepsilon_0 r^2} = \frac{\mathbf{p} \cdot \mathbf{r}}{4\pi\varepsilon_0 r^3}. \tag{1.26}$$

Notice that this last expression, which is in terms of the vectors **p** and **r**, and the magnitude r, makes no reference to a particular choice of coordinate system. The potential is $\mathbf{p} \cdot \mathbf{r}/4\pi\varepsilon_0 r^3$ even if **p** is not pointing along the z-axis, and at all points **r**, whether or not they lie in the x–z plane. In all directions the potential at some distance from the dipole is varying as $1/r^2$, a more rapid fall-off than the $1/r$ dependence around an isolated point charge. Similarly the electric field around the dipole falls off more rapidly than for a single point charge, and the leading term in the expansion in powers of $1/r$ is proportional to $1/r^3$. We can find the magnitude of the field by calculating the gradient of the potential. In Cartesian coordinates, for a dipole pointing in the z-direction, Equation (1.26) applied to points in the x–z plane has the form

$$\phi = \frac{pz}{4\pi\varepsilon_0(x^2 + z^2)^{3/2}}. \tag{1.27}$$

The x- and z-components of the electric field are

$$E_x = -\frac{\partial \phi}{\partial x} = \frac{p}{4\pi\varepsilon_0}\left(\frac{3xz}{(x^2+z^2)^{5/2}}\right)$$

$$= \frac{p}{4\pi\varepsilon_0}\left(\frac{3\cos\theta\sin\theta}{r^3}\right), \tag{1.28}$$

and

$$E_z = -\frac{\partial \phi}{\partial z} = \frac{p}{4\pi\varepsilon_0}\left\{-\frac{1}{(x^2+z^2)^{3/2}} + \frac{3z^2}{(x^2+z^2)^{5/2}}\right\}$$

$$= \frac{p(3\cos^2\theta - 1)}{4\pi\varepsilon_0 r^3}. \tag{1.29}$$

The dipole potential and field are important because they apply to real atoms as well as to idealized point charges. Outside an atom or molecule such as the hydrogen chloride molecule, in which the centres of mass of the positive charge and the negative charge do not coincide, the leading term in the expansion of the potential in powers of $1/r$ has exactly the same form as Equation (1.26) for the dipole. The higher order terms have different relative magnitudes in the dipole and in the molecule, but at some distance from a molecule only its dipole moment is required in order to specify its contribution to the potential. We shall find in Chapter 2 that the higher order terms can be completely neglected when discussing macroscopic electric fields in matter, and that only the average dipole moment of the constituent atoms and molecules needs to be known.

1.5.4 Energy changes associated with the atomic field

The electrostatic potential has very large values near any atomic nucleus. Large amounts of potential energy can therefore be released when the charge carried by a nucleus is redistributed. Such a redistribution occurs in the fission of uranium. The nucleus of the uranium atom splits into two fragments of roughly equal size, which are almost at rest immediately after fission. Both fragments, however, carry a positive charge, and they fly apart because of their electrostatic repulsion. When the fragments are separated by a distance large compared with their radii, they can both be regarded as point charges. If each fragment carries a charge Z times the electronic charge, then the potential energy of the pair of fragments at separation r is $(Ze)^2/4\pi\varepsilon_0 r$. It remains quite a good approximation to represent the fragments by point charges even when they are almost touching, although the potential due to one fragment is then by no means constant over the whole extent of the other. If the uranium atom splits into two equal fragments, each fragment carries a charge of $46e$, and each has a radius of roughly 8×10^{-15} m. When touching, their centres are separated by $1.6 \times$

10^{-14} m, and their potential energy is approximately

$$\frac{(Ze)^2}{4\pi\varepsilon_0 r} = \frac{(46 \times 1.6 \times 10^{-19})^2}{4\pi \times 8.85 \times 10^{-12} \times 1.6 \times 10^{-14}}$$

$$= 3.0 \times 10^{-11} \text{ joules.}$$

The joule is an inconveniently large unit of energy when one is discussing single atoms, and in atomic and nuclear physics the usual unit of energy is the *electron volt*. One electron volt (abbreviated as eV) is the kinetic energy gained by an electron in accelerating through a potential difference of one volt. Since the electronic charge is 1.6×10^{-19} coulombs,

$$1 \text{ eV} = 1.6 \times 10^{-19} \text{ joules.}$$

The potential energy of fission fragments in contact, which is equal to their kinetic energy when far apart, is thus about 1.9×10^8 eV. The measured value of kinetic energy for the fragments formed by fission of uranium is 1.7×10^8 eV: the difference from our estimate occurs mainly because the fragments are not in fact spherical. This is a very large amount of energy, as is to be expected since electrostatic forces are very strong near the atomic nucleus. When one gram atom of uranium (about half a pound) undergoes fission in a reactor, the energy released as fragment kinetic energy is $6 \times 10^{23} \times 3 \times 10^{-11} \simeq 2 \times 10^{13}$ joules, enough to operate a 1000 megawatt power station for an hour or so.

Another example of energy changes associated with the atomic field is the energy of chemical reactions, which is derived from the rearrangement of electrons in atoms and molecules. Atomic radii are about ten thousand times as great as nuclear radii, and because electrons are on the average further apart than particles in the nucleus, their electrostatic interactions are much weaker. Chemical energy changes are millions of times smaller than the energy released in fission. An example of a chemical energy change which can be estimated easily is the heat of formation of an ionic crystal such as sodium chloride. The crystal consists of a regular lattice of positively charged sodium ions and negatively charged chlorine ions. The ions are formed by prising away one electron from each sodium atom and transferring it to a chlorine atom. Energy is required to effect the transfer, amounting to 1.5 eV per ion pair. However, more than 1.5 eV is available because of the mutual attraction of the oppositely charged sodium and chlorine ions. Energy is therefore released in formation of the crystal from its constituent atoms; this energy is called the heat of formation. To calculate the heat of formation of the ionic crystal we use the same approximation as for the fission fragments, and represent each ion by a point charge at its centre. All that remains to be done is to apply the principle of superposition to find the potential energy of an assembly of many ions.

Imagine an isolated charge q_1. When a second charge q_2 is brought up to a distance r_{12} from the first, the potential energy of the system is $q_1q_2/4\pi\varepsilon_0 r_{12}$. To bring up a third charge q_3 requires work to be done against the fields of both q_1 and q_2. If the final position of q_3 is r_{13} from q_1, and r_{23} from q_2, the additional potential energy is

$$\frac{q_3}{4\pi\varepsilon_0}\left\{\frac{q_1}{r_{13}} + \frac{q_2}{r_{23}}\right\}.$$

By building up the whole assembly in this way, a single charge at a time, we see that the total potential energy of the assembly is

$$U = \frac{q_2}{4\pi\varepsilon_0}\left\{\frac{q_1}{r_{12}}\right\} + \frac{q_3}{4\pi\varepsilon_0}\left\{\frac{q_1}{r_{13}} + \frac{q_2}{r_{23}}\right\} + \frac{q_4}{4\pi\varepsilon_0}\left\{\frac{q_1}{r_{14}} + \frac{q_2}{r_{24}} + \frac{q_3}{r_{34}}\right\} + \cdots$$

$$= \frac{1}{4\pi\varepsilon_0}\sum_i q_i \sum_{j<i} \frac{q_j}{r_{ji}}. \tag{1.30}$$

The restriction $j < i$ in the summation makes sure that the interaction between each pair of charges is only counted once. We can express the potential energy in a different way by deliberately counting each interaction twice, and then dividing the answer by two, writing

$$U = \frac{1}{8\pi\varepsilon_0}\sum_i q_i \sum_{j\neq i} \frac{q_j}{r_{ji}}$$

$$= \frac{1}{2}\sum_i q_i\phi_i \tag{1.31}$$

since $\dfrac{1}{4\pi\varepsilon_0}\displaystyle\sum_{j\neq i} \frac{q_j}{r_{ji}}$ is just the potential ϕ_i at the charge q, due to all the other charges q_j.

In the sodium chloride crystal, all the q_i are either $+e$ or $-e$. If the separation between neighbouring sodium and chlorine ions is d, then the potential energy of each neighbouring pair is $-e^2/4\pi\varepsilon_0 d$. Because there are exactly equal numbers of positive and negative ions surrounding a particular ion pair, there is a lot of cancellation in the summation (1.31), and the total potential energy per ion pair becomes

$$U = -\frac{e^2}{4\pi\varepsilon_0 d} \times (\text{constant of order unity}).$$

For sodium chloride, $d = 2.8 \times 10^{-10}$ m, and the constant has the value 1.75, leading to $U = -8.0$ eV. Adding the 1.5 eV needed to transfer an electron from the sodium atom to the chlorine atom, the heat of formation of sodium chloride is found to be 6.5 eV per ion pair, or equivalently, 6.8×10^5 joules per gm molecule. This is within 10 % of the experimentally measured value, and the magnitude of a few electron volts is typical of chemical energy changes[*].

1.5.5 Capacitors

The potential energy in Equation (1.30) depends on the product of charges, and therefore if two distributions of charge are superimposed the total potential energy is not the sum of the potential energies of the two systems separately. This non-linearity means that the potential energy associated with a macroscopic distribution of charge cannot usually be found from the macroscopic charge density and the macroscopic potential, which are simply averages over the corresponding atomic quantities. However, we can at once work out the potential energy of macroscopic charges on conductors in vacuum. Since the macroscopic field is always zero inside a conductor, any energy changes occur because of the movement of charges outside, where there is no distinction between the atomic and the macroscopic field. The only charges are on the surfaces of the conductors, and the summation in Equation (1.31) can be replaced by an integration over surface charge density:

$$U = \frac{1}{2} \int_{\text{all surfaces}} \sigma \, \phi \, dS. \tag{1.32}$$

In electrical circuits, charges are stored on a *capacitor*, which consists of a pair of conductors. When the fields some distance away from the capacitor are negligible, as is usually the case, Gauss' theorem tells us that the total charge carried by the capacitor is zero, and that the two conductors must carry equal and opposite charges. There is no component of electric field lying along the surface of a conductor, since the surface is an equipotential. Suppose the two conductors in a capacitor carry charges $\pm Q$. These charges are distributed with surface charge densities σ_+ and σ_-, which may not be constant over the two conductors. However, the potentials of the conductors are constant, and have values ϕ_+ and ϕ_- say. Substituting in Equation (1.32), the potential energy becomes

$$U = \tfrac{1}{2}\phi_+ \int \sigma_+ \, dS + \tfrac{1}{2}\phi_- \int \sigma_- \, dS.$$

The surface integrals yield simply the total charges $\pm Q$ on the two conductors, and

$$U = \tfrac{1}{2}Q(\phi_+ - \phi_-) = \tfrac{1}{2}QV. \tag{1.33}$$

We have introduced the symbol V for the *potential difference* between the two conductors, to distinguish it from the scalar field ϕ, which is a function of position.

---

[*] The energy of an ionic crystal is discussed in more detail in section 3.8 of Flowers and Mendoza, *Properties of Matter*, Wiley (Manchester Physics 1). Here we are chiefly concerned to establish the relations (1.30) and (1.31) for the potential energy of an assembly of charges, and to find the order of magnitude of the interaction of electronic charges at spacings on an atomic scale.

The expression for the potential energy of the capacitor can be simplified even further. In a given capacitor, Q and V are always proportional to one another, because there is only one way in which surface charge can be distributed so that the macroscopic field is zero everywhere inside the conductors. The pattern of the lines of force between the conductors is determined only by their geometry, and the principle of superposition then ensures that charge and voltage are proportional; if the charge on each conductor is doubled, then the field retains the same pattern, but is everywhere doubled in strength. For any capacitor, we can write

$$Q = CV. \tag{1.34}$$

The constant of proportionality C, which depends only on the geometry of the conductors, is called the *capacitance*. Now the energy stored by the capacitor can be expressed in terms of the constant C and only one of the variables Q or V

$$U = \tfrac{1}{2}CV^2 = \tfrac{1}{2}Q^2/C. \tag{1.35}$$

A simple arrangement of conductors, which is used for example in the tuning circuit of a radio set, is the *parallel plate capacitor*. Two plates, each of area A, are separated by a distance d. When there is a potential difference V between the

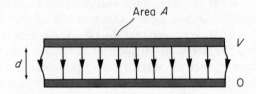

Figure 1.23. A cross-section through a parallel plate capacitor.

plates, field lines pass from one to the other as shown in Figure 1.23. Except at the edges, the magnitude of the field is V/d. Provided that d is small compared with the dimensions of the plates, edge effects are unimportant, and we shall ignore them, assuming the field to be uniform between the plates, and zero elsewhere. It was shown in section 1.4.2 that the field at the surface of a conductor is related to the surface charge density σ by $\sigma = \varepsilon_0 E$, and it follows that there is a uniform charge density $\pm Q/A$ on the inner surface of the plates. Hence

$$Q/A = \varepsilon_0 E = \varepsilon_0 V/d,$$

and the capacitance

$$C = \frac{Q}{V} = \frac{\varepsilon_0 A}{d}. \tag{1.36}$$

The unit of capacitance is the *farad*. The farad is a very large unit, and capacitances are often quoted in *microfarads* or *picofarads*:

$$1 \ \mu\text{F (microfarad)} \equiv 10^{-6} \text{ farads,}$$

$$1 \ \text{pF (picofarad)} \equiv 10^{-12} \text{ farads.}$$

Small parallel plate capacitors have capacitances in the picofarad range. Radio tuning capacitors such as the one illustrated in Figure 1.24 are constructed of stacks of plates which can swivel so that their effective area is variable. Normally,

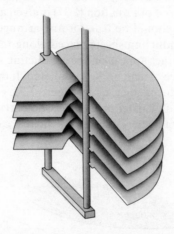

Figure 1.24. A variable tuning capacitor.

of course, the space between the plates is filled with air, but this makes practically no difference, and we can calculate the capacitance as if the capacitor were in vacuum. A typical plate separation is 1 mm, and the maximum area a few hundred square centimetres. At $A = 100 \text{ cm}^2$, the capacitance is

$$C = \frac{\varepsilon_0 A}{d} = \frac{8.85 \times 10^{-12} \times 10^{-2}}{10^{-3}}$$

$$= 88.5 \times 10^{-12} \text{ farads}$$

$$= 88.5 \text{ pF,}$$

and the energy stored when 100 volts is applied across the plates is $U = \frac{1}{2}CV^2 = 4.4 \times 10^{-7}$ joules.

For the parallel plate capacitor, $V = Ed$, and, expressing the energy in terms of the electric field instead of the voltage, we find

$$U = \tfrac{1}{2}CE^2 d^2 = \tfrac{1}{2}\varepsilon_0 E^2 \cdot (Ad).$$

The volume of the region between the plates is (Ad) and the amount of work needed to build up the field **E** can be written as an integral over this volume:

$$U = \int \tfrac{1}{2}\varepsilon_0 E^2 \, d\tau. \tag{1.37}$$

Although we have derived this result by considering the field in a parallel plate capacitor, it is true for any arrangement of conductors in vacuum, provided that the integration extends over the whole of the region where the field is non-zero. A rigorous proof of Equation (1.37) is given at the end of this section, but one can see that it should be true for any arrangement of conductors by imagining very thin conducting sheets placed along the equipotential surfaces of the field. Induced charges appear on the conducting sheets, but the pattern of equipotentials, and hence of field lines, is unaltered. If the equipotential surfaces are chosen close enough together, as in Figure 1.25, each volume element $d\tau$ looks like a parallel plate capacitor, and has potential energy $\tfrac{1}{2}\varepsilon_0 E^2 \, d\tau$. Integration then gives the total energy stored by the real capacitor.

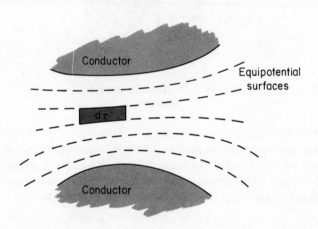

Figure 1.25. The space between conductors can be imagined as being divided into a lot of little parallel plate capacitors like the one occupying the volume $\delta\tau$.

It is sometimes helpful to think of the energy $\tfrac{1}{2}\varepsilon_0 E^2 \, d\tau$ as residing in the volume $d\tau$, that is, to associate an energy density $\tfrac{1}{2}\varepsilon_0 E^2$ with the field. At first this may seem a strange idea, since the whole concept of electric field is an abstract one. However, the energy density has been defined so that it must give the correct value for the total energy, and it is a useful concept, because it has eliminated reference to individual charges.

Remembering that practically realizable fields are limited to about 10^9 volts/metre, the maximum energy density is $\frac{1}{2}\varepsilon_0 \times 10^{18}$, or about 4×10^6 joules/m^3. This is a very small amount compared with the electrostatic energy liberated in fission (2×10^{13} joules from 25 gm of uranium) or in chemical reactions (6×10^5 joules for the heat of formation of a gm mol of salt—about 60 gm). Nevertheless the forces on surface charges can be quite large, and in fact the design of large capacitors is limited by the mechanical stresses they undergo when charged. We can calculate the force between the plates of a charged capacitor indirectly by finding the energy change when they move. If the force attracting the two oppositely charged plates of a parallel plate condenser is F, an amount of work $F\,dx$ is expended in pulling them apart a further distance dx, and the potential energy change is

$$dV = F\,dx.$$

If the area of the plates is A, and they are at separation x, then

$$U = \frac{1}{2}\varepsilon_0 E^2 Ax.$$

The field remains constant when the separation changes, because it is determined only by the charge density on the plates. Hence

$$F = \frac{dU}{dx} = \frac{1}{2}\varepsilon_0 E^2 A \text{ or } \frac{1}{2}\varepsilon_0 E^2 \text{ per unit area.}$$

When the field at its surface is 10^9 volts/metre, the stress on a conductor is 4×10^6 newtons/sq. metre, or about 40 kg wt. per square centimetre.

★ **1.5.6 Energy stored by a number of charged conductors**

The rigorous proof of Equation (1.37) requires some manipulation with vector operators. Consider a number of conductors 1, 2, 3 ... at potentials $\phi_1, \phi_2, \phi_3 \ldots$, and carrying charges on their surfaces of densities $\sigma_1, \sigma_2, \sigma_3 \ldots$. Such an assembly of conductors is illustrated in Figure 1.26. In general the charge densities will not be constant over each surface. The total potential energy of the system of conductors is

$$U = \frac{1}{2}\int_{S_1} \sigma_1\phi_1\,dS_1 + \frac{1}{2}\int_{S_2} \sigma_2\phi_2\,dS_2 + \frac{1}{2}\int_{S_3} \sigma_3\phi_3\,dS_3 + \cdots.$$

The conductors are surrounded by another surface S_0, sufficiently far away that the field is negligible on S_0. The set of surface integrals for each conductor can be replaced by a single integral over the surface $S = S_0 + S_1 + S_2 + S_3 \ldots$ bounding the volume V where the field is non-zero. The outward normals to V are directed towards the *inside* of each conductor, and since the charge density σ is related to the *outward* field at the surface by $\sigma = \varepsilon_0 E$, the potential energy

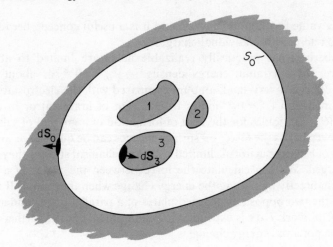

Figure 1.26.

can be written

$$U = -\tfrac{1}{2} \int_{S_1} \varepsilon_0 \phi_1 \mathbf{E}_1 \cdot d\mathbf{S}_1 - \tfrac{1}{2} \int_{S_2} \varepsilon_0 \phi_2 \mathbf{E}_2 \cdot d\mathbf{S}_2 - \cdots$$

$$= -\tfrac{1}{2} \int_S \varepsilon_0 \phi \mathbf{E} \cdot d\mathbf{S}.$$

Using the divergence theorem, the surface integral is converted into a volume integral:

$$U = -\tfrac{1}{2} \int_V \varepsilon_0 \operatorname{div}(\phi \mathbf{E}) \, d\tau.$$

There is an identity

$$\operatorname{div}(\phi \mathbf{E}) \equiv \phi \operatorname{div} \mathbf{E} + \mathbf{E} \cdot \operatorname{grad} \phi$$

$$= \phi \operatorname{div} \mathbf{E} - E^2.$$

The conductors are in vacuum, so there are no charges within V, and $\operatorname{div} \mathbf{E} = 0$. In the volume integral, $\operatorname{div}(\phi \mathbf{E})$ can therefore be replaced by $(-E^2)$, and finally

$$U = \tfrac{1}{2} \int_V \varepsilon_0 E^2 \, d\tau.$$

Notice that the proof holds for any distribution of charges on the conductors, and does not require that the net charge be zero, as it is in the parallel plate capacitor. When there is a net charge on the surfaces $S_1, S_2, S_3 \ldots$, an equal

and opposite charge is always induced on S_0. But by choosing S_0 to be sufficiently far away, the contribution of this induced charge can be made indefinitely small.

PROBLEMS 1

1.1 Three point charges are placed at the vertices of an equilateral triangle with sides one
metre long. The charges are all positive and each of magnitude 1 μC. What is the force
on each charge? What is the magnitude of the electric field (a) at the centre of the
triangle, (b) at the mid-point of one side of the triangle?

1.2 Draw schematically the equipotentials and field lines around two parallel wires each
carrying line charges λ coulomb/m (a) if the charges are of the same sign, (b) if they are
of opposite sign.

1.3 The maximum electric field which can be supported by dry air at atmospheric pressure
is about 10^6 volts/m. What is the maximum potential difference to earth for a conducting sphere of radius 10 cm in air? (Take the distance from earth to the sphere to be
infinite: provided that the distance is large compared to 10 cm, the correction is small.)

1.4 Figure 1.27 shows a cross-section of the cylindrical high-voltage terminal of a van de
Graaff generator, surrounded by an 'intershield' and a pressure vessel, both of which

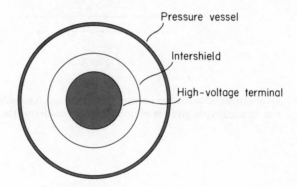

Figure 1.27.

are also cylindrical. The gas in the pressure vessel breaks down in electric fields greater
than 1.6×10^7 volts/m. If the radii of the terminal, intershield and pressure vessel are
1.5 m, 2.5 m and 4 m respectively, what is the highest potential difference that can be
maintained between the terminal and the pressure vessel? (Hint: the intershield must
be maintained at a potential such that breakdown is about to occur on its own outer
surface as well as on the surface of the terminal.)

1.5 Two concentric conducting spherical shells have radii a and b ($a < b$). Work out their
capacitance.

1.6 Estimate the capacitance of two conducting spheres each of radius 1 cm whose centres
are 10 cm apart. Assume that the field between the spheres has the same shape as for
equal and opposite point charges situated at their centres. This is a good approximation for spheres separated by a distance much greater than their radii—the charge
distribution on one sphere is then hardly affected by the field due to the other.

1.7 The uranium nucleus contains 92 protons and has a radius of about 10^{-14} m. Assuming that the positive charge of the protons is uniformly distributed throughout the nucleus, calculate its electrostatic energy in MeV. (Imagine that the nucleus is built up layer by layer. When a sphere of radius r has already been built up, how much work is required to add an additional spherical shell of thickness dr?) If the uranium nucleus splits into two equal fragments each of radius 8×10^{-15} m, how much electrostatic energy is released?

1.8 Two large parallel plates of area A are a distance d apart and are maintained at potentials 0 and V. A third similar plate, carrying a charge q, is isolated from the other two and placed midway between them. What is the potential of this plate?

1.9 Three charges are placed on the z-axis as in Figure 1.28. This arrangement of charges, which is equivalent to two electric dipoles with dipole moment qa a distance a apart, is called an *electric quadrupole*. Calculate the potential at (r, θ, ϕ) for $r \gg a$.

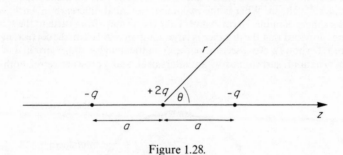

Figure 1.28.

1.10 An electrostatic voltmeter consists of 6 fixed and 5 movable vanes whose size, shape and separation are given in Figure 1.29. By considering the electrostatic energy stored

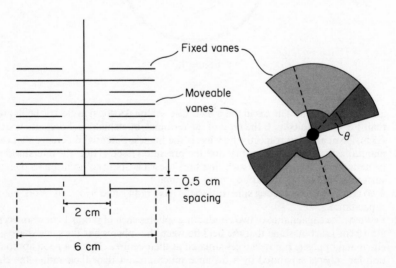

Figure 1.29.

in the voltmeter as a function of θ, find the couple on the movable vanes when a potential difference of 5 kV is maintained between the fixed and movable vanes. (Neglect end effects.)

1.11 The field emission microscope consists of a fine hemispherical tungsten point at the centre of an evacuated sphere; the inside surface of the sphere forms a fluorescent screen. If the screen is maintained at a positive potential with respect to the point, there is a large electric field at the point, and when the magnitude of this field reaches about 10^8 V/m, electrons are drawn from the point even at room temperature; this process is called field emission. The electrons leaving the point follow the field lines, and a magnified image of objects on the point appears on the screen. In a typical microscope the radius of curvature of the point and the screen are 10 nm and 10 cm respectively. What is the magnification of the microscope? What is the minimum screen potential needed to cause field emission?

CHAPTER

2

Dielectrics

The electric field generated by an assembly of many charges can be found by applying the principle of superposition. In practice it is not easy to specify the charge distribution in a piece of matter, since atomic charges are not fixed in position. Any one charge moves about until it is in equilibrium under the action of internal atomic forces and the forces due to external fields. We have already seen that when a metallic conductor is placed in an external field, conduction electrons move until induced charges ensure that the macroscopic field is zero everywhere inside the metal. In an insulator, all the electrons are bound to particular atoms, but external fields still slightly displace the electrons in each atom. The displacement results in the appearance of induced charges which reduce the field in the insulator, though not completely cancelling it. The purpose of this chapter is to discuss the induced charges occurring in insulating materials, or *dielectrics* as they are often called, and to formulate the rules obeyed by the electrostatic field in the presence of dielectrics.

2.1 POLARIZATION

The isolated neutral atom shown in Figure 2.1(*a*) has Z electrons moving around a nucleus carrying a charge $+Ze$. (The number Z, which is called the *atomic number* of the atom, determines its chemical behaviour: each chemical element has a different atomic number, and in the periodic table the elements are listed in the order of increasing atomic number.) In Figure 2.1(*a*) the size of the nucleus is enormously exaggerated; its diameter in a real atom is only about one ten thousandth of the atomic diameter. The electrons are represented as being smeared out in a cloud around the nucleus, since according to quantum

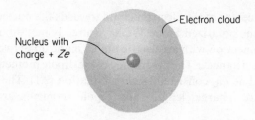

Electron cloud

Nucleus with
charge + Ze

(a) An isolated atom of the element with atomic number Z.

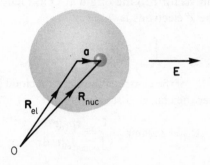

(b) The electric field displaces the electron cloud.

Figure 2.1. Polarization of an atom by an electric field.

mechanical theory this is the most useful way of thinking of the electrons when we are discussing their electrostatic properties.

In the absence of an external field, the nucleus of an atom is at the centre of the electron cloud, where it experiences no net electrostatic force. It is in a position of stable equilibrium, so that if the nucleus were displaced from the centre of the electron cloud a restoring force would arise from the mutual attraction of nucleus and electrons. What happens if the atom is placed in an external electric field pointing towards the right as in Figure 2.1(b)? The field pushes the positively charged nucleus to the right and the negatively charged electrons to the left. As a result, the centre of the electron cloud moves away from the nucleus until the external force is balanced by the internal restoring force. The atom as a whole is electrically neutral, so its centre of mass must remain at rest. Most of the mass of the atom is in the nucleus, which hardly moves at all when the electron cloud is displaced by the external field.

In its new equilibrium position, in which the centre of the electron cloud no longer coincides with the nucleus, the atom is said to be *polarized*.

In Figure 2.1(b) the electron cloud has been displaced a distance a. The vector

$$\mathbf{p} = Ze\mathbf{a} \tag{2.1}$$

pointing from the centre of the electron cloud towards the nucleus, is called the *dipole moment* of the polarized atom. We have already met dipole moments in section 1.5.3 in connection with an electric dipole made up of two point charges $\pm q$ separated by a distance a. The magnitude of the dipole moment of the electric dipole was defined as qa, consistently with Equation (2.1). The dimensions of dipole moment are [charge][length] and dipole moments are measured in coulomb metres.

The dipole moment of a polarized atom can be expressed concisely in terms of its atomic charge density. If the charge density of the electron cloud is $\rho_{el}(\mathbf{r})$ referring the position vector $\mathbf{r}$ to the origin at O in Figure 2.1(b), then the total charge carried by the Z electrons is

$$\int_{atom} \rho_{el}(\mathbf{r})\, d\tau = -Ze.$$

The position vector $\mathbf{R}_{el}$ of the centre of the electron cloud (i.e. the mean position vector, averaged over the charge distribution) is

$$\mathbf{R}_{el} = \text{mean } \mathbf{r} = \frac{\int_{atom} \mathbf{r}\rho_{el}(\mathbf{r})\, d\tau}{\int_{atom} \rho_{el}(\mathbf{r})\, d\tau}$$

$$= -\frac{1}{Ze} \int_{atom} \mathbf{r}\rho_{el}(\mathbf{r})\, d\tau.$$

Similarly, writing the charge density of the nucleus as $\rho_{nuc}(\mathbf{r})$, the position vector $\mathbf{R}_{nuc}$ of the centre of the nucleus is

$$\mathbf{R}_{nuc} = +\frac{1}{Ze} \int_{atom} \mathbf{r}\rho_{nuc}(\mathbf{r})\, d\tau.$$

Substituting in Equation (2.1),

$$\mathbf{p} = Ze\mathbf{a} = Ze(\mathbf{R}_{nuc} - \mathbf{R}_{el})$$

$$= \int_{atom} \mathbf{r}(\rho_{nuc}(\mathbf{r}) + \rho_{el}(\mathbf{r}))\, d\tau \qquad (2.2)$$

$$= \int_{atom} \mathbf{r}\rho_{atomic}(\mathbf{r})\, d\tau.$$

where $\rho_{atomic}(\mathbf{r})$ is the total atomic charge density including the contributions of both the nucleus and the electron cloud.

When an insulator is placed in an electric field, all its atoms become polarized. For simplicity, consider a dielectric material consisting only of the element with atomic number Z, and containing N atoms per unit volume. In a uniform electric field, the centre of the electron cloud in each atom moves the same distance away from the nucleus, in a direction opposite to the electric field. Each atom therefore acquires a dipole moment of magnitude Zea in the direction of the field*. The displacement of electrons leads to the appearance of induced charges on the surface of the dielectric. We will now relate the dipole moments of the individual atoms to the density of the surface charge—the *polarization charge* as it is called.

The appearance of a surface charge density means that a macroscopic volume at the surface carries a net charge. But the macroscopic charge density *inside* a uniformly polarized dielectric remains zero. Imagine a small cubical box inside the dielectric before it is polarized, as shown in section at the top of Figure 2.2. The box is small on the macroscopic scale, but large enough to contain many atoms (the box should really contain far more atoms than have actually been drawn). Some atoms are partly inside and partly outside the box, but on average such a box contains no net charge. When an electric field is switched on, pointing from left to right, each atom becomes polarized, and all the electron clouds move a distance a to the left. The shaded parts of the electron clouds pass through the sides of the box. Exactly as much charge leaves the left-hand side of the box as enters the right-hand side, and the box remains electrically neutral. Notice that this is only true for *uniform* polarization. If the polarization is non-uniform, different amounts of charge pass through opposite faces of the cube; we shall treat this problem in section 2.3.

Even with uniform polarization, neutrality is not preserved at the dielectric surface. The box in the third part of Figure 2.2 encloses an area δS on the left-hand surface of the dielectric. When the dielectric is polarized, the shaded charge enters the box, but no charge leaves it. The volume of the shaded zone is $a\delta S$ and the average charge density of the electrons is $(-NZe)$, so the box acquires a net charge $(-NZea)\delta S$. In other words, the electric field induces a polarization charge of surface density $\sigma_p = -NZea = -Np$. Similarly, the box on the right-hand surface loses electrons and acquires a net positive charge, corresponding to a surface charge density $(+Np)$.

We can write the charge density on both surfaces in the same form by using vector notation. The dipole moment $\mathbf{p}$ is a vector pointing in the same direction as the field $\mathbf{E}$. Referring again to Figure 2.2, we see that if $\delta\mathbf{S}$ is an outward normal to the dielectric surface, in both cases the polarization charge on the area δS can be written as

$$\sigma_p\,\delta S = N\mathbf{p}\cdot\delta\mathbf{S},$$

or

$$\sigma_p\,\delta S = \mathbf{P}\cdot\delta\mathbf{S}, \qquad (2.3)$$

where

$$\mathbf{P} = N\mathbf{p}. \qquad (2.4)$$

---

* The magnitude and direction of the induced dipole moments in some crystalline materials depend on the orientation of the crystal axes with respect to the field direction. Such materials are called *anisotropic* dielectrics, in contrast to *isotropic* dielectrics which have the same response to an electric field whatever their orientation, and for which induced dipole moments always point along the field direction. In this book we shall only consider isotropic dielectrics.

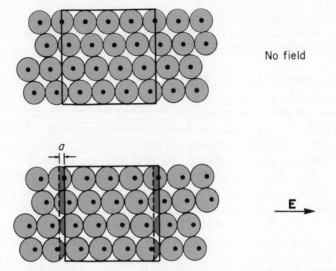

No field

A small volume embedded in the dielectric remains neutral when the atoms are polarized by the uniform field **E**.

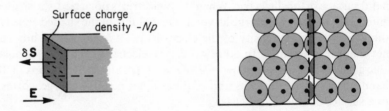

If the field points from left to right, a volume enclosing part of the left-hand surface *gains* electrons and acquires a net negative charge...........

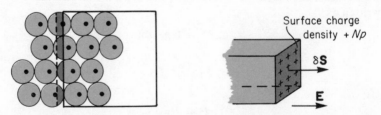

while a volume enclosing part of the right-hand surface *loses* electrons and acquires a net positive charge.

Figure 2.2. The surfaces of a polarized dielectric carry a net charge, even though all the atoms in the dielectric remain electrically neutral.

The vector **P**, which represents the *dipole moment per unit volume* of the dielectric, is called the *polarization*.

Equation (2.3) holds generally, even for surfaces which are not normal to the electric field. In Figure 2.3 the surface element δS is at an angle θ to the field, and its projected area along the field is $\delta S \cos \theta$. The net charge acquired by the box after polarization is

$$NZea \, \delta S \cos \theta = P \, \delta S \cos \theta = \mathbf{P} \cdot \delta \mathbf{S} \text{ as before.}$$

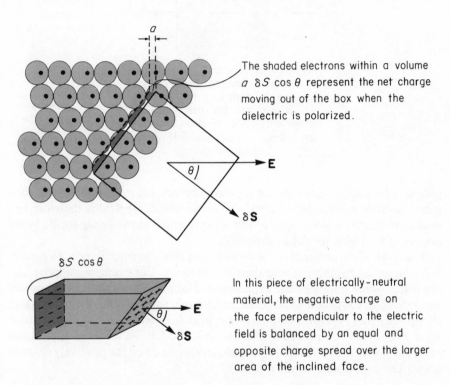

The shaded electrons within a volume $a \, \delta S \cos \theta$ represent the net charge moving out of the box when the dielectric is polarized.

In this piece of electrically-neutral material, the negative charge on the face perpendicular to the electric field is balanced by an equal and opposite charge spread over the larger area of the inclined face.

Figure 2.3. The polarization charge density is reduced on a surface inclined to the electric field.

2.2   RELATIVE PERMITTIVITY AND ELECTRIC SUSCEPTIBILITY

Polarization charges induced on the surface of a dielectric material make a contribution to the macroscopic electric field inside the material. Their contribution is called the 'depolarizing field', because the sign of the induced charge always ensures that the field just inside the dielectric surface is less than the field just outside. This is illustrated in Figure 2.4, which shows a slab of dielectric material in a uniform field. The depolarizing field points away from the positive

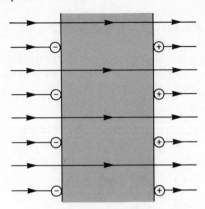

Figure 2.4. Electric field lines in a slab of polarized dielectric. Lines end on negative polarization charges and start on positive ones. The density of lines is reduced inside the dielectric, corresponding to a smaller field inside than outside.

polarization charges towards the negative charges, i.e. from right to left, thus partly cancelling the external field. The net macroscopic field is therefore less inside the dielectric than outside: this is represented in the figure by the lower density of field lines inside the dielectric.

As a result of the presence of the depolarizing field, the capacitance of a capacitor is increased by filling it with dielectric material. For example, suppose that the plates of the parallel plate capacitor in Figure 2.5 have an area A and are separated by a distance d. When the capacitor is in vacuum, and there are surface charge densities $\pm\sigma$ on the plates, the field between them is of magnitude $E_0 = \sigma/\varepsilon_0$. (This was proved in section 1.4.2, and follows from the application of Gauss' law to the surface S which encloses an area δS on the positively charged plate.) The capacitance in vacuum is

$$C_0 = \frac{\text{charge}}{\text{potential difference}} = \frac{A\sigma}{E_0 d} = \frac{\varepsilon_0 A}{d},$$

the result already given in Equation (1.36).

Now keeping the same charge densities $\pm\sigma$ on the conducting plates, fill the space between them with a slab of dielectric. Induced charges appear on the dielectric, causing a depolarizing field, and the magnitude of the macroscopic field is reduced by a factor ε, say, becoming

$$E = E_0/\varepsilon = \sigma/\varepsilon\varepsilon_0. \tag{2.5}$$

It is found experimentally that for fields below the field at which breakdown occurs, the factor ε is a constant depending only on the nature of the dielectric

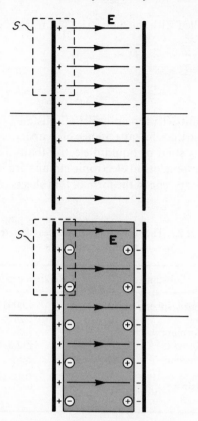

Figure 2.5. When a slab of dielectric is inserted into a parallel plate capacitor, the electric field between the plates is reduced. The diagram shows the field to be halved in the presence of the dielectric, corresponding to $\varepsilon = 2$.

and not on the size and shape of the capacitor. Charge and voltage on the capacitor remain proportioned to one another, and the capacitance has a new value

$$C = \frac{\text{charge}}{\text{potential difference}} = \frac{A\sigma}{Ed} = \frac{\varepsilon\varepsilon_0 A}{d}.$$

The dimensionless constant ε is called the *relative permittivity* of the dielectric material*.

The relative permittivity ε of a dielectric material is usually found by measuring the capacitance C_{air} of a capacitor in air and the capacitance C of the same

* The relative permittivity is often called the *dielectric constant*, especially in books which do not use SI units.

capacitor when it is filled with the dielectric. The ratio of capacitances

$$C/C_{air} = \varepsilon/\varepsilon_{air},$$

and hence

$$\varepsilon = \varepsilon_{air} C/C_{air}.$$

Values of relative permittivity for a number of substances are given in Table 2.1. One sees from the table that the capacitance of a capacitor is several times larger when it is filled with a solid or liquid dielectric than in vacuum. Many of the capacitors used as components in electronic circuits are in fact made with solid dielectrics such as polystyrene in the form of thin sheets coated with conducting films.

Table 2.1. The relative permittivities of materials
at room temperatures.

Substance	Relative permittivity
Air at atmospheric pressure	1.00059
CCl_4	2.24
Transformer oil	2.2
Paraffin wax	2–2.5
Polyethylene	2.3
Nylon	3.5
Porcelain	6
Mica	7

The use of solid dielectrics gives a bonus to the value of the capacitance because the relative permittivity is larger than one, and also enables much smaller spacings to be maintained between the conductors than is possible with air-filled capacitors. By rolling up the thin sheets large capacitances can be squeezed into a small volume. Typically a 1 μF capacitor capable of sustaining 50 volts across its conductors is about 1 cm in diameter and a few cm long. In contrast the capacitance of air-filled capacitors is not usually more than a few hundred pF.

The fact that relative permittivity is found to be constant means that the polarization **P** is proportioned to the field **E**. By making another application of Gauss' law we can see that this is so for the parallel plate capacitor after it has been filled with dielectric. The surface S in Figure 2.5 now encloses a negative polarization charge as well as the positive charge $\sigma \, \delta S$ on the conducting plate. The magnitude of the polarization charge within S is given by Equation (2.3) as $(-P \, \delta S)$, and the total charge within S is therefore $(\sigma - P) \, \delta S$. By Gauss' law,

$$E \, \delta S = \frac{1}{\varepsilon_0}(\sigma - P) \, \delta S,$$

and substituting $\sigma = \varepsilon\varepsilon_0 E$ from Equation (2.5),

$$E\, \delta S = \frac{1}{\varepsilon_0}(\varepsilon\varepsilon_0 E - P)\, \delta S,$$

leading to

$$P = (\varepsilon - 1)\varepsilon_0 E.$$

Since the vectors $\mathbf{P}$ and $\mathbf{E}$ point in the same direction, this equation can be re-written as

$$\mathbf{P} = (\varepsilon - 1)\varepsilon_0 \mathbf{E}$$

or   (2.6)

$$\mathbf{P} = \chi_E \varepsilon_0 \mathbf{E},$$

where the dimensionless constant of proportionality

$$\chi_E = \varepsilon - 1 \qquad (2.7)$$

is called the *electric susceptibility* of the material.

Equation (2.6) always holds for isotropic dielectrics below their breakdown fields and not just in parallel plate capacitors. We can use this equation to work out the dipole moment $\mathbf{p}$ of an individual molecule, since the polarization $\mathbf{P} = N\mathbf{p}$. Let's do this for carbon tetrachloride, which at $20\,^{\circ}\text{C}$ has relative permittivity 2.24 and density $1.60\,\text{gm/cm}^3$. The molecular weight of CCl_4 is 156, so that 156 gm contains 6.02×10^{23} molecules, and the molecular density is

$$N = 6.02 \times 10^{23} \times 1.60/156 = 6.20 \times 10^{21} \text{ molecules/cm}^3,$$

$$\text{i.e. } 6.20 \times 10^{27} \text{ molecules/m}^3.$$

In a field of 10^7 volts/metre (near the highest value obtainable in a liquid before breakdown occurs), the dipole moment of a single molecule is

$$\mathbf{p} = \mathbf{P}/N = \chi_E \varepsilon_0 \mathbf{E}/N$$

(2.8)

$$= \frac{1.24 \times 8.84 \times 10^{-12} \times 10^7}{6.20 \times 10^{27}} \simeq 1.8 \times 10^{-32} \text{ coulomb metres.}$$

There are 74 electrons in each CCl_4 molecule, and if the average electron displacement is a, the dipole moment of the whole molecule is $p = 74\,ae$. In a field of 10^7 volts/metre,

$$a = \frac{p}{74e} \simeq \frac{1.8 \times 10^{-32}}{74 \times 1.6 \times 10^{-19}} \simeq 1.5 \times 10^{-15} \text{ m};$$

only about one hundred thousandth of the diameter of an atom. It is because of the smallness of the displacement that polarization effects are proportional to the magnitude of the electric field. The amount of polarization is determined by the balance between the external forces trying to displace electrons and the internal restoring forces. If electrons were moved away from their equilibrium position through distances comparable with the size of an atom, the restoring forces would certainly include higher powers of the displacement than the first. But the higher order terms are quite negligible for the small displacements caused by polarization, even in the highest external fields. The induced dipole moment of a molecule is always proportional to the magnitude of the electric field in which the molecule is situated.

2.2.1 The local field

For a uniformly polarized dielectric, Equation (2.8) establishes a connection between the average dipole moment of a single molecule and the macroscopic electric field. However, the macroscopic field, which has been smoothed out over a region large enough to contain many molecules, is not necessarily the same as the *local field* polarizing one molecule. The local field E_{local} acting on a particular molecule is the field generated by all charges outside the molecule: in other words E_{local} is the same as the atomic field except that the field generated by the charges making up the molecule in question has been subtracted away.

As well as the fields due to charges outside the dielectric and to polarization charges on its surface, the local field includes contributions from nearby neutral molecules. In section 1.5.3 the potential at a distance r from an electric dipole was expanded as a series in $(1/r)$. The leading term depends on the dipole moment $\mathbf{p}$, and is given in Equation (1.26) as

$$\phi(\mathbf{r}) = \frac{p \cos \theta}{4\pi\varepsilon_0 r^2} = \frac{\mathbf{p} \cdot \mathbf{r}}{4\pi\varepsilon_0 r^3}.$$

Exactly the same expression applies for a molecule with dipole moment $\mathbf{p}$. Since this potential is proportional to $(1/r)^2$, the field associated with the dipole moment is proportioned to $(1/r)^3$, and falls off rapidly with distance. The field outside a molecule also has terms in higher powers of $(1/r)$, with magnitudes which depend on the shape of the electron clouds in the molecule. The higher order fields have an even shorter range than the dipole field, but they are sometimes important for closely neighbouring molecules.

At the site of one particular molecule in a liquid or solid dielectric, the vector sum of the short-range fields of the neighbouring molecules is usually different from zero. The local field E_{local} is then equal to the macroscopic field (i.e. the slowly-varying field due to charges more than a few atomic diameters away) *plus* the contribution from nearby molecules. In a gas, on the other hand, molecules spend most of their time so far apart that the short-range fields are negligible, and the local field E_{local} is almost the same as the macroscopic field $\mathbf{E}$.

When applied to a gas, Equation (2.8) thus relates the induced dipole moment $\mathbf{p}$ of a single molecule to the actual field $\mathbf{E}_{local} = \mathbf{E}$ acting on the molecule:

$$\mathbf{p} = (\chi_E/N)_{gas}\varepsilon_0\mathbf{E}_{local}$$

or

$$\mathbf{p} = \alpha\varepsilon_0\mathbf{E}_{local}, \tag{2.9}$$

where

$$\alpha = (\chi_E/N)_{gas}. \tag{2.10}$$

The dimensionless constant α, called the *molecular polarizability*, is a property of the individual molecule: it measures the resistance of the molecule to displacement of its electron cloud. Equation (2.9) therefore holds with the same value of α even in liquids and solids, for which short-range fields are important, and in which $\mathbf{E}_{local}$ is not the same as the macroscopic field. We shall discuss the problem of estimating the local field in liquids in section 2.2.3, but first must digress to explain the dielectric behaviour of molecules which have a permanent dipole moment, even when they are not placed in an external field.

2.2.2 Polar molecules

All chemical bonds are formed by the sharing of electrons between atoms. Often the sharing is unequal, and some atoms gain electron density at the expense of their neighbours. The alkali halides in the gaseous state, for example, form diatomic molecules which are almost ionic in character—just as the sodium chloride crystal discussed in section 1.5.4 contains a lattice of ions each carrying charge $\pm e$, so the diatomic NaCl molecule consists of an Na^+ ion bound to a Cl^- ion by electrostatic attraction. Such a molecule has a permanent dipole moment of much larger magnitude than the induced dipole moments we have been considering, since the induced dipole moments are caused by moving electrons through a very small fraction of the interatomic spacing.

The electron clouds of the sodium chloride and water molecules are sketched in Figure 2.6. Although the water molecule is not ionic, there is substantial electron transfer from hydrogen to oxygen, and the molecule has a permanent dipole moment almost as large as that of an alkali halide. Yet in the absence of an external electric field, there is no polarization in water, that is, the dipole moment per unit volume in a large-scale sample is zero. This is because thermal motion ensures the random orientation of the molecular dipole moments. However, water does become strongly polarized when an external field is switched on, and the relative permittivity of water at room temperature is 80, much larger than the values listed in Table 2.1 for non-polar molecules. The high relative permittivity occurs because the molecular dipole moments tend to line up along the direction of an external field. To see why this should be so, we shall first investigate the behaviour of a single isolated dipole.

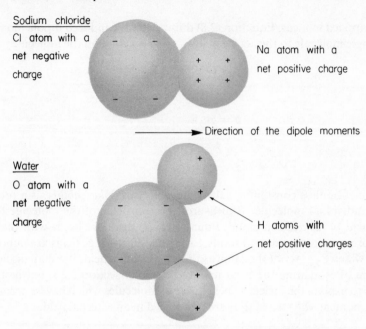

Sodium chloride
Cl atom with a
net negative
charge

Na atom with a
net positive charge

Direction of the dipole moments

Water
O atom with a
net negative
charge

H atoms with
net positive charges

Figure 2.6. Sodium chloride and water molecules both have a permanent
dipole moment.

The dipole in Figure 2.7 consists of charges $\pm q$ separated by a distance a. The dipole moment **p** is of magnitude $p = qa$, and points along the line joining the charges, at an angle θ to a uniform external field **E**. There is no net force on the dipole, but a couple $qaE \sin \theta$ acts in the sense required to line up the dipole parallel to the field. Work must be done against this couple to turn the dipole,

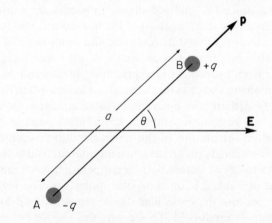

Figure 2.7. A dipole in an external field.

whose potential energy therefore depends on its orientation. The potential energy of the dipole is

$$U = \sum_i q_i \phi_i = q(\phi_B - \phi_A),$$

where ϕ_B and ϕ_A are the potentials of the field $\mathbf{E}$ at the positions of $+q$ and $-q$ respectively. (There should strictly be an additional term to allow for the potential energy of the two charges in one another's fields. But for fixed separation this energy is constant, and we don't have to worry about it when working out what happens when the dipole turns in the external field.) The potential difference $(\phi_B - \phi_A)$ is given by

$$(\phi_B - \phi_A) = -\int_A^B \mathbf{E} \cdot d\mathbf{l} = -aE\cos\theta$$

and

$$U = -qaE\cos\theta = -\mathbf{p} \cdot \mathbf{E}. \tag{2.11}$$

This result holds for polar molecules as well as for the idealized dipole made up of point charges. Taking the origin to be at the centre of a particular molecule, the potential $\phi(\mathbf{r})$ can be expanded in a Taylor series as

$$\phi(\mathbf{r}) = \phi_0 + \mathbf{r} \cdot (\nabla\phi)_0 + \cdots$$
$$= \phi_0 - \mathbf{r} \cdot \mathbf{E}_0 + \cdots,$$

where ϕ_0 and $\mathbf{E}_0$ are the potential and field at the origin. Neglecting the higher order terms in the expansion, the potential energy of the molecule is

$$\int_{mol} \rho_{atomic}\phi\, d\tau = \phi_0 \int_{mol} \rho_{atomic}\, d\tau - \left\{ \int_{mol} \mathbf{r}\rho_{atomic}\, d\tau \right\} \cdot \mathbf{E}$$

$$= -\mathbf{p} \cdot \mathbf{E}.$$

The susceptibility of a gaseous polar dielectric

The local field acting on a molecule in a gas is almost the same as the external field $\mathbf{E}$. For a molecule with dipole moment $\mathbf{p}$, Equation (2.11) shows how the potential energy of the molecule varies with its orientation. The energy has a minimum at $\cos\theta = 1$, when the dipole is aligned along the field. An amount of energy $2pE$ must be expended to reverse the dipole moment and point it in the direction opposite to the field. How much energy is this? If one valence electron in the molecule has been transferred from one atom to another, the resulting dipole moment is roughly

$$e \times (\text{atomic spacing}) \simeq 1.6 \times 10^{-19} \times 10^{-10} = 1.6 \times 10^{-29} \text{ coulomb m.}$$

Typical polar molecules have dipole moments of this order of magnitude, and in a field $\mathbf{E}$ of 10^6 volts/m (about as big as can be achieved in a gaseous dielectric),

$$2pE \simeq 2 \times 1.6 \times 10^{-29} \times 10^6 \text{ joules}$$

$$= 2 \times 10^{-4} \text{ eV}.$$

At ordinary temperatures this is much less than the energy associated with thermal motion. The average kinetic energy $\frac{3}{2}kT$ of each molecule in a gas is about 0.04 eV at room temperature. The gas molecules are continually knocked about in collisions, changing their orientations and directions of motion, and transferring kinetic energy from one to another. Because the potential energy $-\mathbf{p} \cdot \mathbf{E}$ is so small compared with the kinetic energy, the random thermal motion is hardly affected by the presence of the field $\mathbf{E}$, and molecules are still to be found in all orientations. However, there is a slight preponderance of molecules with dipoles lined up along the field, the direction in which their potential energy is a minimum. To calculate the net polarization caused by this alignment, we must know the probability of finding a molecule at any particular orientation to the field.

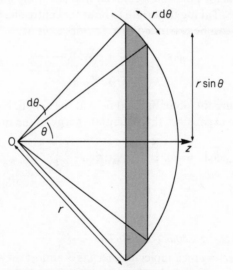

Figure 2.8. The shaded area represents the band between θ and $(\theta + d\theta)$ on the surface of a sphere of radius r. The area of the band is $dS = 2\pi r^2 \sin\theta\, d\theta$, and hence the solid angle $d\Omega = dS/r^2 = 2\pi \sin\theta\, d\theta$.

Before the electric field is switched on, all orientations of the molecules in the gaseous polar dielectric are equally probable. The number of molecules whose dipole moments have directions which lie within a solid angle $d\Omega$ is simply proportional to $d\Omega$. Now the solid angle included by the range of orientations in the cone between angles θ and $(\theta + d\theta)$ to the z-axis is $d\Omega = 2\pi \sin \theta \, d\theta$ (see Figure 2.8). When there is no electric field the density of molecules with dipole moments pointing in directions in the range $d\theta$ is

$$N \, d\Omega/4\pi = \tfrac{1}{2}N \sin \theta \, d\theta,$$

where N is the total number of molecules per unit volume. After switching on an electric field **E** directed along the z-axis, a molecule with dipole moment at θ acquires an additional potential energy $U = -pE \cos \theta$. The probability distribution at thermal equilibrium is now modified by the presence of a Boltzmann factor* $\exp(-U/kT)$. If in the presence of the field, the number of molecules per unit volume with dipole moments in the range $d\theta$ is $N(\theta) \, d\theta$, then

$$N(\theta) \, d\theta \propto \exp(-U/kT) \, d\Omega$$

or

$$N(\theta) \, d\theta = A \exp(-U/kT) \times 2\pi \sin \theta \, d\theta.$$

The constant of proportionality A is determined by the requirement that the total number of molecules per unit volume at all orientations is still N. For polar molecules U/kT is normally very small, and the Boltzmann factor can be expanded in powers of U/kT with the retention of only the first term: $\exp(-U/kT) \simeq 1 - U/kT = 1 + pE \cos \theta/kT$, leading to

$$N(\theta) \, d\theta = 2\pi A \sin \theta \left(1 + \frac{pE \cos \theta}{kT}\right) d\theta.$$

The total number of molecules per unit volume is

$$N = \int_0^\pi N(\theta) \, d\theta = 2\pi A \int_0^\pi \sin \theta \left(1 + \frac{pE \cos \theta}{kT}\right) d\theta = 4\pi A.$$

Hence

$$A = N/4\pi \quad \text{and} \quad N(\theta) = \tfrac{1}{2}N \sin \theta \left(1 + \frac{pE \cos \theta}{kT}\right). \tag{2.12}$$

In Equation (2.12) the term in $\cos \theta$ is small, and there are almost the same numbers of molecules parallel to the field ($\cos \theta = 1$) and opposite to the field ($\cos \theta = -1$). Nevertheless, in a volume large enough to contain many molecules, there is a net dipole moment, which by symmetry must be in the direction of the field. For a dipole moment pointing in the direction θ, the component along the field is $p \cos \theta$. To find the net dipole moment per unit volume, we must add the components along the field for all the molecules, that is, we must integrate

---

* The rôle of the Boltzmann factor in equilibrium probability distributions is discussed in section 4.3.1 of the Manchester Physics series volume on *Properties of Matter*, by Flowers and Mendoza.

$p \cos \theta$ over the probability distribution. The dipole moment per unit volume is

$$P = \int_0^\pi N(\theta)p \cos \theta \, d\theta$$

$$= \frac{N}{2} \int_0^\pi \sin \theta \left(1 + \frac{pE \cos \theta}{kT}\right) \cos \theta \, d\theta \qquad (2.13)$$

$$= \frac{Np^2E}{3kT}.$$

All the molecules also acquire an induced dipole moment, which points along the field whatever may be the direction of the permanent dipole moment. If the molecular polarizability is α, there is an additional induced polarization $N\alpha\varepsilon_0 E$. Finally, we have the result that the net polarization of the gaseous polar dielectric is

$$\mathbf{P} = \left(N\alpha\varepsilon_0 + \frac{Np^2}{3kT}\right)\mathbf{E},$$

and its susceptibility

$$\chi_E = N\left(\alpha + \frac{p^2}{3\varepsilon_0 kT}\right). \qquad (2.14)$$

By measuring susceptibilities at different temperatures, it is possible to distinguish between permanent and induced dipole moments. In Figure 2.9, χ_E is plotted against $1/T$ for HCl gas at a fixed density of $1/22.4$ moles/litre (i.e. the density found at N.T.P.). As expected from Equation (2.14), the plot is a straight line. The intercept at $1/T = 0$ represents the susceptibility due to induced polarization at this density, while from the slope the permanent dipole moment can be deduced to be 3.6×10^{-30} coulomb metres. To achieve this dipole moment, charges $\pm e$ would be needed to be separated by a little more than 2×10^{-11} m, or about one-fifth of a typical atomic diameter.

The susceptibilities and dipole moments of a number of gases are listed in Table 2.2. It is interesting to observe that carbon disulphide has no dipole moment although, like water, it contains two identical atoms bound to a common partner. There is electron transfer across the C—S bonds in carbon

Table 2.2. Dipole moments of polar molecules.

Substance	Susceptibility of gas at N.T.P.	Dipole moments in coulomb metres
HCl	0.0046	3.6×10^{-30}
Ethane (C_2H_6)	0.0015	0
Ethyl alcohol	0.0061	5.7×10^{-30}
H_2O	0.0126	6.2×10^{-30}
CS_2	0.0029	0

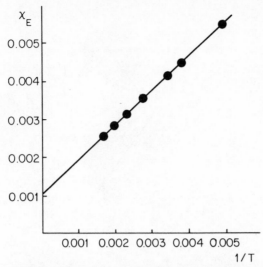

Figure 2.9. The temperature dependence of the susceptibility of HCl gas at a density 1/22.4 mole/litre. (Data from Zahn, *Phys. Rev.*, **24**, 400 (1924)).

disulphide, and dipole moments associated with each bond. But the atoms in the CS_2 molecule lie on a straight line, and the dipole moments exactly cancel one another, whereas in water the two O—H bonds make an angle of 105° (Figure 2.10). By glancing at a table of electric susceptibilities one can learn something about the shape of molecules!

2.2.3 Non-polar liquids

The molecules in a solid or liquid dielectric are so close together that they are affected by one another's short-range fields, and the average field $\mathbf{E}_{local}$ acting on an individual molecule is not the same as the macroscopic field $\mathbf{E}$. However, if the molecules do not interact chemically, they have the same shape as in the gaseous state. The polarizability α of the molecules is unchanged, since it depends only on their internal structure, and each one acquires an average dipole moment $\alpha\varepsilon_0\mathbf{E}_{local}$. If the molecular density is N, the dipole moment per unit volume, i.e. the polarization, is

$$\mathbf{P} = N\alpha\varepsilon_0\mathbf{E}_{local}.$$

Comparing with the definition of susceptibility (given in Equation (2.6))

$$\mathbf{P} = \chi_E\varepsilon_0\mathbf{E},$$

we see that

$$\chi_E = N\alpha\frac{\mathbf{E}_{local}}{\mathbf{E}}. \tag{2.15}$$

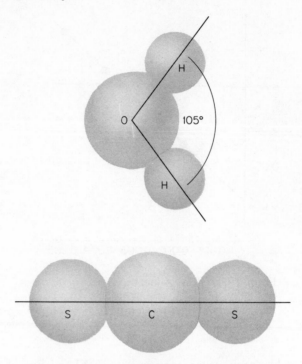

Figure 2.10. The water molecule has a permanent dipole
moment, but the linear CS_2 molecule has none.

Thus if we know $\mathbf{E}_{\text{local}}/\mathbf{E}$ we can calculate the susceptibility of a solid or liquid
dielectric in terms of its molecular polarizability.

Generally it is difficult to estimate $\mathbf{E}_{\text{local}}$ for the fixed molecules in solids.
Liquids are more manageable, because the short-range fields may have a small
average effect as neighbouring molecules continually change their relative
positions and orientations. Even in a liquid, dielectric behaviour is complicated when polar molecules are present, since very large fields are generated by
the permanent dipole moments. (Estimate the field at a distance of a few atomic
diameters away from a water molecule—it is far, far bigger than any macroscopic field!).

But in a non-polar liquid the short-range fields are not so strong, and it is
not too bad an approximation to assume that an individual molecule is surrounded by a continuous medium in which atomic structure has been smeared
out. Like the real liquid, the continuous medium is polarized with a dipole
moment per unit volume $\mathbf{P}$ so that polarization charges appear on any surface
of the medium. Imagine that the molecule in question is removed, leaving a
cavity embedded in the uniformly polarized medium. The field in the cavity is
the sum of the macroscopic field $\mathbf{E}$ and the field generated by polarization charges

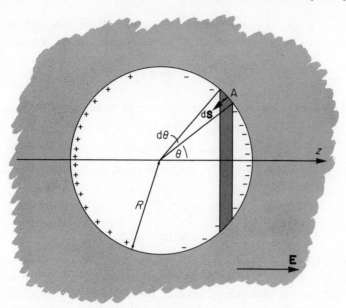

Figure 2.11. The field in a spherical cavity in a uniformly polarized
liquid.

on its surface—and this sum is the field E_{local} which would act on a molecule
filling the cavity. Let us assume that the cavity is a sphere of radius R, and as in
Figure 2.11 take the z-axis to be in the direction of $\mathbf{E}$. At the point A in the figure,
where the normal to the surface of the cavity is at an angle θ to $\mathbf{E}$, the polarization charge on an element of surface dS is $\sigma_p(\theta)\, dS$, say, where

$$\sigma_p(\theta)\, dS = \mathbf{P} \cdot d\mathbf{S} = \varepsilon_0 \chi_E \mathbf{E} \cdot d\mathbf{S}$$

$$= -\varepsilon_0 \chi_E E \cos \theta \, dS.$$

The minus sign arises because the outward normal to the dielectric is the inward
normal to the cavity. By symmetry the net field generated at the centre of the
cavity by the polarization charges is directed along the z-axis. At the centre the
z-component of the field due to the charge $\sigma_p(\theta)\, dS$ is

$$\frac{\sigma_p(\theta)\, dS \cos \theta}{4\pi\varepsilon_0 R^2}.$$

The area of the band on the surface of the cavity lying between θ and $(\theta + d\theta)$
is $2\pi R^2 \sin \theta \, d\theta$, and hence the field due to the band is

$$-\frac{\sigma_p(\theta) \cos \theta}{4\pi\varepsilon_0 R^2} \times 2\pi R^2 \sin \theta \, d\theta = \tfrac{1}{2}\chi_E E \cos^2 \theta \sin \theta \, d\theta.$$

The field at the centre of the cavity generated by all the polarization charge is thus in the same direction as the macroscopic field **E** and is of magnitude

$$\tfrac{1}{2}\chi_E E \int_0^\pi \cos^2 \theta \sin \theta \, d\theta = \tfrac{1}{3}\chi_E E.$$

The field in the cavity is uniform (this is proved in section 3.6 of the next chapter), and so our estimate of the local field is

$$\mathbf{E}_{\text{local}} \simeq (1 + \tfrac{1}{3}\chi_E)\mathbf{E}.$$

Substituting this value for $\mathbf{E}_{\text{local}}$ in Equation (2.15) we find

$$\chi_E = N\alpha \frac{\mathbf{E}_{\text{local}}}{\mathbf{E}}$$

$$\simeq N\alpha(1 + \tfrac{1}{3}\chi_E). \tag{2.16}$$

This approximation, which is known as the Clausius–Mossotti formula, cannot be expected to be very accurate since it is based on such a crude representation of the neighbouring molecules. However, the Clausius–Mossotti formula does demonstrate that polarization in the liquid dielectric causes the local field to be larger than the macroscopic field, and makes a reasonable estimate of the size of the increase.

The molecular polarizability can be derived from the susceptibility of a gas (Equation (2.10)):

$$\alpha = (\chi_E/N)_{\text{gas}}$$

Comparison of the susceptibility of liquids and gases therefore gives a measure of the local field in the liquid, since

$$\frac{\mathbf{E}_{\text{local}}}{\mathbf{E}} = \frac{1}{\alpha}\left(\frac{\chi_E}{N}\right)_{\text{liq}} = \left(\frac{N}{\chi_E}\right)_{\text{gas}} \times \left(\frac{\chi_E}{N}\right)_{\text{liq}}$$

$$= \left(\frac{\rho}{\chi_E}\right)_{\text{gas}}\left(\frac{\chi_E}{\rho}\right)_{\text{liq}},$$

where ρ is the density. Some experimental values are listed in Table 2.3, which compares the value of $\mathbf{E}_{\text{local}}/\mathbf{E}$ derived from this expression with the Clausius–Mossotti prediction $(1 + \tfrac{1}{3}\chi_E)$. The table shows that the Clausius–Mossotti approximation is not bad even when the local field is as much as 50% greater than the macroscopic field.

2.3 MACROSCOPIC FIELDS IN DIELECTRICS

While discussing the atomic basis of dielectric behaviour, it was adequate to restrict attention to uniform polarization. But in practice dielectrics are often found in situations where the polarization is not uniform, either because the

Table 2.3. Electric susceptibilities of gases and liquids.

Substance	Gas			Liquid		E_{local}/E $= \left(\dfrac{\rho}{\chi_{E}}\right)_{\text{gas}} \left(\dfrac{\chi_{E}}{\rho}\right)_{\text{liq}}$	$1 + \frac{1}{3}(\chi_E)_{\text{liq}}$
	$(\chi_E)_{\text{gas}}$	$\rho_{\text{gas}} \times 10^3$	State	$(\chi_E)_{\text{liq}}$	ρ_{liq}		
He	0.0000684	0.179	2.3 K	0.0559	0.147	1.00	1.02
O_2	0.000523	1.43	80 K	0.507	1.19	1.165	1.169
CO_2	0.000985	1.96	100 atm	0.610	0.975	1.24	1.20
CS_2	0.0029	3.40		1.64	1.29	1.48	1.55

ρ is the density in g/cm^3. The data for gases are at atmospheric pressure.

dielectric is non-uniform, or because the electric field varies with position even inside a uniform dielectric. For example, the familiar coaxial cable used to connect a television set to its aerial consists of a central conductor separated from an outer cylindrical conductor by a layer of dielectric material, as shown in Figure 2.12. If a potential difference is set up across the conductors, what is the polarization and the electric field in the dielectric?

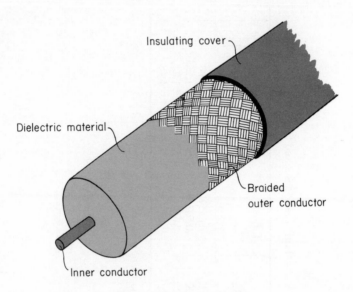

Figure 2.12. A coaxial cable.

To answer questions like this, we need to know how induced polarization charges behave in non-uniform fields. In section 2.3.1 it will be shown that in non-uniform fields, as well as induced charges appearing on the surface of a dielectric, there may also be an induced polarization charge distributed throughout the volume of the dielectric. This looks at first sight to be a very awkward complication. Fortunately, by introducing a new vector field called the *electric displacement* we shall be able to account for the effects of the induced polarization charges *automatically*, without even working out the polarization charge density. The electric displacement vector is discussed in section 2.3.2, and then in section 2.3.3 we go on to explain how electrostatic fields change at the boundaries between different dielectric media.

2.3.1   The volume density of polarization charge

Imagine that the cube of volume $\delta x\,\delta y\,\delta z$ sketched in Figure 2.13 is inside an electrically neutral dielectric. The cube is small on the macroscopic scale, yet still large enough to contain many atoms. Suppose that the cube is given a

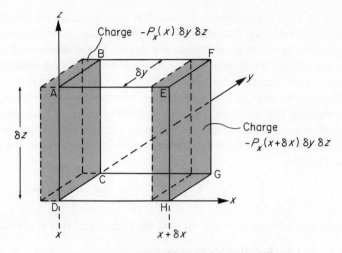

Figure 2.13. In a non-uniformly polarized dielectric, different amounts of charge pass through opposite faces of a cube embedded in the dielectric.

polarization **P** which has a component $P_x(x)$ in the x-direction at the face ABCD. Assume for the moment that $P_x(x)$ is positive. When the atoms are polarized, electrons move leftwards out of the box at ABCD, and (using the same argument as in the case of uniform polarization discussed in section 2.1) they carry a charge $-P_x(x)\,\delta y\,\delta z$. If the polarization has a slightly different value $P_x(x + \delta x)$ at the face EFGH, a charge $-P_x(x + \delta x)\,\delta y\,\delta z$ enters the cube here. The net charge entering the cube through ABCD and EFGH is thus*

$$-\{P_x(x + \delta x) - P_x(x)\}\,\delta y\,\delta z = -\frac{\partial P_x}{\delta x}\,\delta x\,\delta y\,\delta z.$$

The same expression holds if $P_x(x)$ is negative, since the electrons then move in the opposite direction. There may be similar contributions from components of **P** in the y and z directions, and the total polarization charge acquired by the cube is

$$\left\{-\frac{\partial P_x}{\partial x} - \frac{\partial P_y}{\partial y} - \frac{\partial P_z}{\partial z}\right\}\,\delta x\,\delta y\,\delta z.$$

In other words there is a macroscopic polarization charge density

$$\rho_p = -\frac{\partial P_x}{\partial x} - \frac{\partial P_y}{\partial y} - \frac{\partial P_z}{\partial z} = -\,\text{div}\,\mathbf{P}. \tag{2.17}$$

* It is not rigorously correct to replace $P_x(x + \delta x) - P_x(x)$ by $(\partial P_x/\partial x)\,\delta x$, because δx can never be allowed to tend to zero—it must always be large compared with atomic sizes. There are corrections to Equation (2.15) which are small if the fractional change in polarization is small over a distance of one atomic spacing. In practice these corrections are completely negligible.

Equation (2.17) is similar to Gauss' law which relates the electric field to the total charge density. Note however the minus sign, which arises because of the convention that dipole moments are drawn in the direction from negative to positive charge. Field lines of the vector **P** terminate on polarization charges, starting from negative charges and ending on positive charges.

The appearance of the polarization charge density is consistent with the requirement that the total polarization charge carried by a piece of dielectric should be zero. The surface charge on an element dS on the surface of the dielectric is $\mathbf{P} \cdot d\mathbf{S}$, and, summing surface and volume charges, the total polarization charge is

$$\int_S \sigma_p \, dS + \int_V \rho_p \, d\tau = \int_S \mathbf{P} \cdot d\mathbf{S} - \int_V \operatorname{div} \mathbf{P} \, d\tau.$$

The divergence theorem tells us that the right-hand side is identically zero.

2.3.2   The electric displacement vector

In the general case, dielectric material may not be electrically neutral even when unpolarized. If the dielectric carries a charge density ρ_f of 'free' charges, representing a net surplus or deficit of electrons in the atoms of the dielectric then the total charge density is

$$\rho = \rho_f + \rho_p. \tag{2.18}$$

It must be emphasized that both terms in this equation represent real physical charge. Nevertheless, we shall find the distinction between free charge and polarization charge to be a useful one, particularly when we come to consider the energy associated with fields in dielectric material. The macroscopic electric field **E** is related to the total charge density, and in matter Gauss' law becomes

$$\operatorname{div} \mathbf{E} = \rho/\varepsilon_0 = \frac{1}{\varepsilon_0}(\rho_f + \rho_p)$$

$$= \frac{1}{\varepsilon_0}(\rho_f - \operatorname{div} \mathbf{P}).$$

Rearranging this equation we can write

$$\varepsilon_0 \operatorname{div} \mathbf{E} + \operatorname{div} \mathbf{P} = \rho_f.$$

It can easily be verified, by writing out the equation in full in terms of the components of the vectors, that

$$\varepsilon_0 \operatorname{div} \mathbf{E} + \operatorname{div} \mathbf{P} = \operatorname{div}(\varepsilon_0 \mathbf{E} + \mathbf{P});$$

hence

$$\operatorname{div}(\varepsilon_0 \mathbf{E} + \mathbf{P}) = \rho_f,$$

or

$$\boxed{\text{div } \mathbf{D} = \rho_f.}$$

(2.19)

Here we have introduced the new vector field

$$\boxed{\mathbf{D(r)} = \varepsilon_0 \mathbf{E(r)} + \mathbf{P(r)}.}$$

(2.20)

The vector field $\mathbf{D}$ has the same dimensions as $\mathbf{P}$, namely dipole moment per unit volume, and it is called the *electric displacement*. Since $\mathbf{P} = \chi_E \varepsilon_0 \mathbf{E}$, the electric displacement can also be written as

$$\mathbf{D} = (1 + \chi_E)\varepsilon_0 \mathbf{E}$$
$$= \varepsilon\varepsilon_0 \mathbf{E}.$$

(2.21)

Equation (2.19) is still really Gauss' law, now modified in such a way that the effects of polarization charge are automatically included. (Gauss' law looks a bit different in that it contains a factor $1/\varepsilon_0$ on the right-hand side. This is simply because we have chosen to give $\mathbf{D}$ and $\mathbf{E}$ different dimensions.) Gauss' law can be expressed alternatively in the integral form $\int_S \mathbf{E} \cdot d\mathbf{S} = (1/\varepsilon_0)\int_V \rho \, d\tau$. Similarly there is an integral relation for $\mathbf{D}$ which is equivalent to Equation (2.19). The divergence theorem tells us that

$$\int_S \mathbf{D} \cdot d\mathbf{S} \equiv \int_V \text{div } \mathbf{D} \, d\tau.$$

Substituting for div $\mathbf{D}$ from Equation (2.19)

$$\boxed{\int_S \mathbf{D} \cdot d\mathbf{S} = \int_V \rho_f \, d\tau}$$

(2.22)

or in words,

Flux of $\mathbf{D}$ out of a closed surface S = Total free charge enclosed within S.

Unlike the electric field $\mathbf{E}$ (which is the force acting on unit charge) or the polarization $\mathbf{P}$ (the dipole moment per unit volume), the electric displacement $\mathbf{D}$ has no clear physical meaning. The only reason for introducing it is that it enables one to calculate fields in the presence of dielectrics without first having to know the distribution of polarization charges.

Let us illustrate this by applying Equation (2.22) to the example of the co-axial cable shown in section in Figure 2.14. The annular region between the

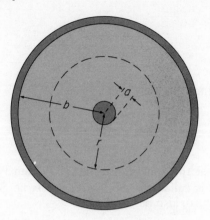

Figure 2.14. Cross-section of a coaxial cable.

inner and outer conductors is filled with a uniform dielectric material of relative permittivity ε. The two conductors make up a capacitor, and when a
potential difference is maintained between them, induced charges appear on the
conductor surfaces. Suppose that there is a surface charge density σ on the
inner conductor, which has a radius a, so that the total charge on a one metre
length is $Q = 2\pi a\sigma$. The electric displacement is obviously cylindrically symmetrical, and its magnitude depends only on the distance r from the axis. The
flux of $\mathbf{D}$ out of a cylindrical surface S of radius r and length one metre is

$$\int_S \mathbf{D} \cdot d\mathbf{S} = 2\pi r D(r) = 2\pi a\sigma.$$

Hence

$$D(r) = a\sigma/r,$$

and from equation (2.21)

$$E(r) = \frac{a\sigma}{\varepsilon\varepsilon_0 r}.$$

The potential difference between the inner and outer conductor is

$$V = -\int_b^a \mathbf{E} \cdot d\mathbf{l} = -\int_b^a \frac{a\sigma}{\varepsilon\varepsilon_0 r} \, dr = \frac{a\sigma}{\varepsilon\varepsilon_0} \ln\left(\frac{b}{a}\right),$$

and the capacitance per unit length of the cable is

$$C = \frac{Q}{V} = 2\pi\varepsilon\varepsilon_0/\ln\,(b/a) \simeq 75 \text{ pF/metre}$$

for a typical coaxial cable with polythene as the dielectric ($\varepsilon = 2.3$) and $a =$
0.5 mm, $b = 2.5$ mm. The capacitance per unit length is an important property

of a coaxial cable, which one needs to know when discussing its performance in the transmission of high-frequency signals.

The electric field in the cable varies as $1/r$, just as it would do if there were no dielectric between the conductors. Whenever the relative permittivity is the same throughout the whole of the region where the field is non-zero, Equation (2.19) simplifies to

$$\text{div } \mathbf{E} = \rho_f/\varepsilon\varepsilon_0.$$

Apart from the factor ε, this is the equation satisfied by the same distribution of free charge ρ_f in vacuum. The presence of the dielectric does not alter the shape of the electric field, but simply reduces its magnitude by the factor ε.

2.3.3   Boundary conditions for D and E

When the space near a set of charges contains dielectric, but is not completely filled by a single uniform dielectric material, then the electric field no longer has the same form as in vacuum. Suppose, for example, that a slab of dielectric, which carries no free charges, is placed in a parallel plate capacitor, but that the thickness b of the slab is less than the distance a between the capacitor plates. When there is a potential difference between the plates, polarization charges appear at the free surface of the dielectric. There is therefore a discontinuity in the electric field: by Gauss' law, the fluxes of the field $\mathbf{E}$ through opposite faces of the box S shown in Figure 2.15 are not equal, since the box encloses polarization charges. But the flux law for the electric displacement $\mathbf{D}$ (Equation

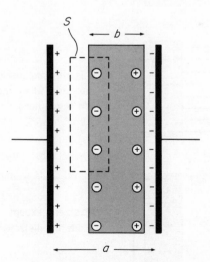

Figure 2.15. There is no change in **D** across the dielectric surface enclosed by the box S.

(2.22)) involved only *free charge*. There is no free charge inside the box, and the fluxes of **D** across its opposite faces are equal. In other words, the vector **D** obeys the simple *boundary condition* that **D** is continuous across a surface perpendicular to **D**, if there is no free charge on the surface.

It is now easy to work out the electric field distribution in the capacitor and hence derive its capacitance. The electric displacement **D** is uniform throughout the capacitor. In the region where there is no dielectric material, the electric field is $E_1 = D/\varepsilon_0$. The dielectric has a uniform polarization **P**, and inside it the electric field is $E_2 = D/\varepsilon\varepsilon_0$. The potential difference between the plates is thus

$$V = E_1(a - b) + E_2 b = E_1\left(a - b + \frac{b}{\varepsilon}\right).$$

The charge densities on the conducting plates are $\pm\varepsilon_0 E_1$, and if their area is A, the plates carry total charges $\pm Q = \pm\varepsilon_0 E_1 A$. The capacitance is therefore $C = Q/V = \varepsilon_0 A/(a - b + b/\varepsilon)$. The fields **D**, **E** and **P** in the capacitor are illustrated in Figure 2.16.

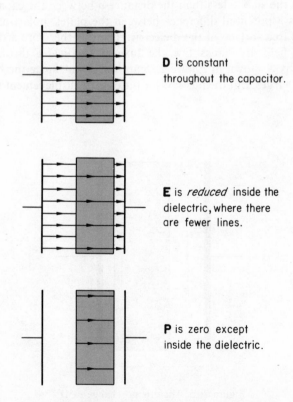

D is constant throughout the capacitor.

E is *reduced* inside the dielectric, where there are fewer lines.

P is zero except inside the dielectric.

Figure 2.16. Lines of **D**, **E** and **P** in a parallel plate capacitor which is partially filled with dielectric material.

The parallel plate capacitor is a specially simple case, but we can apply similar arguments to find the boundary conditions at any dielectric boundary. Imagine a disc enclosing part of the boundary surface between two media of relative permittivity ε_1 and ε_2, as indicated in Figure 2.17. The thickness of the disc is allowed to tend to zero, so that the only contributions to the outward flux

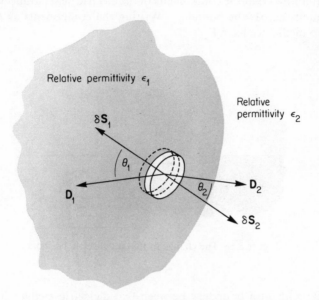

Relative permittivity ϵ_1

Relative permittivity ϵ_2

Figure 2.17. The change of **D** at a dielectric boundary.

of **D** from the disc come from its flat faces. These faces have areas represented by the outward normals $\delta\mathbf{S}_1$ and $\delta\mathbf{S}_2$. If there is no free charge inside the disc, the total outward flux of **D** is zero:

$$\mathbf{D}_1 \cdot \delta\mathbf{S}_1 + \mathbf{D}_2 \cdot \delta\mathbf{S}_2 = 0.$$

Writing $D_{1\perp}$ and $D_{2\perp}$ for the components of $\mathbf{D}_1$ and $\mathbf{D}_2$ perpendicular to the boundary, and bearing in mind that $\delta\mathbf{S}_1$ and $\delta\mathbf{S}_2$ are pointing in opposite directions, it follows that $D_{1\perp} = D_{2\perp}$, i.e. $D_\perp$ is continuous across the boundary. (At a boundary carrying a surface charge density σ_f, there is a discontinuity in $D_\perp$ of magnitude σ_f.)

A different boundary condition applies to the components of the fields parallel to the boundary; in this case it is the component of the electric field which is continuous. Consider the rectangular loop shown in Figure 2.18 with sides $\delta\mathbf{l}_1$ and $\delta\mathbf{l}_2$ lying close to the boundary between media with relative permittivities ε_1 and ε_2. Since the electric field is conservative, no work is done in taking a test charge around the loop. Making the short sides vanishingly small, the work done on unit charge is

$$\oint \mathbf{E} \cdot \mathbf{dl} = -\mathbf{E}_1 \cdot \delta \mathbf{l}_1 - \mathbf{E}_2 \cdot \delta \mathbf{l}_2 = 0.$$

This is true for any $\delta \mathbf{l}_1$ parallel to the boundary, whatever its direction along the surface. It follows that the components of the electric field parallel to the surface is continuous across the boundary. Writing the components as $E_{1\parallel}$ and $E_{2\parallel}$ in the two media, we have $E_{1\parallel} = E_{2\parallel}$.

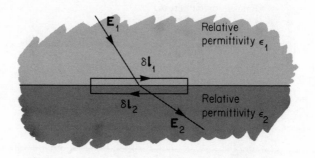

Figure 2.18. The change of **E** at a dielectric boundary.

Summarizing, at a boundary between two dielectric media which carry no surface charges,

$$D_{1\perp} = D_{2\perp} \quad \text{or} \quad D_\perp \text{ is continuous,} \tag{2.23}$$

$$E_{1\parallel} = E_{2\parallel} \quad \text{or} \quad E_\parallel \text{ is continuous.} \tag{2.24}$$

When a boundary is neither parallel nor perpendicular to the electric field, the direction of the field changes across the boundary. Labelling the angles to the normal as in Figure 2.17, the boundary conditions become

$$D_\perp \text{ continuous:} \quad \varepsilon_1 E_1 \cos \theta_1 = \varepsilon_2 E_2 \cos \theta_2,$$

$$E_\parallel \text{ continuous:} \quad E_1 \sin \theta_1 = E_2 \sin \theta_2.$$

$$\text{Hence} \quad \varepsilon_1 \cot \theta_1 = \varepsilon_2 \cot \theta_2.$$

This equation is reminiscent of Snell's law in optics. The resemblance is no accident. Optical refraction can be explained in terms of the boundary conditions applying to changing electric and magnetic fields; in Chapter 11 we shall derive Snell's law, and relate refractive index to the relative permittivity.

2.4   ENERGY IN THE PRESENCE OF DIELECTRICS

What is the potential energy of charges in the presence of dielectric material? This is a rather tricky problem, and before dealing with a general system of charges, we shall start by working out the energy stored by a single capacitor. The capacitor is initially uncharged, and a potential difference V is built up between the plates by transferring a charge $Q = CV$ from one to the other. Energy—supplied for example by a battery—is needed to move the charge, and at the moment when the potential difference is V', energy $V' \, dQ'$ is required to transfer a further charge dQ'. Now $dQ' = C \, dV'$, and the total energy stored is

$$U = \int_0^V V' \, dQ' = \int_0^V CV' \, dV' = \tfrac{1}{2}CV^2 = \tfrac{1}{2}QV. \qquad (2.25)$$

This expression has the same form as Equation (1.33) for the energy of a vacuum capacitor. The actual value of the energy is of course affected by the presence of dielectric material, since C depends on the relative permittivity. Nevertheless, only Q, the free charge on the capacitor plate, appears in Equation (2.25) which contains no explicit reference to polarization charges. This is because it is only the free charge which is directly moved across the potential difference by external forces.

In a parallel plate capacitor of area A and plate separation d, the energy stored when it is filled with dielectric is $\tfrac{1}{2}CV^2 = \tfrac{1}{2}\varepsilon_0\varepsilon(A/d)V^2$. The volume of the capacitor is (Ad), and therefore the energy density is

$$\tfrac{1}{2}\varepsilon_0\varepsilon\left(\frac{V}{d}\right)^2 = \tfrac{1}{2}\varepsilon_0\varepsilon E^2 = \tfrac{1}{2}DE.$$

Just as with fields in vacuum, we can regard the energy density $\tfrac{1}{2}DE$ as residing in the field. As before, it is plausible that the energy density is $\tfrac{1}{2}DE$ in any electrostatic field, because we can imagine that the field is split up into a large number of parallel plate capacitors by thin conductors placed along closely spaced equipotentials. The total electrostatic energy stored in a volume V is

$$U = \tfrac{1}{2}\int_V \mathbf{D} \cdot \mathbf{E} \, d\tau. \qquad (2.26)$$

This result is proved rigorously in the next section.

2.4.1   Some further remarks about energy

To find the energy stored by a general system of charges, we use the same technique as was applied to the capacitor in the previous section. We begin with unpolarized dielectric with no free charges, and then assemble the free charges by bringing them up from infinity. External work is done in bringing up the free charges, which automatically accounts for the potential energy of any polarization charges which may appear. The energy required to assemble a set of charges is $U = \frac{1}{2} \sum_i q_i \phi_i$ (Equation (1.31)), where ϕ_i is the potential at the position of q_i. If the free charges are distributed with surface charge density σ_f on a number of conducting surfaces, and with volume charge density ρ_f in the region V bounded by the conductors, the sum is replaced by an integral, and

$$U = \frac{1}{2} \int_V \rho_f \phi \, d\tau + \frac{1}{2} \int_S \sigma_f \phi \, dS. \tag{2.27}$$

Here, as in the analogous problem of the energy density in vacuum which was treated in section 1.4.4, S includes all conducting surfaces on which there are surface charges and also a distant surface enclosing the whole system.

On the conducting surfaces the outward normal to S is the inward normal to the conductor and by considering the flux of $\mathbf{D}$ out of the small box (illustrated in Figure 2.19) on the surface we find $\sigma_f \, \delta S = -\mathbf{D} \cdot \delta \mathbf{S}$. The volume density of free charge is given by $\rho_f = \operatorname{div} \mathbf{D}$. Substituting in the integrals, the potential energy becomes

$$U = \frac{1}{2} \int_V \phi \operatorname{div} \mathbf{D} \, d\tau - \int_S \phi \mathbf{D} \cdot d\mathbf{S}$$

$$= \frac{1}{2} \int_V \phi \operatorname{div} \mathbf{D} \, d\tau - \frac{1}{2} \int_V \operatorname{div} (\phi \mathbf{D}) \, d\tau.$$

Using the vector identity $\operatorname{div} (\phi \mathbf{D}) \equiv \phi \operatorname{div} \mathbf{D} + \mathbf{D} \cdot \operatorname{grad} \phi$, and putting $\mathbf{E} = -\operatorname{grad} \phi$, we find that the potential energy is indeed given by Equation (2.26):

$$U = \frac{1}{2} \int_V \mathbf{D} \cdot \mathbf{E} \, d\tau.$$

A word of caution must be added here. In assembling the system of charges piece by piece to arrive at Equation (2.27), we have tacitly assumed the polarization to be proportional to the electric field. Except in some very uncommon materials this is always true, but where non-linear effects do occur, the external work done may depend on the precise path by which the final state of the system is approached. Energy is of course always conserved, but sometimes it is not possible to make a unique distinction between potential energy and heat. We shall meet the equivalent problem in magnetism, where non-linear materials are of practical importance.

Even with materials in which polarization is proportional to the field, when calculating the potential energy one must be careful to follow the recipe given above, and assemble the free charges when the matter is already in place.

Suppose that we reverse the order for the parallel plate capacitor, first charging it up, and then afterwards inserting the dielectric while maintaining the potential

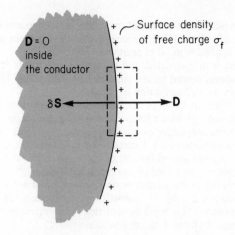

Figure 2.19. At a conducting surface, **D** is related to the surface density of free charge.

difference constant across the plates. In vacuo an amount of work $\frac{1}{2}CV^2 = \frac{1}{2}\varepsilon_0 AV^2/d$ is done in raising the potential difference across the plates to V. When the dielectric is inserted, an additional charge $(\varepsilon - 1)\varepsilon_0 AV/d$ must be added to keep the potential difference at the value V, requiring additional work $(\varepsilon - 1)\varepsilon_0 AV^2/d$.

The total amount of external work is

$$\tfrac{1}{2}\varepsilon_0 AV^2/d + (\varepsilon - 1)\varepsilon_0 AV^2/d = \tfrac{1}{2}\varepsilon\varepsilon_0 AV^2/d + \tfrac{1}{2}(\varepsilon - 1)\varepsilon_0 AV^2/d,$$

more than the amount $\frac{1}{2}\varepsilon\varepsilon_0 AV^2/d$ expended when the dielectric was in place before the free charges were moved. The difference $\frac{1}{2}(\varepsilon - 1)\varepsilon_0 AV^2/d$ occurs because there is a force on the dielectric, and work is done on it as it enters the capacitor. To be definite, let us assume that the capacitor plates are square and of side a. When the dielectric is inserted a distance x into the capacitor, extra work $\frac{1}{2}(\varepsilon - 1)(\varepsilon_0 aV^2/d)\,dx$ is done on the dielectric if it is moved a further distance dx, and the force is therefore $\frac{1}{2}(\varepsilon - 1)\varepsilon_0 aV^2/d$. The force is independent of x because it only acts at the edge of the plates, where the field is non-uniform. The same argument applied to a general field shows that the force per unit volume acting on a piece of dielectric is $\frac{1}{2}\operatorname{grad}(\varepsilon_0 E^2 - \mathbf{D}\cdot\mathbf{E})$. This is a force which attracts dielectric material towards regions where the electric field is high.

PROBLEMS 2

2.1 Two identical capacitors are connected in parallel, charged by a battery of voltage V, and then isolated from the battery. One of the capacitors is then filled with a material of relative permittivity ε. What is the final potential difference across its plates?

2.2 A capacitor is placed in a tank, and the following sequence of events occur:
  (i) a battery of voltage V is connected across the capacitor terminals;
  (ii) the tank is filled with oil of relative permittivity ε;
  (iii) the battery is disconnected, and the tank is afterwards drained.
What is the energy stored by the capacitor at each stage? How do you reconcile the energy changes with conservation of energy?

2.3 A 1 μF capacitor is made up of thin sheets of polycarbonate (the transparent foil used as a wrap in cooking) which have a conducting film deposited on them. The foil is rolled up to form a cylinder of diameter 1 cm and length 2 cm. The relative permittivity of polycarbonate is 2.3. Estimate the thickness of the polycarbonate sheet. At the maximum recommended voltage of 50 volts across the terminals of the capacitor, what is the force per unit area compressing the dielectric?

2.4 At room temperature the relative permittivity of water is 80. The dipole moment of a water molecule is 6.2×10^{-30} coulomb metres. What is the average value of E_{local}/E for a water molecule? (In working out this problem, neglect the contribution to the relative permittivity from induced dipole moments.)

2.5. Two molecules each have a dipole moment $\mathbf{p}$ pointing along the line joining their centres, as shown in Figure 2.20. How does the force between the molecules vary with

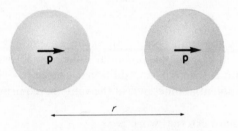

Figure 2.20.

their separation r? If two water molecules are oriented in this way, what is the potential energy due to their dipole–dipole interaction when their centres are at their average separation of 3.1×10^{-10} m? (Dipole moment of a water molecule $= 6.2 \times 10^{-30}$ coulomb m.)

2.6 Imagine that the dipoles in Figure 2.20 have their centres fixed, but are both free to rotate and take up any orientation. Think about the couples the dipole moments can exert on one another, and sketch the orientations in which they are in stable or unstable equilibrium.

2.7 A slab of dielectric of relative permeability ε fills the space between $z = \pm a$ in the x–y plane, and contains a uniform density of free charge ρ_f per unit volume. Find $\mathbf{E}$, $\mathbf{D}$ and $\mathbf{P}$ as functions of z. What is the surface density of polarization charge on the surface of the dielectric?

2.8 A slab of dielectric of thickness a and relative permittivity ε is placed in a uniform external field $\mathbf{E}$ whose field lines make an angle θ with normals to the surface of the slab. What is the density of polarization charge on the surface of the slab? Neglect end effects.

CHAPTER

Electrostatic calculations

3.1 INTRODUCTION

Information about the electric field may be required in a variety of different physical situations. For example, we may want to know where breakdown is most likely to occur in a piece of high-voltage equipment; or what path an electron follows as it travels through an electron microscope; or the value of the capacitance between a pair of conductors. To find the answer to each of these problems, we must first work out the electric field. Often the electric field cannot be expressed in terms of a finite number of elementary functions, and the best that can be done then is to look for approximations. Even when a manageable solution to the problem exists, there is no universal way of finding out what it is. One must know (or guess) the functional form of the solution, and fit it to the problem in hand, in much the same way as one guesses the result of integrating a function by comparison with standard forms of integral, and then checks by differentiation. Some of the techniques which can be used to lead to solutions of electrostatic problems are outlined in this chapter. The examples discussed are not simply intended as mathematical exercises, but have been chosen to illustrate important properties of the electric field.

3.2 ELECTRIC FIELDS WITH A SIMPLE SYMMETRY

When fields have a very simple symmetry, they can often be calculated directly from the integral form of Gauss' law (Equation (1.12)). For example, the electric displacement between the conductors of a coaxial cable has already been found in this way. This was possible because the cylindrical symmetry of the cable itself implies that the displacement $\mathbf{D}$ is also cylindrically symmetrical, that is to say, that it depends only on the radial distance r from the axis of the cable. Consideration of the flux of **D** out of a cylindrical surface then shows immediately that **D** is proportional to $1/r$, as we found in section 2.3.2. Since the dielectric in the cable is uniform, the electric field **E** is also proportional to $1/r$.

The same procedure works for problems with plane and spherical symmetry. One imposes the limitation that **D** has the appropriate symmetry, and then applies Gauss' law over a surface on which D is constant. We have already used this technique to find the relation between electric field and charge in a parallel plate capacitor and to find the atomic field around a spherically symmetrical ion.

3.3   POISSON'S EQUATION

Unfortunately, in most practical problems in which the electric field needs to be known, there is no simple symmetry. It is then usually better to begin with the differential form of Gauss' law rather than the integral form. We shall discuss the differential form in this section, illustrating its properties with very simple examples.

Let us start by applying Gauss' law in its differential form to the coaxial cable, in which we already know the electric field. Assume that the dielectric material between the conductors is uniform, and has relative permittivity ε. The outer conductor, of radius b, is at zero potential, and the inner conductor, of radius a, is held at a potential V. We write Gauss' law in its differential form (Equation (1.16))

$$\operatorname{div} \mathbf{E} = \rho/\varepsilon_0, \tag{3.1}$$

where ρ is the total charge density, including any polarization charges.

The electric field must obey this equation everywhere between the conductors, and it must also satisfy the *boundary conditions* of the problem. The boundary conditions are that the potential is constant over the surface of the conductors, and has the values 0 and V on the outer and inner conductors respectively. Because the boundary conditions refer to the potential, it is convenient to express Gauss' law in terms of the potential as well. The field is related to the potential by $E = -\operatorname{grad} \phi$, and hence by substituting in Equation (3.1),

$$\boxed{\operatorname{div} \operatorname{grad} \phi = \nabla^2 \phi = -\rho/\varepsilon_0.} \tag{3.2}$$

This equation is known as *Poisson's equation*. The *Laplacian operator* div grad, written concisely as ∇^2, is a second order differential operator. Now the gradient (grad ϕ) of a scalar function of position ϕ is the vector with components $(\partial\phi/\partial x, \partial\phi/\partial y, \partial\phi/\partial z)$ in Cartesian coordinates. The divergence of this vector is $(\partial^2\phi/\partial x^2 + \partial^2\phi/\partial y^2 + \partial^2\phi/\partial z^2)$, and thus in Cartesian coordinates the Laplacian operator is div grad $\equiv \nabla^2 = \partial^2/\partial x^2 + \partial^2/\partial y^2 + \partial^2/\partial z^2$. The forms of the Laplacian operator in cylindrical and spherical polar coordinates are given in Appendix B.

The problem of finding the electrostatic field has now been reduced to the solution of Poisson's equation, a second order differential equation. A formal solution to Poisson's equation has already been written down in section 1.5.2, where the principle of superposition was invoked to show that

$$\phi(\mathbf{r}) = \frac{1}{4\pi\varepsilon_0} \int_{\text{all space}} \frac{\rho(\mathbf{r}') \, d\tau'}{|\mathbf{r} - \mathbf{r}'|} + \frac{1}{4\pi\varepsilon_0} \int_{\text{all surfaces}} \frac{\sigma(\mathbf{r}') \, dS'}{|\mathbf{r} - \mathbf{r}'|}.$$

Here the potential at the point with position vector $\mathbf{r}$ is given in terms of the volume charge density ρ and surface charge density σ at the point with position vector $\mathbf{r}'$. This expression for the potential is quite general, and therefore it is a solution to Poisson's equation. But it is not of much use when dielectrics and conducting materials are present, because induced surface charges and polarization charges then appear, of a magnitude which can only be evaluated when the field or potential is already known. The polarization charges must be accounted for by introducing the electric displacement $\mathbf{D}$, and writing Gauss' law in terms of the free charge density as in Equation (2.19):

$$\text{div } \mathbf{D} = \text{div } (\varepsilon\varepsilon_0 \mathbf{E}) = \rho_f.$$

When the space between conductors is filled with a uniform dielectric of constant relative permittivity, as it is in the coaxial cable, this equation simplifies to

$$\text{div } \mathbf{E} = \rho_f / \varepsilon\varepsilon_0.$$

Again $\mathbf{E}$ can be replaced by $(-\text{grad } \phi)$, leading to

$$\boxed{\nabla^2 \phi = -\rho_f / \varepsilon\varepsilon_0,} \tag{3.3}$$

an equation still retaining the form of Poisson's equation. This equation is the one we usually want to use for solving electrostatic problems, since it makes no reference to polarization charges or to induced charges on conductors—the presence of conductors only imposes the condition that the potential ϕ must be constant on any conducting surface.

In the coaxial cable there is no free charge between the inner and outer conductors, and Equation (3.3) simplifies to $\nabla^2 \phi = 0$. This important special case of Poisson's equation is called *Laplace's equation*. The appropriate co-ordinates for the cable are the cylindrical polar coordinates (r, θ, z). In cylindrical polar coordinates r is the perpendicular distance from (r, θ, z) to the z-axis, as shown in Figure 3.1. (The variable r must not be confused with the distance to the origin, which we label r in *spherical* polar coordinates.) The angle θ is the azimuthal angle of the position vector, with respect to an x-axis drawn through the origin. In cylindrical polar coordinates the Laplacian operator has the form

$$\nabla^2 = \frac{1}{r} \frac{\partial}{\partial r}\left(r \frac{\partial}{\partial r}\right) + \frac{1}{r^2} \frac{\partial^2}{\partial \theta^2} + \frac{\partial^2}{\partial z^2}.$$

Since the coaxial cable has cylindrical symmetry, there is no θ dependence, and

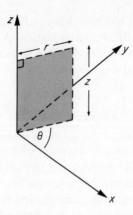

Figure 3.1. Cylindrical polar coordinates.

for a long cable we may neglect end effects, so that there is no z-dependence either. Laplace's equation then reduces to

$$\frac{1}{r}\frac{d}{dr}\left(r\frac{d}{dr}\right)\phi(r) = 0.$$

Hence

$$r\frac{d\phi}{dr} = A, \quad \text{a constant,}$$

and

$$\phi(r) = A \ln(r) + C, \text{ where } C \text{ is another constant of integration.}$$

Choosing the zero of potential to be at the radius b of the outer conductor, the boundary conditions when a potential V is maintained across the cable are $\phi(a) = V$ and $\phi(b) = 0$, where a is the radius of the inner conductor. The values of the constants of integration A and B which satisfy these conditions lead to

$$\phi(r) = \frac{-V}{\ln(b/a)}(\ln(r) - \ln(b)). \tag{3.4}$$

This is in agreement with the result found in section 2.3.2 from the integral form of Gauss' theorem.

3.3.1 The uniqueness theorem

The potential $\phi(r)$ given in Equation (3.4) obeys Laplace's equation in the region $a < r < b$, and also satisfies the conditions imposed at the boundaries of this region. Obviously $\phi(r)$ is the *only* function which has these properties, because we found it by integrating Laplace's equation step by step in a systematic manner. Unfortunately, systematic integration is not generally possible. However, suppose that in some region we have managed to find a solution to

Poisson's equation or Laplace's equation, having specified values on the boundaries of the region. Then this solution is the *only possible* solution with these boundary values.

It is easy to prove this *uniqueness theorem* by considering first of all the potential inside a cavity in a piece of conducting material. The potential on the boundary of the cavity is constant—at a value V_0 say. If there is no charge in the cavity, the potential inside it must obey Laplace's equation, $\nabla^2 \phi = 0$. One solution of this equation, which satisfies the boundary condition, is that $\phi = V_0$ throughout the cavity. This is indeed the only solution. For suppose there were some other solution ϕ_1, then ϕ_1 must have at least one maximum or minimum in the cavity, since $\phi_1 = V_0$ on the walls, but has a different value somewhere inside. But solutions of Laplace's equation cannot have maxima or minima. At a maximum, for example, $\partial^2 \phi / \partial x^2 < 0$, $\partial^2 \phi / \partial y^2 < 0$ and $\partial^2 \phi / \partial z^2 < 0$, which contradicts the original requirement that

$$\partial^2 \phi / \partial x^2 + \partial^2 \phi / \partial y^2 + \partial^2 \phi / \partial z^2 = 0.$$

(Think of a hill in a two-dimensional relief map of ϕ, as drawn in Figure 3.2. At the top of the hill both $\partial^2 \phi / \partial x^2$ and $\partial^2 \phi / \partial y^2$ are less than zero. Similarly at a

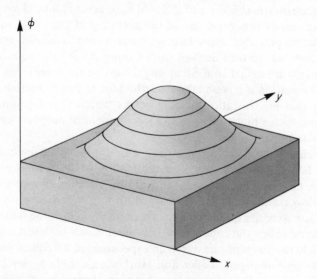

Figure 3.2. A relief map representing a 'hill' in a two-dimensional potential.

maximum in three dimensions, $\partial^2 \phi / \partial z^2$ is also less than zero.) It follows that there is no other solution than $\phi = V_0$ in the cavity.

The proof of the uniqueness theorem can now be extended to show that the theorem holds for Poisson's equation as well as for Laplace's equation, and for any boundary conditions. Imagine that ϕ_1 and ϕ_2 are two solutions of Poisson's equation with the same boundary values. The difference $\phi = \phi_1 - \phi_2$ must be zero on the boundary, and must obey Laplace's equation $\nabla^2\phi = 0$ everywhere. One solution to Laplace's equation is $\phi = 0$ everywhere, and we have proved that this must be the *only* possible solution. It follows that there cannot be *different* solutions ϕ_1 and ϕ_2, and that the uniqueness theorem holds for Poisson's equation also.

3.3.2 Electrostatic shielding

Since the potential is constant inside a hollow conductor, the electric field is zero. Sometimes it is useful to put apparatus in a region shielded from electric fields by being enclosed in a conducting box. A very precisely field-free region has been constructed at Stanford University, where the free fall of the electron has been observed inside an evacuated conducting box. The gravitational force $m_e g$ on an electron is extremely small, and if it is not to be swamped by the electrostatic force due to a stray electric field of strength E, say, then $eE < m_e g$, or $E < 10^{-10}$ volts/metre. In an unshielded part of the atmosphere, there is a vertical electric field of about 100 volts/metre near the surface of the earth, larger than the permissible stray fields in the cavity by a factor of the order of 10^{12}. The electrostatic shielding of a closed conductor is indeed very effective, and its effectiveness is a good test of the accuracy of Laplace's equation for the macroscopic potential. Now Laplace's equation was derived from Gauss' theorem, which was in turn derived from Coulomb's inverse square law. The three statements are in fact equivalent ways of saying the same thing, and a test of Laplace's equation is a roundabout method for testing Coulomb's law. The observation that electrostatic shielding works means that Coulomb's law is certainly valid with a high precision and for all practical purposes can be taken as exactly true.

The experiment with falling electrons was not designed in order to test Coulomb's law, although it is a good illustration of the accuracy of the law. The law can be tested even more sensitively by searching for an electric field inside a hollow conductor which is charged up by a high-frequency voltage. If Coulomb's law were not quite exact, then a changing field of the same frequency would be found inside the conductors. It is much easier to detect a small high-frequency field than a static one, yet in an experiment at a frequency of a million cycles per second Williams, Faller and Hill* were unable to detect any field within the conductor. If the law of force between charges is not quite inversely proportional to r^2, but varies as $1/r^{2+q}$, their results show that q must be less than 10^{-16}.

Of course electrostatic shielding can only be carried out on a laboratory scale, with macroscopic pieces of conductor. It is worth mentioning at this point that there are other experiments, which we shall not describe, which afford rigorous

---

* E. R. Williams, J. E. Faller and H. Hill, Phys. Rev. Letters *26*, 721 (1971).

tests of Coulomb's law on the atomic scale. At very small distances quantum mechanical effects become important, but after making allowances for them, no deviation from the theory based on the inverse square law has ever been observed, even at distances as small as 10^{-14} cm. Coulomb's law can be regarded as exact for atomic as well as for macroscopic fields.

3.4   BOUNDARIES BETWEEN DIFFERENT REGIONS

The electric field inside a 'closed region' is generated by charges which may lie inside or outside the region. The field can be derived from a potential which must satisfy Poisson's equation. Because of the uniqueness theorem, the potential inside the region is determined once we know the charge distribution and the value of the potential on the boundary. For example, the potential must be constant inside a closed conductor containing no charges, although the potential outside may have all kinds of shapes, provided that it is constant on the conducting surface itself. When a situation arises in which there are several regions, each with different properties, we can find the potential in each region separately by solving Poisson's equation or Laplace's equation, subject to a consistent set of boundary conditions. As an example of this procedure, we shall solve Laplace's equation for a coaxial cable containing two materials of different relative permittivity. The insulation between the conductors is made up of two cylindrical sleeves of relative permittivities ε_1 and ε_2, as shown in Figure 3.3. We shall discover that it can be advantageous to make high voltage cables in this way, because if $\varepsilon_1 > \varepsilon_2$, such a cable can sustain a greater potential difference between the conductors without breakdown than a cable of the same size with a single dielectric. Suppose that both dielectric materials break down in an electric field greater than some critical value E_b. As before, the radii of the inner and outer conductors are a and b, and the radius of the interface

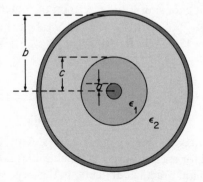

Figure 3.3. Cross-section of a co-axial cable containing two different dielectric materials.

between the two dielectrics we shall call c. What choice of the radius c will allow the cable to carry the maximum possible voltage across its conductors, and what is this voltage?

Boundary conditions are not always most conveniently expressed in terms of the potential, and in this example they are in terms of the field, since the condition for breakdown refers to the field. When the inner dielectric is on the point of breakdown, the field must be E_b at the surface of the inner conductor, and the field E_1 in the region $a < r < c$ is

$$E_1(r) = E_b a/r.$$

We have already shown in section 2.3.3 that at a boundary between dielectrics the perpendicular component of the electric displacement is continuous. In the cable the electric field is perpendicular to the boundary, and the field E_2 in the region with relative permittivity ε_2 therefore satisfies the condition

$$\varepsilon_2 E_2(c) = \varepsilon_1 E_1(c) = \frac{\varepsilon_1 E_b a}{c}.$$

Clearly the highest possible voltage occurs across the cable when *both* dielectrics are on the point of breakdown, and

$$E_2(c) = E_b \frac{\varepsilon_1 a}{\varepsilon_2 c} = E_b,$$

or

$$c = \varepsilon_1 a/\varepsilon_2.$$

The field in the outer region is also proportional to $1/r$, and

$$E_2(r) = E_b c/r.$$

Now we know the electric field in the whole of the cable when it is on the point of breakdown, and can evaluate the potential difference between the inner and outer conductors (again choosing V to be zero on the outer conductor).

$$V = -\int_a^b \mathbf{E} \cdot d\mathbf{r}$$

$$= -\int_b^c E_b \frac{c}{r} dr - \int_c^a E_b \frac{a}{r} dr$$

$$= E_b \{ c \ln (b/c) + a \ln (c/a) \}.$$

This technique of making high-voltage cables with two dielectrics has occasionally been used for special purposes. Such cables are not in general use, however, because the main practical difficulty in the large scale manufacture of high-voltage cables is in the avoidance of small pockets of air within the dielectric. Since the relative permittivity is low in the air pockets, the electric field is high inside them, and they are sites at which breakdown is likely to be initiated.

3.5. ELECTROSTATIC IMAGES

The field around a long conducting wire must be cylindrically symmetrical if the wire is far away from other charges or conductors. The magnitude $E(r)$ of the field at a distance r from the axis of the wire can therefore be found by applying the integral form of Gauss' law to a cylindrical surface of radius r. If the radius of the wire is a, and it carries a charge λ per unit length, then the flux of E out of unit length of the cylindrical surface is

$$2\pi r E(r) = \lambda/\varepsilon_0,$$

and thus

$$E(r) = \frac{\lambda}{2\pi\varepsilon_0 r}, \quad \text{if } r > a.$$

Integrating this equation we find that the potential is

$$\phi(r) = \frac{-\lambda}{2\pi\varepsilon_0} \ln(r) + \text{constant}$$

$$= \frac{\lambda}{2\pi\varepsilon_0} \ln\left(\frac{b}{r}\right),$$

choosing the potential to be zero* at $r = b$. This expression is only valid for $r > a$, since inside the conductor the potential has the constant value $(\lambda/2\pi\varepsilon_0) \ln(b/a)$. But outside $r = a$ the potential is independent of a, and would be the same if the conductor were collapsed onto the axis to form a line charge λ/unit length. The fictitious line charge is called the 'image' of the real surface charges on the conductor.

For a single conductor, we do not learn anything new by introducing the image charge. The image method is helpful, however, in finding the field around two parallel cylindrical conductors carrying equal and opposite charges. We shall prove shortly that the equipotential surfaces around the conductors are cylindrical, as drawn in Figure 3.4a. The equipotentials are not coaxial with the conductors; the electric field, and hence also the charge density, are not uniform on the conducting surface. Nevertheless the conductors can be represented by the image system shown in Figure 3.4(b), in which the surface charge is collapsed onto two lines carrying charges $\pm \lambda$ per unit length. Let us calculate the potential generated by these images. A section through the parallel pair of conductors is shown in Figure 3.5.

The conductors are each of radius a, and their centres at A and D on the x-axis are at distances d from the origin. Image charges $\pm\lambda$/unit length lie along the conductors, crossing the x-axis at C and B respectively. The distances

---

* When deriving this potential the line charge is assumed to be infinitely long, and hence carrying an infinite charge. It is therefore not possible to choose the potential to be zero at infinity. The best that can be done is to take the potential as zero at some radius b, which may be as large as we please.

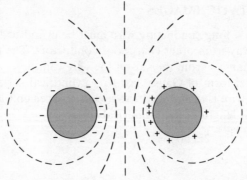

(a)   Charge is distributed non-uniformly on the surfaces of the conductors.

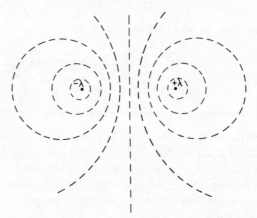

(b)   The image charges are collapsed onto lines.

Figure 3.4. Equipotentials around a pair of parallel cylindrical conductors,
and the equivalent image system.

$\pm p$ of the line charges from the origin can be chosen so that the surfaces of the conductors lie on equipotentials of the image charges. Choosing the potential to be zero at infinity* and on the y-axis, the potential at the point P in the x–y plane is

$$\phi_{\mathrm{P}} = \frac{\lambda}{2\pi\varepsilon_0} \ln (1/r_1) - \frac{\lambda}{2\pi\varepsilon_0} \ln (1/r_2)$$

$$= \frac{\lambda}{4\pi\varepsilon_0} \ln \left(\frac{r_2^2}{r_1^2} \right).$$

---

* The total charge carried by the pair of line charges is zero, and no logarithmic infinities arise if the potential is chosen to be zero at an infinite distance from the z-axis.

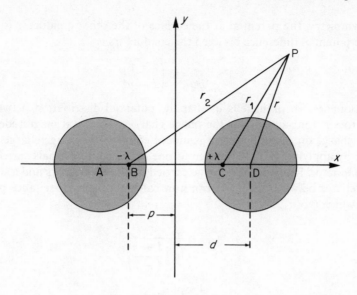

Figure 3.5. Working out the potential due to the image charges.

Now express this potential in terms of the distance from P to the axis of the conductor D; r is the distance PD, and θ the angle PD makes with the x-axis. In the triangle BPD,

$$r_2^2 = (d + p)^2 + r^2 + 2r(d + p)\cos\theta$$
$$= 2(d + p)(d + r\cos\theta) + r^2 + p^2 - d^2.$$

Similarly in triangle CPD,

$$r_1^2 = (d - p)^2 + r^2 + 2r(d - p)\cos\theta$$
$$= 2(d - p)(d + r\cos\theta) + r^2 + p^2 - d^2.$$

Hence

$$\phi_P = \frac{\lambda}{4\pi\varepsilon_0}\ln\left\{\frac{2(d + p)(d + r\cos\theta) + r^2 + p^2 - d^2}{2(d - p)(d + r\cos\theta) + r^2 + p^2 - d^2}\right\}.$$

If the position of the image charge is such that $p^2 = (d^2 - a^2)$, then at $r = a$

$$\phi(a) = \frac{\lambda}{4\pi\varepsilon_0}\ln\left(\frac{d + p}{d - p}\right),$$

independent of θ, and the surface of the conductor with axis through D is an equipotential of the image charge.

By symmetry, the potential at the surface of the other conductor is $-\phi(a)$, and the potential difference between the conductors is

$$V = \frac{\lambda}{2\pi\varepsilon_0} \ln \left| \frac{d + p}{d - p} \right|.$$

The uniqueness theorem tells us that the potential distribution between the conductors is the same as for the image charges. The total magnitude of the surface charge on each conductor is indeed equal to the image charge, as can be seen by applying Gauss' theorem to one of the equipotentials surrounding each conductor. Knowing the charge carried by the conductors and the potential difference between them, we can now calculate their capacitance per unit length, which is

$$C = \frac{\lambda}{V} = \frac{2\pi\varepsilon_0}{\ln \left| \dfrac{d + p}{d - p} \right|}$$

$$\simeq \frac{4\pi\varepsilon_0}{\ln \left(\dfrac{2d}{a} \right)} \quad \text{if } d \gg a$$

$$\simeq 15 \text{ pF/metre for } 2d/a = 6.$$

We shall need to know this capacitance in Chapter 9 when discussing the transmission of electrical signals. It is of the same order of magnitude as the capacitance per unit length of a typical coaxial cable, which we worked out in section 2.3.2.

The method of electrostatic images has here led us to a neat solution of a complex problem, but as a technique for solving Poisson's equation outside conductors, it only works backwards. If we are lucky enough to find a charge distribution which has equipotentials to fit the conducting surfaces, then this distribution can be used as an image of the real surface charges. There is however no systematic way of finding a distribution of image charges equivalent to an arbitrary system of conductors.

★ **3.6 POTENTIAL DISTRIBUTIONS IN THE PRESENCE OF DIELECTRICS**

At the surface of a conductor, the potential meets the straightforward requirement that it must be constant. The boundary conditions at the surface of a dielectric are more complicated, and potential problems are correspondingly more difficult to solve when dielectric materials are present. One of the few examples for which the field has a simple form is a dielectric sphere placed in a uniform external field $\mathbf{E}_{ext}$.

We shall show that the dielectric is uniformly polarized, and that outside the dielectric the original field $\mathbf{E}_{ext}$ is modified by a field which can be represented by a point dipole at the centre of the sphere. The uniform polarization is a special property of the spherical shape of the dielectric: in general, dielectric material is not uniformly polarized when placed in a uniform external field.

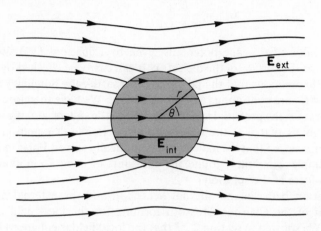

Figure 3.6. A uniform external field is distorted by a dielectric
sphere.

For the spherical dielectric, the field inside has a magnitude E_{int}, different from E_{ext}. Taking the potential to be zero at the centre of the sphere, and choosing the z-axis in the direction of the field as in Figure 3.6, the potential inside the sphere at a point with coordinates (r, θ) is

$$\phi_{int} = -zE_{int} = -r\cos\theta E_{int}. \tag{3.6}$$

If the susceptibility of the dielectric is χ_E, it has a dipole moment per unit volume $\mathbf{P} = \chi_E \varepsilon_0 \mathbf{E}_{int}$, and the total dipole moment of the whole sphere is

$$\mathbf{p} = \tfrac{4}{3}\pi a^3 \chi_E \varepsilon_0 \mathbf{E}_{int},$$

where a is the radius of the sphere. In Section 1.5.3 it was shown that the potential due to a point dipole with dipole moment $\mathbf{p}$ pointing in the z-direction is

$$\frac{p\cos\theta}{4\pi\varepsilon_0 r^2} = \frac{\chi_E a^3 \cos\theta E_{int}}{3r^2}.$$

The potential outside the sphere is the sum of that due to the external field and the dipole potential, i.e.

$$\phi_{ext} = \frac{\chi_E a^3 \cos\theta E_{int}}{3r^2} - r\cos\theta E_{ext}. \tag{3.7}$$

At the surface of the sphere, where $r = a$, this potential matches the internal potential provided that

$$a \cos \theta E_{int} = \tfrac{1}{3}\chi_E a \cos \theta E_{int} - a \cos \theta E_{ext}$$

or

$$E_{int} = \frac{E_{ext}}{1 + \tfrac{1}{3}\chi_E}. \tag{3.8}$$

With this value of E_{int}, the potentials given by Equations (3.6) and (3.7) are continuous at the surface of the dielectric. It can easily be checked by differentiating these expressions that the potentials also satisfy the boundary conditions that $D_\perp$ and $E_\parallel$ are continuous across the surface. We have therefore found one possible form for the potential inside and outside the sphere, and the uniqueness theorem tells us that it is the only solution of Poisson's equation. (Again it must be emphasized that we have not systematically solved Poisson's equation in this example, but simply found the constants of integration after *assuming* the correct form of the solution.)

The relation between the internal field E_{int}, and the external field E_{ext} is reminiscent of the Clausius–Mossotti formula for the local field in a liquid dielectric. We showed in section 2.2.3 that the local field is enhanced by a factor $(1 + \tfrac{1}{3}\chi_E)$ over the macroscopic field. In both cases the factor arises in the same way from the polarization charges of a uniformly-polarized material, induced over the surface of a sphere. But it must be pointed out that the local field calculated in the Clausius–Mossotti formula is *not* the same as the field in a spherical cavity. The local field is calculated from an atom placed in a uniformly polarized dielectric, and part of the field causing the polarization is due to the atom in question.

If the atom is not actually present in the dielectric, and there is a real cavity, the field and potential around the cavity are not uniform. The field in the cavity is less than the local field in a liquid which is uniformly polarized, as illustrated in Figure 3.7.

The field in a cavity is a little more awkward to work out than the field in a dielectric sphere, but again it is uniform and its value is

$$E_{cavity} = \left(\frac{1 + \chi_E}{1 + \tfrac{2}{3}\chi_E} \right) E_{ext}.$$

The enhancement factor

$$\frac{1 + \chi_E}{1 + \tfrac{2}{3}\chi_E}$$

is not much different from $(1 + \tfrac{1}{3}\chi_E)$ when χ_E is small, because the presence of the cavity then causes only a small distortion of the neighbouring field.

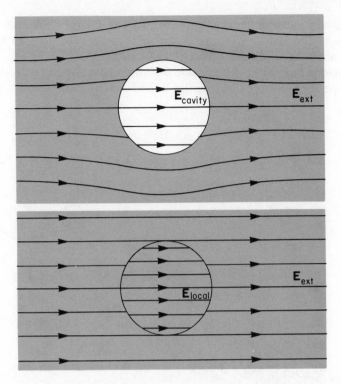

Figure 3.7. The distinction between the field in a cavity and the
local field.

3.7 ELECTROSTATIC LENSES

In most practical situations where one needs to know how the electric field
varies with position, Laplace's equation has no exact solution in terms of simple
functions. For example, the beam leaving the electron gun in a cathode ray tube
passes a number of electrodes, as illustrated in Figure 3.8. The electrodes are
made of conducting material, and each is maintained at a constant potential
with respect to the cathode (the cathode is the heated electrode which is the
source of the electrons). Around the electrodes there is a constant pattern of lines
of force of the electric field. We need to know this pattern of lines of force before
we can solve the equation of motion of an electron and find its path through the
electron gun. In such a complicated geometry, there is no possibility of finding
an exact solution to Laplace's equation, and we must be content with an
approximation.

Later on we shall briefly describe one of the methods of calculating approximate numerical values of the potential in complicated geometries. At present,
however, we shall concentrate on the qualitative behaviour of the electron gun.

We shall make some rather crude approximations, but as so often in physics, it is instructive to build up a rough picture of the way something works before worrying about the details in a special application.

The drawing in Figure 3.8 was made from an actual electron gun used in an oscilloscope. The electrons leaving the cathode are already constricted into a

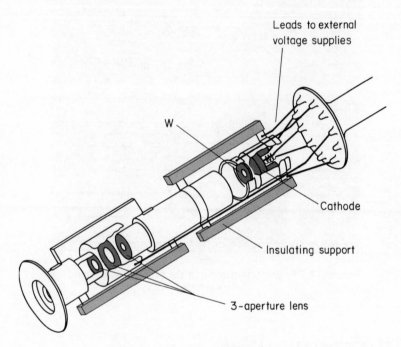

Leads to external
voltage supplies

W

Cathode

Insulating support

3-aperture lens

Figure 3.8. An electron gun.

narrow beam when they pass through the aperture in the electrode W, which is held at a negative potential with respect to the cathode. The remaining three electrodes are used to focus the beam onto a fine spot on the screen. Let us restrict attention to one of these focusing electrodes. What does the potential look like near its aperture? If we start by having an electrode with *no* aperture, with uniform but different fields on either side of it, then the equipotential surfaces are planes. Now punch a circular hole in the electrode. The electric field must change smoothly, and the field lines which are close together on the high field side of the aperture move apart on the low field side. The equipotential surfaces, always normal to the field lines, bulge through the aperture as shown in Figure 3.9.

Near the aperture, the electric field has a radial component directed away from the axis. An electron, arriving at the aperture on a path parallel to the axis, experiences a force deflecting it towards the axis. There is also an electric field along the axis, but in practical electron guns the electron already has an energy of about one keV when it reaches the focusing electrodes, and it is not a bad approximation to assume that the axial velocity of the electron is constant. We

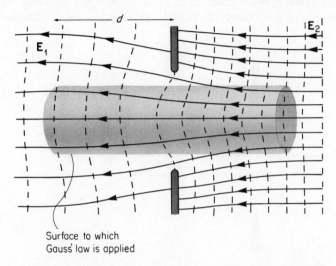

Surface to which
Gauss' law is applied

Figure 3.9. The pattern of field lines and equipotentials near an aperture.

only need to consider the radial field when estimating the deflection of the electron.

Let us follow the trajectory of an electron which approaches parallel to the axis, and is then deflected towards the axis as it passes through the aperture. Such a trajectory is illustrated by the thick line in Figure 3.10, for an electron initially at a distance r from the axis. We shall assume that the radial field is negligible at points further away than a distance d from the aperture. Taking the origin of cylindrical coordinates at the centre of the aperture, the electron is in a radial field E_r between the planes $z = \pm d$, and it then experiences a radial force $F = -eE_r$; the minus sign indicates that the force is directed towards the axis. An electron with axial velocity v moves a distance dz in a time $dt = dz/v$, and acquires a radial momentum $F\, dt = -eE_r(dz/v)$. The total radial momentum acquired by the electron in passing through the aperture is

$$\int_{-\infty}^{\infty} F\, dt = -\frac{e}{v}\int_{-d}^{d} E_r\, dz,$$

neglecting the change of the electron's axial velocity as it moves through the aperture. Provided that the deflection is small, we can estimate the integral on the right-hand side of this equation, without knowing the detailed form of E_r, by applying Gauss' theorem to the cylindrical surface of radius r and length $2d$ shown shaded in Figure 3.9.

At the end of the cylinder, the field is uniform and pointing along the axis, with magnitudes E_1 and E_2 on either side of the aperture. The outward flux at the two flat surfaces is therefore $\pi r^2(E_1 - E_2)$. Only the radial component E_r contributes to the flux out of the curved surface: the area of a circular strip of

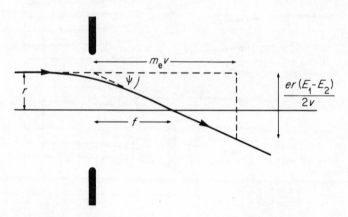

Figure 3.10. An electron trajectory through the aperture.

length dz on the surface is $2\pi r\, dz$, and the total flux out of the curved surface is $2\pi r \int_{-d}^{d} E_r\, dz$. The whole cylindrical surface encloses no charge, and according to Gauss' law the total outward flux is zero;

i.e.
$$\pi r^2(E_1 - E_2) + 2\pi r \int_{-d}^{d} E_r\, dz = 0.$$

Hence
$$\int_{-d}^{d} E_r\, dz = -\tfrac{1}{2}r(E_1 - E_2),$$

and substituting in Equation (3.8), the radial momentum acquired by the electron in passing through the aperture is $+er(E_1 - E_2)/2v$. The angle of deflection ψ of the electron towards the axis, as shown in Figure 3.10, is given by

$$\psi \simeq \tan \psi = \frac{-(\text{radial momentum})}{(\text{axial momentum})} = \frac{er(E_2 - E_1)}{2m_e v^2}.$$

In optics, when a ray passes through a thin lens of focal length f, at a distance r from the axis, the ray is deflected through an angle $\psi = r/f$. Because the electron deflection in passing through the aperture is also proportional to r, the aperture is acting like a lens of focal length

$$f = \frac{r}{\psi} = \frac{2m_e v^2}{e(E_2 - E_1)} = \frac{4 \times \text{(energy of electron in eV)}}{(E_2 - E_1)}. \tag{3.9}$$

A single aperture therefore acts as an electron lens. However, just as cameras are made with compound optical lenses, so an electron gun usually has a number of focusing apertures. A set of three apertures as in Figure 3.8 is commonly used: one advantage of this arrangement is that there can be field-free regions on either side of the three-aperture-lens, with the outer electrodes kept at a fixed potential. The focal length of the lens can be varied simply by changing the potential of the inner electrode. We can make a rough estimate of the performance of the lens in Figure 3.8 by assuming it to be made up of three thin lenses with focal lengths given by Equation (3.9). Suppose that the separation of each pair of electrodes is 1 cm, that the outer electrodes are both at a potential of 1 kV and that the inner electrode is held at 250 V. Application of the thin lens formula to each aperture in turn shows that 1 keV electrons travelling parallel to the axis are brought to a focus 12 cm from the central electrode, i.e. the three-aperture system acts as a lens with focal length 12 cm.

3.8 RELAXATION

The distortion of the electric field caused by an aperture spreads over a distance comparable with the size of the aperture. For a series of apertures, there is no region between the electrodes where the field is nearly uniform when the dimensions of the apertures become comparable with the electrode separation. The assumption made in the last section that each aperture can be treated independently as a thin lens is then not a good one. It does not matter very much using a rather bad approximation to find the focal length of the compound lens, because the electrode voltages are normally variable. One always in practice finds the smallest spot on an oscilloscope by twiddling the focus control knob; in the electron gun in Figure 3.8, the focus control adjusts the potential of the central electrode of the triplet lens. But when designing a lens, it is important to know how much aberration there is, and to check that the best focus will indeed yield a small spot. To investigate the aberration, we must know the potential accurately in the lens system, in order to be able to calculate small changes in focal length for different electron paths through the lens.

No general solution to Laplace's equation exists for a system of several apertures, and all that we can do is to generate *numerical* solutions. The numerical method we shall describe is called the *relaxation method*. To start with, consider a problem in which there is no variation in the potential in one direction, which we shall take to be the z-direction. Laplace's equation in Cartesian coordinates then reduces to

$$\frac{\partial^2 \phi}{\partial x^2} + \frac{\partial^2 \phi}{\partial y^2} = 0.$$

We shall consider a square region in the x–y plane, bounded on two sides by equipotentials at 0 and 10 volts, and on the other two sides by boundaries along which the potential changes linearly. (The potential obviously varies linearly everywhere between the equipotentials, but this trivial problem serves to illustrate the numerical techniques.) Now mark out a grid of points in the region where we want to know the potential, as in Figure 3.11. Each point is a distance h from its four nearest neighbours. Concentrate attention for the moment on the point labelled G. If the grid is fine enough, it will be a good approximation to express the potential at the neighbouring points C, F, H and K as a Taylor series about the potential ϕ_G at G, retaining only the second-order terms. Thus

$$\phi_F = \phi_G - h\frac{\partial \phi}{\partial x}\bigg|_G + \tfrac{1}{2}h^2\frac{\partial^2 \phi}{\partial x^2}\bigg|_G + \cdots$$

$$\phi_H = \phi_G + h\frac{\partial \phi}{\partial x}\bigg|_G + \tfrac{1}{2}h^2\frac{\partial^2 \phi}{\partial x^2}\bigg|_G + \cdots$$

$$\phi_C = \phi_G - h\frac{\partial \phi}{\partial y}\bigg|_G + \tfrac{1}{2}h^2\frac{\partial^2 \phi}{\partial y^2}\bigg|_G + \cdots$$

$$\phi_K = \phi_G + h\frac{\partial \phi}{\partial y}\bigg|_G + \tfrac{1}{2}h^2\frac{\partial^2 \phi}{\partial y^2}\bigg|_G + \cdots.$$

Adding the four equations, and using Laplace's equation to eliminate the second order terms, we find

$$\phi_C + \phi_F + \phi_H + \phi_K = 4\phi_G,$$

or

$$\phi_G = \tfrac{1}{4}(\phi_C + \phi_F + \phi_H + \phi_K). \tag{3.10}$$

This equation is not exact, because we have ignored the third and higher order terms in the Taylor expansion. However, these terms can be made as small as we please by choosing the grid sufficiently fine.

Now suppose that we can find a set of values for the potential at the grid points which satisfy the boundary conditions, and for which each value is the mean of the four nearest neighbour values, as required by Equation (3.10). Such a set of values is a good approximation to the exact potential at the points of the grid. To find this set of values, we must solve a large number of simultaneous equations—in fact there is one equation of the form (3.10) for each point of the grid. These equations can be solved by an iterative procedure. One starts with a guess at the potential distribution. For example, if in the grid in Figure 3.11 the values of the potential on the boundary are as labelled, the exact solution is obviously a potential varying linearly from 0 to 10 volts from left to right. Let us make a rather poor initial guess at this potential, assuming it to be zero everywhere except on the boundary. A lot of the points on the grid do not now satisfy Equation (3.10). However, we can systematically adjust them, starting for example at point A. The first guess for the potential at A is $\phi_A = 0$, but a better guess is

$$\phi_A = \tfrac{1}{4}(0 + 2 + \phi_E + \phi_B).$$

Putting $\phi_E = 0$ and $\phi_B = 0$ leads to $\phi_A = 0.5$. Similarly

$$\phi_B = \tfrac{1}{4}(4 + \phi_A + \phi_F + \phi_C).$$

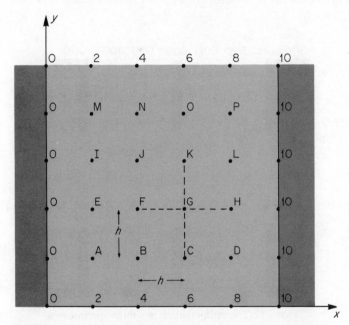

Figure 3.11. An equally spaced mesh of points used in finding an approximate solution to Laplace's equation.

Using the new guess of $\phi_A = 0.5$, and putting $\phi_F = 0$, $\phi_C = 0$, we find $\phi_B = 1.125$.

After evaluating the new approximation to the potential at each point, the whole operation is carried out over and over again until Equation (3.10) is satisfied at every point on the grid. In the present example, the method converges to within 1 % of the exact solution only after 18 iterations. The relaxation method is thus very laborious, and it also suffers from the disadvantage, common to all numerical methods, that the solution is only applicable to one set of boundary conditions. If some of the boundary potentials in Figure 3.11 are changed, the whole iterative procedure has to be carried out again. Nevertheless the relaxation method is useful because of the following features:

(i) the potential is evaluated in a *systematic* manner for an arbitrary set of boundary conditions.

(ii) each step in the calculation is straightforward, and although the iteration may be lengthy, it is ideally suited to calculation on a digital computer.

The relaxation method can be applied to problems in three dimensions, or to problems in which Cartesian coordinates are not appropriate. In all cases, a Taylor expansion leads to approximate relations between the potentials at neighbouring points (though the relations are not always as simple as Equation (3.10)). This technique of dividing up a region into a finite grid is nowadays the

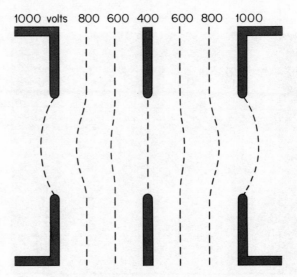

1000 volts 800 600 400 600 800 1000

Figure 3.12. The equipotentials in a three-aperture lens.

most important method of finding approximate solutions to Laplace's equation, since digital computers are fast enough, and have enough capacity, to cope with the large number of variables which are needed in two or three dimensions.

The relaxation method has been used to compute the potential distribution around the three aperture lens of an electron gun, in which the separation of the electrodes is equal to the diameter of the apertures*. When the outer electrodes

* The relaxation method does *not* work near edges and corners. In the computed solution shown in Figure 3.12 the electrodes are thick, and the apertures are rounded. This of course corresponds more closely to reality than does the idealized thin aperture.

are at 1 kV, this lens system has a focal length of 12 cm when the inner electrode is at 400 volts. The value of 250 volts, derived earlier using the thin lens formulae, is a poor approximation for the practical lens with wide apertures. The potential distribution is drawn in Figure 3.12.

3.9 TWO-DIMENSIONAL POTENTIAL PROBLEMS

When discussing the potential around a pair of parallel wires in section 3.5, we assumed that the potential does not vary along the direction of the conductors. Of course this is not true near the ends of the conductors, but provided that they are long by comparison with their separation, end effects do not much alter the quantities like capacitance which one needs to know. In problems of this nature, in which it is a good approximation to neglect the variation of potential in one direction, if we align the z-axis along this direction, Laplace's equation reduces to

$$\frac{\partial^2 \phi}{\partial x^2} + \frac{\partial^2 \phi}{\partial y^2} = 0.$$

The potential $\phi(x, y)$ is a function of the two variables x and y only, and is therefore called a two-dimensional potential.

The two-dimensional version of Laplace's equation has some special properties which make possible a very elegant method of solution using complex variables. Let us start by considering a uniform field of 1 volt/metre between the plates of a parallel plate capacitor. The lines of force and the equipotentials are drawn in Figure 3.13(a), neglecting any end effects at the edges of the plates. If we regard this figure as an Argand diagram, we can label any point on it with the complex variable $\mathbf{z} = x + \mathrm{j}y$. Choosing the zero of potential to be along the x-axis, the potential inside the capacitor is $v(\mathbf{z}) = y = \mathscr{I}(\mathbf{z})$, the imaginary part of $\mathbf{z}$. Now we can get another solution of Laplace's equation, though one corresponding to different boundary conditions, if we interchange the role of lines of force and equipotentials, as shown in Figure 3.13(b). This second solution, again for a field of 1 volt/metre, and now with the zero of potential along the y-axis, is $u(\mathbf{z}) = x = \mathscr{R}e(\mathbf{z})$, the real part of $\mathbf{z}$. Both functions $u(\mathbf{z})$ and $v(\mathbf{z})$ obviously satisfy Laplace's equation.

What happens to the field lines and equipotentials if we twist the conducting plates so that they point towards the origin, as shown in Figure 3.14(a)? Lines of force and equipotentials are always perpendicular to one another and we can still interchange them to construct the potential diagram in Figure 3.14(b). The lines of force now point towards the origin; this diagram represents the familiar situation of an image which is a negative line charge at the origin. The equipotentials are circular arcs and the potential is best expressed in polar coordinates (r, θ), in which the complex variable is $\mathbf{z} = r \exp(\mathrm{j}\theta)$. If we make the line charge $\lambda = -4\pi\varepsilon_0$ per unit length, and choose the potential to be zero at

$r = 1$, then the potential is

$$U(\mathbf{z}) = \ln r.$$

The equipotentials in Figure 3.14(a) are straight lines through the origin, and with a suitable choice of scale and potential zero, the potential can be written as

$$V(\mathbf{z}) = \theta.$$

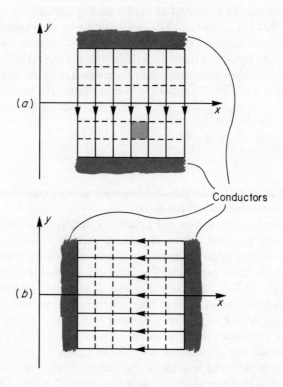

Figure 3.13. A complementary pair of field and
potential lines.

    The potentials u and U are related in an interesting way. By moving the conductors in Figure 3.13(a) to their position in Figure 3.14(a) we have, as it were, stretched the lines of force, but in such a way that the shape of small sections of the potential distribution is unaltered. The small square shown shaded in Figure 3.13(a) is transformed into the shaded area in Figure 3.14(a). This area is almost square, and if we drew the lines of force and equipotentials close enough together, could be made as nearly square as we please. Only the size of the square is changed. Such a transformation, in which the shape (or *form*) of small areas is preserved, is called a *conformal transformation*. It can be shown that

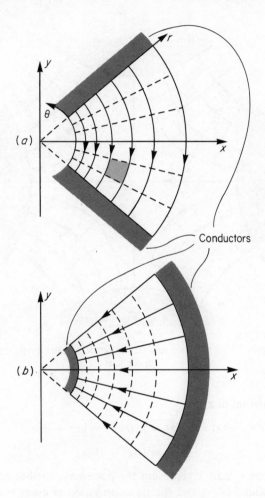

Figure 3.14. The fields from Figure 3.13 after a
conformal transformation.

when a potential function undergoes a conformal transformation the transformed function obeys Laplace's equation, and therefore also represents a potential distribution. Furthermore, pairs of potential distributions like $U(\mathbf{z})$ and $V(\mathbf{z})$, related by the interchange of equipotentials and lines of force, are always the real and imaginary parts of a complex function*, just as u and v are the real and imaginary parts of the complex variable $\mathbf{z}$ itself. Functions of a complex variable can be formed in the same way as functions of a real variable. If we write the complex variable in the form $\mathbf{z} = r \exp{(j\theta)}$, then we can write

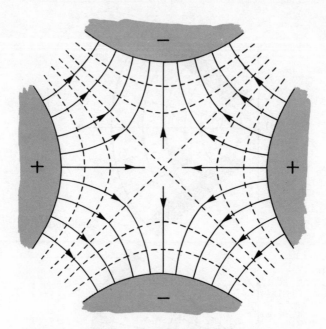

Figure 3.15. Field lines and equipotentials in an electric quadrupole lens.

the natural logarithm of $\mathbf{z}$ as

$$f(\mathbf{z}) = \ln{(\mathbf{z})} = \ln{(r\,e^{j\theta})} = \ln r + j\theta$$

$$= U(\mathbf{z}) + jV(\mathbf{z}).$$

The real functions U and V represent the potential distributions appearing in Figures 3.14(a) and 3.14(b), and the transformation of every point $\mathbf{z}$ in Figure 3.14 to a corresponding point $\ln{(\mathbf{z})}$ recovers the rectangular grid of field lines and equipotentials of Figure 3.13.

Since any function generates two potential functions, it is easy to invent lots of two-dimensional solutions to Laplace's equation. But, as with the method of images, only a few of the solutions have practical applicability. We shall give

---

* The proof that all reasonably well-behaved functions of a complex variable generate conformal transformations, and that the real and imaginary parts of complex functions obey Laplace's equation, can be found in books on complex variables, and we shall not discuss the proof here. See for example, Copson's 'Theory of Functions of a Complex Variable'.

one further simple example to illustrate the power of the method, generating potentials from the complex function $\mathbf{z}^2$.

Now

$$\mathbf{z}^2 = (x + \mathrm{j}y)^2 = x^2 - y^2 + 2\mathrm{j}xy.$$

The real functions $U(\mathbf{z}) = x^2 - y^2$ and $V(\mathbf{z}) = 2xy$ both satisfy Laplace's equation, and the curves defined by $U = $ constant and $V = $ constant can be taken as equipotentials and lines of force. Both sets of curves are rectangular hyperbolae, and in Figure 3.15 the equipotentials and field lines are drawn for a system of conductors on curves $U = $ constant. Notice that as for the potential near an aperture, the potential is proportional to the square of the distance from each axis. The electrodes act as a lens for charged particles moving along the z-direction: positively charged particles are focused towards the y-axis, but away from the x-axis. Because there are four electrodes, the lens is called a quadrupole lens. Electrostatic quadrupole lenses are not much used, but magnetic quadrupole lenses, in which magnetic lines of force have the same configuration as the lines of force of the electric field in Figure 3.15, are frequently used in charged-particle optics. Magnetic quadrupole lenses are discussed in section 4.7.2.

SUMMARY OF ELECTROSTATICS

The key definitions and relations of electrostatics are set out here, numbered as in their first appearance in the text.

Coulomb's Law states that the force between two charges q_1 and q_2 is inversely proportional to the square of the distance between them

$$\mathbf{F}_{21} = \frac{1}{4\pi\varepsilon_0} \frac{q_2 q_1}{r_{21}^3} \mathbf{r}_{21}. \tag{1.2}$$

The *electric field* at a point with position vector $\mathbf{r}$ is the force per unit charge on an imaginary (and vanishingly small) test charge placed at $\mathbf{r}$.

$$\mathbf{E}(\mathbf{r}) = \frac{1}{4\pi\varepsilon_0} \int_{\text{all space}} \frac{(\mathbf{r} - \mathbf{r}')\,\rho(\mathbf{r}')\,\mathrm{d}\tau'}{|\mathbf{r} - \mathbf{r}'|^3} + \frac{1}{4\pi\varepsilon_0} \int_{\text{all surfaces}} \frac{(\mathbf{r} - \mathbf{r}')\,\sigma(\mathbf{r}')\,\mathrm{d}S'}{|\mathbf{r} - \mathbf{r}'|^3}. \tag{1.8}$$

Gauss' Theorem states that the flux of $\mathbf{E}$ out of a closed surface S enclosing a volume V is $(1/\varepsilon_0)$ times the total charge within V

$$\int_S \mathbf{E} \cdot \mathrm{d}\mathbf{S} = \frac{1}{\varepsilon_0} \int_V \rho \, \mathrm{d}\tau. \tag{1.12}$$

Gauss' theorem can be expressed equivalently in differential form by the equation

$$\operatorname{div} \mathbf{E} = \rho/\varepsilon_0. \tag{1.16}$$

The *potential difference* between two points A and B is defined as the work done per unit charge in moving a small test charge from A to B:

$$\phi(\mathbf{r_B}) - \phi(\mathbf{r_A}) = -\int_A^B \mathbf{E} \cdot d\mathbf{l}. \tag{1.19}$$

The differential form of this equation is

$$\mathbf{E} = -\text{grad } \phi. \tag{1.21}$$

The *potential energy* of an assembly of point charges q_i is the work done to bring them up one by one from infinity to their final positions. This amount of work is

$$U = \tfrac{1}{2} \sum_i q_i \phi_i. \tag{1.31}$$

Potential energy can also be expressed in terms of the field by associating an *energy density* $\tfrac{1}{2}\varepsilon_0 E^2$ with the field. The total energy of a system of charges generating a field $\mathbf{E}$ is

$$U = \int \tfrac{1}{2}\varepsilon_0 E^2 \, d\tau. \tag{1.37}$$

The *capacitance* of a pair of conductors is the ratio of the charge stored on them to the voltage difference between them:

$$C = Q/V. \tag{1.34}$$

The capacitance of a pair of conductors in free space is determined only by their geometry. When the space between the conductors is completely filled with a homogeneous dielectric material, the capacitance increases by a factor ε, the *relative permittivity* of the material. The increase in capacitance arises because the dielectric material becomes polarized and in a field $\mathbf{E}$ it acquires a dipole moment per unit volume, or *polarization*

$$\mathbf{P} = (\varepsilon - 1)\varepsilon_0\mathbf{E}. \tag{2.6}$$

The factor $(\varepsilon - 1)$ has a special name. It is called the *electric susceptibility*:

$$\chi_E = \varepsilon - 1. \tag{2.7}$$

When discussing dielectric materials, it is useful to introduce a new vector field, the *electric displacement* $\mathbf{D}$ defined by

$$\mathbf{D} = \varepsilon_0\mathbf{E} + \mathbf{P}. \tag{2.20}$$

The electric displacement satisfies a relation rather like Gauss' theorem for the field $\mathbf{E}$, but referring only to the *free charge density* ρ_f:

$$\text{div } \mathbf{D} = \rho_f. \tag{2.19}$$

At boundaries between regions containing materials of different dielectric constant, the *boundary conditions* are

$$D_\perp \text{ continuous} \tag{2.23}$$

$$E_\parallel \text{ continuous.} \tag{2.24}$$

Within a region containing uniform dielectric with relative permittivity ε, the potential satisfies *Poisson's equation*:

$$\nabla^2 \phi = -\rho_f/\varepsilon\varepsilon_0. \tag{3.3}$$

Where no free charges are present, Poisson's equation reduces to *Laplace's equation*:

$$\nabla^2 \phi = 0.$$

PROBLEMS 3

3.1 A silicon rectifier consists of a disc of silicon $\frac{1}{2}$ mm thick with a thin gold film on one face. The silicon has been 'doped' by the addition of donor atoms with a density $N_D = 10^{12}/\text{cm}^3$; each donor atom contributes one electron which is free to move through the silicon. When the rectifier is back-biased, the mobile electrons are swept away, but other electrons in the gold are unable to surmount the 'surface barrier'. A depletion layer builds up in which there are no mobile electrons, leaving a positive charge density $N_D e$ due to the unbalanced charge on the donor atoms. Write down Poisson's equation within the depletion layer, and solve it to find the minimum voltage needed to make the depletion layer extend through the whole thickness of the silicon disc. (Relative permittivity of silicon = 12.)

3.2 The potential distribution in a thermionic diode is quite different from that in the silicon diode. The space between the electrodes is evacuated, and if the electrodes are parallel plates, the field between them is nearly uniform when the electron density in the diode is small, as occurs when the anode is at a high potential with respect to the cathode. But at low voltages, electrons move slowly near the cathode, and their charge density becomes high: the diode becomes *space-charge limited* when the electric field at the cathode is zero. Find the current density in a space-charge limited parallel plate diode with plate separation d when there is a voltage V across the plates. (Tackle the problem in the following way: (i) express the electron velocity in terms of the potential between the plates; (ii) this gives the charge density in terms of the current density; (iii) solve Poisson's equation with the conditions $\phi = 0$ and V at the electrodes, $\partial\phi/\partial x = 0$ at the cathode.)

3.3 A proton is at rest 1 cm below a conducting plate. Does the electrostatic attraction towards the plate overcome the gravitational force acting on the proton? (Hint: what image charge can reproduce the surface of the plate as an equipotential?)

3.4 A telephone wire of diameter 1 mm is suspended 10 m above the ground. Estimate its capacitance to earth per unit length. (Images again.)

3.5 A point charge is situated at a distance a from the centre of a conducting sphere of radius R as shown in Figure 3.16. Show that the image charge q' satisfies the condition that the surface of the sphere be an equipotential if $q' = -qR/a$ and $b = R^2/a$. What is the force attracting q towards the sphere if the sphere carries no net charge?

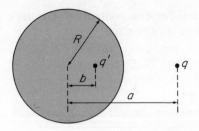

Figure 3.16.

3.6 Two conducting spheres, each of radius 2 cm have their centres 10 cm apart. If the breakdown field in air is 30 kV/cm, estimate the breakdown voltage across the sphere gap. (Sphere gaps of this kind are often used as an overvoltage protection in high voltage equipment.)

3.7 A sheet of charge of density σ/unit area lies midway between two earthed conducting plates. What are the surface charge densities induced on the plates? (You cannot use images for this one, because an infinite number of sheets of image charge are required. Use Gauss' law.)

3.8 A hemispherical bulge protrudes from the otherwise flat plate of a parallel plate capacitor. Away from the bulge there is a uniform field in the z-direction. Choosing the origin as in Figure 3.17, show that the addition of a potential proportional to $\cos\theta/r^2$ leads to a solution of Poisson's equation in the neighbourhood of the bulge. Hence show that the field at the top of the bulge is three times as great as it is above a completely flat plate.

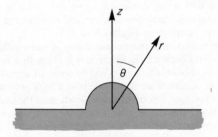

Figure 3.17.

3.9 Convince yourself that when a pair of conductors is moved to follow a conformal transformation, the capacitance between them remains the same. The wedge-shaped capacitor in Figure 3.18 can be converted into a parallel plate capacitor by the transformation of each point z in the figure into ln (z). By working out the transformed coordinates of the edges of the plates find an equivalent parallel plate capacitor, and hence show that the capacitance per unit length of the wedge-shaped capacitor is $\varepsilon_0 \ln 2/\theta$. (End effects are neglected.)

3.10 (*Estimate of capacitance from an approximate field pattern.*) In the capacitor described in Problem 3.9, the field lines are actually arcs of circles. However, if the wedge is narrow, these arcs are not much different from straight lines. Assuming that the field

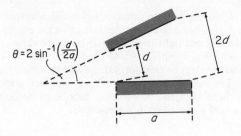

Figure 3.18.

lines are indeed straight, estimate the electric field as a function of position along the capacitor when there is a voltage V across the plates (neglect end effects). Then work out the total field energy stored by the capacitor and derive an approximate capacitance.

3.11 (*Solution of Laplace's equation by a Fourier expansion.*) Figure 3.19 represents a cross-section through a cylindrical annulus made of conducting material. The four quadrants

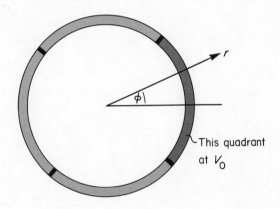

Figure 3.19.

are insulated from one another, and the right-hand quadrant is maintained at a potential V_0 while the others are earthed. What is the potential distribution within the cylinder? This problem can be solved by expanding the potential as a Fourier series in the angular variable ϕ. Since the potential must be symmetrical about $\phi = 0$, we may put

$$V(r, \phi) = A + \sum_{m=1}^{\infty} A_m \cos m\phi \, f_m(r).$$

Show that for $V(r, \phi)$ to be a solution of Laplace's equation which is finite at the origin, $f_m(r) = r^m$. Find the Fourier coefficients which fit the boundary conditions of the problem at the conductor radius $r = a$, and then from the field at $r = a$ derive the charge density on the conducting surface. Hence calculate the total charge per unit length ($-Q$) on the left-hand quadrant, deducing that the capacitance per unit length Q/V_0 between the left- and right-hand quadrants has the value $\varepsilon_0 \ln 2/\pi$.

It can be proved using complex variable theory that any conducting cylinder with symmetrical quadrants can be transformed into the circular cylinder by a conformal transformation. The capacitance per unit length between opposite quadrants in a long cylinder is therefore $\varepsilon_0 \ln 2/\pi$, whatever may be its size and shape. A practical capacitor of this kind is described in Appendix A.

Steady currents and magnetic fields

The previous chapters were concerned with the forces between stationary charges and their description in terms of the electrostatic field and potential. We now consider what happens when charges are in motion. As well as the force represented by the electric field, a moving charge may experience an additional force, called a magnetic force, when it is in the presence of other moving charges or permanent magnets. Magnetic forces are also conveniently described in terms of a vector field—the magnetic field—from which we can derive the force on a moving charge at any point.

A steady electric current in a conducting wire arises from the continuous motion of charge along the wire, and a moving charge near a current-carrying wire may therefore experience a magnetic force. In this chapter we shall first explain the rules governing the flow of current in conductors, and then go on to discuss the magnetic fields caused by steady currents in conductors.

4.1 ELECTROMOTIVE FORCE AND CONDUCTION

4.1.1 Current and resistance

Some of the electrons inside conductors are not bound to particular atoms but are free to move from atom to atom. In a metal we may picture these free electrons, the conduction electrons, as behaving more or less like gas molecules in a box, moving about at random with large speeds, and colliding with atoms in the metallic lattice.

In the absence of an applied electric field the mean electron velocity, averaged over any volume which is very large on a microscopic scale, will be zero. If an electric field $\mathbf{E}$ is maintained in the conductor the electrons acquire an average velocity $\mathbf{v}$ in the direction opposite to the field. This is because, between collisions with atoms in the conductor, the electrons are acted upon by a force $-e\mathbf{E}$. The mean drift velocity $\mathbf{v}$ of the electrons is thus equal to one half of the average extra velocity given to electrons in between collisions.

A conductor carrying current can be, and usually is, electrically neutral. The number of free electrons is balanced by an equal number of positively charged atoms. These atoms are also subjected to an electric force, but they are unable to move, being held in their lattice sites by far stronger interatomic forces. In liquid and gaseous conductors there may exist moving carriers with charge of either sign. We restrict the discussion here to solid conductors.

The electron flow in a solid conductor can be described at every point by the *current density*, $\mathbf{j}$. The current density is a vector function of position; at any point the electrons flow in the direction of their mean drift velocity at that point. Because the charge on the electron is by convention negative, the direction of the vector $\mathbf{j}$ is opposite to the direction of the velocity $\mathbf{v}$. The magnitude of the current density is the net amount of charge crossing unit area of a surface perpendicular to the mean drift velocity. Hence, if the number of free electrons per unit volume in a conductor is N, the current density is given by

$$\mathbf{j} = -Ne\mathbf{v}. \qquad (4.1)$$

The *current* I through any surface S is the net charge that passes through the surface per second. It is given by the surface integral of the current density

$$I = \int_S \mathbf{j} \cdot d\mathbf{S}. \qquad (4.2)$$

In a conductor carrying a steady current the surface integral has the same value for any surface that cuts across the flow of charge. For example, when a wire joins the two terminals of a battery, the current flowing into the wire at one terminal equals the current flowing out at the other terminal. The current in a conductor is thus the total amount of charge that enters or leaves the conductor per second. Current has the dimensions [charge] [time]$^{-1}$, and is measured in *amperes* (A). Current density has the dimensions [charge] [time]$^{-1}$ [area]$^{-1}$, and is measured in A m^{-2}.

Let us estimate the mean drift speed of the electrons in a typical conductor carrying a current. If a current I flows in a metal wire of uniform cross-sectional area A the current density $\mathbf{j}$ will be constant and parallel to the length of the wire. The current I is thus given by equation (4.2)

$$I = NevA. \qquad (4.3)$$

In a good conductor there is usually about one free electron per atom; hence in a copper wire for example we obtain $N \simeq 8 \times 10^{28}$ m^{-3}. The electronic charge e is equal to 1.6×10^{-19} C, and so for a wire of cross-sectional area 1 mm^2 carrying a current of 1 A,

$$v = \frac{I}{NAe} = \frac{1}{8 \times 10^{28} \times 10^{-6} \times 1.6 \times 10^{-19}} \simeq 10^{-4} \, \text{m s}^{-1}.$$

This calculation illustrates that the drift velocities of electrons in a solid are very small. The random thermal velocities upon which the mean drift velocity is superposed are always very much larger.

For many materials it is found that in the steady state the current density is proportional to the electric field. For such materials

$$\boxed{\mathbf{j} = \sigma \mathbf{E},} \tag{4.4}$$

where σ is constant for the material at a given temperature, and is called its *electrical conductivity*. Materials that obey Equation (4.4) are said to be ohmic, and Equation (4.4) is called *Ohm's Law*.

Consider a homogeneous conductor of uniform cross-sectional area A in which a steady current I is flowing. For all surfaces S cutting the wire

$$I = \int_S \mathbf{j} \cdot d\mathbf{S}$$

$$= \sigma \int_S \mathbf{E} \cdot d\mathbf{S}. \tag{4.5}$$

If a potential difference of V volts is maintained across the ends of the conductor by an external source, the electric field $\mathbf{E}$ will be uniform and

$$E = V/l,$$

where l is the length of the conductor. Equation (4.5) now becomes

$$I = \sigma A E$$

$$= \sigma A V/l.$$

The ratio of the voltage to the current in the wire

$$\frac{V}{I} = \frac{l}{\sigma A} \tag{4.6}$$

$$= R, \tag{4.7}$$

where R is called the *resistance* of the conductor. R is measured in *ohms*, (symbol Ω). The conductivity σ is measured in $(\Omega \, \text{m})^{-1}$.

In the more general case of a conductor of arbitrary shape, the electric field inside is not constant when a current I is flowing. However, as long as Equation

(4.4) is obeyed inside the material the current I remains proportional to the voltage V across the ends of the conductor, and the resistance is still defined by Equation (4.7). Ohm's law is not always valid. Gases obey Ohm's law at low electric fields, but not at high fields; some materials at low temperatures deviate from Ohm's law to the extent that their electrical conductivities become infinite. Even for ohmic materials the conductivity of a material is usually not a constant but is very dependent on temperature, and this is especially true for semi-conductors like silicon and germanium. Pure germanium has a conductivity of $2.5 \, (\Omega m)^{-1}$ at 300 K, but at 1000 K the conductivity has risen to 3×10^4 $(\Omega \, m)^{-1}$. Table 4.1 gives the electrical conductivities of several common materials at 293 K, and illustrates the enormous range over which the values of the

Table 4.1. The conductivities of some common materials at 293 K.

Material	Conductivity $(ohm \, m)^{-1}$
Copper	5.9×10^7
Gold	4.1×10^7
Germanium (pure)	2.2
Saturated NaCl solution	23.0
Glass	10^{-10}–10^{-14}
Fused quartz	1.3×10^{-18}
Wood	10^{-8}–10^{-11}

conductivities spread. The reciprocal of the conductivity of a material is called its *resistivity*. The resistivity is denoted by ρ, and is measured in ohm metres. We adopt the commonly used symbols ρ and σ for the resistivity and the conductivity, although we have already used them for volume and surface charge densities. It is always clear from the context which meaning the symbols have.

We have seen in Chapter 1 that if charges are placed on an isolated conductor the electric field inside becomes zero, and the charge is distributed over the surface of the conductor. For a steady current to flow in a single conductor a potential difference must be maintained between its two ends. The agent that does this is called a source of *electromotive force*. If a source maintains a potential difference V volts between the ends of a conductor its electromotive force (e.m.f.) is equal to V volts. When the conductor is connected to the terminals of the source of e.m.f., a steady current flows around the completed circuit formed by the conductor and the source.

There are several types of source of e.m.f. The most widely used are batteries and alternators. We shall not explain the principles of operation of different sources of e.m.f., but simply point out that they all convert some other kind of energy into electrical energy. A battery, for example, uses stored chemical energy to drive a current around a circuit. The chemical energy reappears as heat dissipated in the resistors. This heating arises because part of the energy the electrons acquire by virtue of the work done on them by the applied electric field is given up to the atoms with which they collide.

The power dissipated in a resistor can easily be worked out. Suppose that a steady current I flows in a wire which has a potential difference V between its ends. If, in time dt, a charge dq passes through any surface that cuts through the wire, then

$$I = dq/dt.$$

In the steady state a charge dq leaves the wire at one end and the same amount of charge enters it at the other end. The net effect is that during the time dt a charge dq has lost an amount $V\,dq$ of potential energy, and

$$\text{power dissipated in wire} = V\frac{dq}{dt}$$

$$= VI = I^2R = V^2/R. \tag{4.8}$$

Before closing this section we make a cautionary remark about the above discussion of currents in conductors. The picture of conduction electrons moving freely in conductors is an oversimplification, although it gives a good qualitative description of many of the properties of electric currents. The proper description of metallic conduction requires the use of quantum mechanics, and we will not consider it in this book.

4.1.2 The calculation of resistance

The calculation of the resistance between two electrodes on a piece of conducting material of arbitrary shape is not always easy, even if the material is uniform and obeys Ohm's law. We will outline the technique used in the general case by considering a simple example.

We will calculate the resistance between the inner and outer conductors of the cable shown in Figure 4.1. The inner wire is of radius a and lies along the axis of the outer conducting sheath of radius b, which is earthed. The space between the conductors is filled with a material which is not a perfect insulator, but has a conductivity σ. What is the resistance between the inner and outer conductors of a length L of the cable? This resistance may be of practical significance since it determines the leakage current to ground when the cable is used to transmit a voltage along the inner wire. To determine this resistance we have to calculate the current I that flows from the inner wire to the outer sheath, over a length L of the cable, when the inner wire is at a potential V. Now

$$I = \int_{S_1} \mathbf{j} \cdot d\mathbf{S}, \tag{4.9}$$

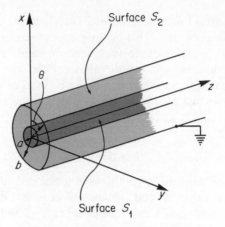

Figure 4.1. A cable consisting of coaxial cylindrical conductors. The space in between them is filled with an imperfect insulator.

where **j** is the current density in the insulator at the surface S_1 of a length L of the inner wire. (The current could be obtained by taking the surface integral of the current density over any surface between the conductors that encloses the inner wire. We choose surface S_1 merely for the sake of convenience.) From Equation (4.4)

$$\mathbf{j} = \sigma\mathbf{E},$$

where **E** is the electric field over the surface S_1. This field is perpendicular to the metal surface and by symmetry has a constant magnitude $E(a)$. Equation (4.9) can now be written

$$I = 2\pi a L \sigma E(a), \tag{4.10}$$

and the problem reduces to the determination of the field **E** in terms of the quantities V, a and b.

When a source of e.m.f. causes a steady current to flow, the electric field outside the source is everywhere static. The field **E** can thus be written in terms of the electrostatic potential ϕ according to Equation (1.21),

$$\mathbf{E} = - \operatorname{grad} \phi. \tag{4.11}$$

We now need an equation from which we can determine the potential everywhere within the material. Combining Equations (4.4) and (4.11) gives

$$\mathbf{j} = -\sigma \operatorname{grad} \phi,$$

or

$$\operatorname{div} \mathbf{j} = -\sigma \operatorname{div} \operatorname{grad} \phi = -\sigma \nabla^2 \phi.$$

The divergence of the current density at any point is a measure of the flux of the vector **j** that flows out over the surface of a very small volume surrounding the point. If the current is steady as much charge enters any small volume within the material as leaves it, and the divergence of **j** is everywhere zero. Hence the equation

$$\nabla^2 \phi = 0, \tag{4.12}$$

is obeyed everywhere within the conducting medium. Equation (4.12) is Laplace's equation, the same equation for the electrostatic potential as we derived in regions in which no current flows and there is no volume density of free charge. In the present problem we require a solution to Equation (4.12) which obeys the boundary conditions that the potential has the constant value V at all points on the surface S_1, and the constant value zero at all points on the surface S_2. This problem has already been solved in section 3.3. For a very long cable the potential ϕ does not vary in the z-direction. Since the cable is cylindrically symmetrical ϕ is independent of the angle θ shown on Figure 4.1, and depends only on the distance r from the z-axis. The required solution of Laplace's equation is

$$\phi = A \ln (r) + C, \tag{4.13}$$

where the constants A and C are determined from the boundary conditions to be

$$A = V/\ln \left(\frac{a}{b}\right)$$

$$C = -V \ln(b)/\ln \left(\frac{a}{b}\right).$$

From Equations (4.10) and (4.11) we find

$$I = -2\pi a L \sigma V/a \ln \left(\frac{a}{b}\right),$$

and hence,

$$R = \ln \left(\frac{b}{a}\right)/2\pi\sigma L. \tag{4.14}$$

A typical insulator has a conductivity of about $10^{-12}\,(\Omega\,\text{m})^{-1}$. The leakage resistance of 100 m of a cable whose inner wire has a radius of 1.5 mm and outer wire a radius of 10 mm, if filled with such an insulator, would be about $3 \times 10^9\,\Omega$.

This example of the calculation of resistance illustrates the analogy between the solution of steady current problems and the solution of electrostatic problems in free space. In each case a solution of Laplace's equation is required, with the boundary conditions that the potentials over metallic surfaces are constant. In the calculation of resistance the metallic surfaces are the electrodes through which the current enters and leaves the resistive material. In the analogous electrostatic problem the metallic surfaces act as the two conductors of a capacitor.

We emphasize that the calculation of resistance made above is valid only when Ohm's law is obeyed by the material. The steps between Equations (4.11) and (4.12) require that Ohm's law holds, and this is not always so. For example, the law breaks down for many materials at high electric fields.

4.2   THE MAGNETIC FIELD

4.2.1   The Lorentz force

As well as the forces exerted on charges by electric fields, there may be additional *magnetic forces* experienced by moving charged particles. Magnetic forces always act perpendicularly to the direction of motion of the charged particles. For example, if two parallel metal wires are brought near to each other and steady currents are set up in them, there is a force between the two as shown in Figure 4.2. If the currents are in the same direction the force is attractive; if the currents are in opposite directions the force is repulsive. The force

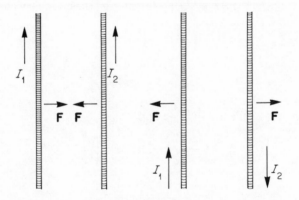

Figure 4.2. The forces between parallel wires carrying currents. Currents in the same direction cause attraction; currents in opposite directions cause repulsion.

between the wires is not the same as the electrostatic force since the wires contain equal amounts of positive and negative charge and are electrically neutral.

Permanent magnets also cause forces to act on wires carrying currents. If a current is passed through a wire suspended between the poles of a permanent magnet the wire experiences a force perpendicular to the direction of the current and is deflected. If the direction of the current is reversed the deflection of the

wire is reversed. A wire with a steady current flowing in it and a permanent magnet both have the same effect on another wire carrying a current. In order to
describe these effects we introduce the *magnetic field* **B** (sometimes called the
magnetic induction field).

The current I_1 in the wire shown on Figure 4.2 gives rise to a field $\mathbf{B}_1$ which
describes the forces on the moving charges constituting the current I_2 in the
other wire. The current I_2 can be seen from Equation (4.3) to be proportional to
the product of the mean speed of the electrons and the electronic charge.
Experiments show that the force on the wire carrying the current I_2 is proportional to I_2, for a fixed current in the first wire. Hence, in a constant magnetic
field, the magnitude of the force on a moving charge is proportional to the
product of the charge q and the speed v of the charge.

We may do more experiments to determine how the force on a moving charge
depends on its velocity and the magnetic field in which it moves. Let us make the
reasonable assumption that the field **B** in the middle of the gap between the
parallel pole faces of a large permanent magnet is uniform and in a direction perpendicular to the pole faces, as shown in Figure 4.3. Experiments with charged

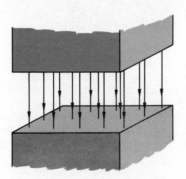

Figure 4.3. A uniform magnetic
field between the pole faces of a
permanent magnet. We represent the magnetic field in strength
and direction with field lines,
just as for the electrostatic field.

particles moving in the gap show that the particles always experience a force **F**
which is perpendicular both to the field direction and to the velocity **v** of the
particles. For example the electron beam of a cathode ray tube placed in the gap
is deflected parallel to the pole faces. The magnitude of the force is found to be
proportional to the speed of the particle, its charge q, and to the sine of the
angle between the vectors **v** and **B**. These experiments are illustrated schematically in Figure 4.4. Their results can be summarized by writing

$$\mathbf{F} \propto q\mathbf{v} \wedge \mathbf{B}.$$

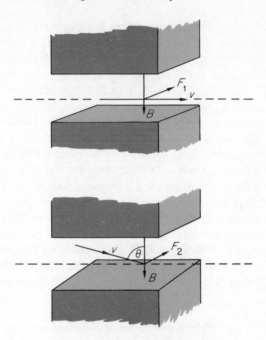

Figure 4.4. The force on a positively charged particle moving in a uniform magnetic field. If the particle moves parallel to the pole faces, as shown in the top diagram, the magnitude of the force $\mathbf{F}_1$ is proportional to the product vB. When the velocity $\mathbf{v}$ of the particle makes an angle θ with the direction of the field $\mathbf{B}$ the magnitude of the force $\mathbf{F}_2$ is $vB \sin \theta$.

The units of the magnetic field $\mathbf{B}$ are defined by choosing the constant of proportionality to be unity, leading to

$$\boxed{\mathbf{F} = q\mathbf{v} \wedge \mathbf{B}.}$$ (4.15)

This equation is known as the *Lorentz force law*. The direction of the force is illustrated in Figure 4.5. In a Cartesian coordinate system the force $\mathbf{F}$ has components:

$$F_x = q(v_y B_z - v_z B_y)$$
$$F_y = q(v_z B_x - v_x B_z)$$
$$F_z = q(v_x B_y - v_y B_x).$$

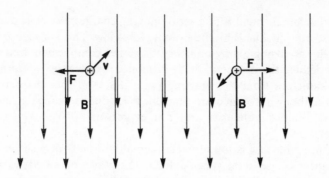

Figure 4.5. The direction of the force on a positively charged
particle moving in a magnetic field.

The definition (4.15) shows that the dimensions of the field **B** are [mass] [charge]$^{-1}$[time]$^{-1}$. The unit of magnetic field is such that a charge of one coulomb moving with a speed of 1 m s^{-1} perpendicularly to unit field experiences a force of one newton. This unit is called the *tesla* (T). The strengths of magnetic fields are often given in an alternative unit, called the *gauss* (G), which is ten thousand times smaller than the tesla,

$$1\ \mathrm{G} \equiv 10^{-4}\ \mathrm{T}.$$

A magnetic field of one gauss is comparable with the Earth's magnetic field. More precisely the field at the surface of the Earth varies from about 0.3 gauss at the equator to about 0.6 gauss at the poles. Typical iron-cored electromagnets can produce steady fields of up to about 2 T, while air-cored electromagnets can generate fields of 100 T for a short time over volumes of a few cubic cm. On a microscopic scale fields as large as 10^4 T have been found to occur near the nuclei of certain atoms.

If an electric field **E** is present as well as a magnetic field **B**, a charge q experiences an additional force $q\mathbf{E}$ and the total force is

$$\mathbf{F} = q\mathbf{E} + q\mathbf{v} \wedge \mathbf{B}. \tag{4.16}$$

This equation defines the fields **E** and **B** in free space, where the force on a moving test charge can be measured. The test charge must be small enough not to affect the charge and current distributions that give rise to the fields, otherwise the measured force will not give those fields present without the test charge.

The way we have attributed steady magnetic fields to steady currents is analogous to the way we ascribed steady electric fields to stationary charges. The electrostatic field was defined in Equation (1.4) via the inverse square law. A similar approach to the discussion of steady magnetic fields due to currents would be to define the field **B** of a set of moving charges. We choose not to do this, but instead to define **B** by the Lorentz force law. This is because the magnetic field due to a single moving charge is rather complicated, and it is very difficult to deduce the magnetic fields generated by complex patterns of currents in conductors from this starting point. Later in this chapter we shall derive laws enabling us to calculate magnetic fields due to steady currents. These laws will eventually enable us to derive an expression for the magnetic field of a moving charge.

The Lorentz force law is found experimentally to be true even for particles moving at speeds close to the speed of light. The charge on a body is thus independent of its speed. All observers in uniform relative motion assign the same value to the charge on a body. Electric charge is therefore relativistically invariant just as the rest mass of a body is invariant in relativistic mechanics. We shall find later, in Chapter 14, that the laws of electromagnetism which we will derive require no modification to satisfy the principle of special relativity.

4.2.2   Magnetic field lines

The magnitude and direction of a magnetic field can be represented on a diagram by field lines, in the same way as an electrostatic field can be shown by lines of force. The direction of the magnetic field at any point is given by the direction of the field line at the point and in a sense given by an arrow on the line; the magnitude of the field is proportional to the density of field lines near the point.

One approach to the determination of the properties of the magnetic field, and hence its relationship to the currents that produce it, is to determine experimentally the fields produced by simple current distributions. The measurement of the Lorentz force on a moving charge provides (at least in principle) a method for determining magnetic fields. The direct measurement is usually not practical, but an indirect measurement, based on an effect known as the *Hall effect*, is often used*. We will outline the theory of the Hall effect meter before discussing the magnetic fields that arise in a few simple examples.

In a Hall effect meter there is a current I flowing in a thin conducting rectangular bar, as shown in Figure 4.6. Suppose that the current, from left to right, corresponds to the net motion of electrons from right to left with mean drift velocity **v**. If there is a uniform field **B** acting over the bar in the direction shown, the electrons will experience a force **F** equal to $-e\mathbf{v} \wedge \mathbf{B}$ which tends to move them over towards the face P of the bar. Electrons will be forced towards this face until a charge distribution is set up which generates an electric field **E** just big enough to prevent more electrons from accumulating on the side near face P. The electric force $-e\mathbf{E}$ on the electrons is then equal and opposite to that due to the magnetic field,

$$-e\mathbf{E} = e\mathbf{v} \wedge \mathbf{B}.$$

---

* Other methods of measuring magnetic fields are discussed in section 6.4.

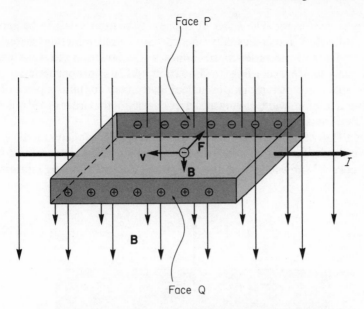

Face P

Face Q

Figure 4.6. If a current flows in a conducting bar placed in a magnetic
field, charges are produced on the faces P and Q as shown. This
charge distribution gives rise to a uniform transverse electric field.

The occurrence of the transverse electric field is known as the *Hall effect*. The
electric field in the bar has constant magnitude vB, and is in the direction from
face Q to face P. If l is the distance between the faces P and Q, the potential
V_{PQ} between the faces is

$$V_{PQ} = El = vBl.$$

The current I is given by Equation (4.3)

$$I = NevA,$$

where N is the number of conduction electrons per unit volume, and A is the
cross-sectional area of the bar. The potential difference between the faces is thus

$$V_{PQ} = \frac{Il}{NeA}B. \qquad (4.17)$$

A measurement of the potential difference across the bar thus gives the magnitude of the field **B** if the other quantities in Equation (4.17) are known.

   Let us examine a common metal like copper in order to see whether it would
be a suitable candidate for use in a Hall effect meter. If the bar were 0.5 cm wide
($l = 0.5$ cm), and 0.1 cm thick ($A = 5 \times 10^{-6}$ m^2), and carried a current of 1 A,
Equation (4.17) predicts a value of the potential difference V_{PQ} equal to about
10^{-7} V for a field of one tesla, assuming that there is one conduction electron

for every copper atom. The effect in copper is thus too small to be generally useful. Hall effect meters usually incorporate semiconducting materials in which there are of the order of 10^7 times less conduction electrons per unit volume, and hence according to Equation (4.17) a proportionately larger potential difference. A typical commercial instrument contains a piece of semiconductor a few mm wide, having a carrier density of the order of 10^{15} electrons per cm^3, and passing a current of a few mA.

Using a Hall effect meter consisting of a thin slice of semiconductor we may now investigate the magnetic field lines around some simple current distributions. Figure 4.7 shows the field due to a steady current in a long straight wire.

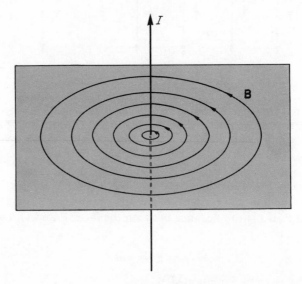

Figure 4.7. The magnetic field of a current in a long straight wire.

The field is cylindrically symmetrical, and its magnitude is inversely proportional to the perpendicular distance from the wire. The direction of the field at any point is perpendicular to both the wire and the perpendicular from the point to the wire. The field lines are circles, and the direction of the arrows on the circles is given by the *right-hand screw rule*. This rule says that the circles go round in the direction the threads of a right-hand screw would rotate if it were screwed in the direction of the current.

Figure 4.8 shows the magnetic field lines from a small loop of wire carrying a current. (The leads carrying the current into and out of the loop are close together so that their magnetic effects cancel. They are not drawn.) The field from the small current loop is axially symmetrical, i.e. it is independent of the

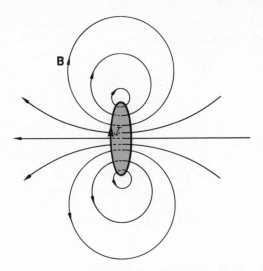

Figure 4.8. The magnetic field of a current in a
small circular loop of wire.

azimuthal angle with respect to the axis of the coil. This means that a map of
the field lines would look the same for any plane containing the axis of the coil.
At distances from the loop large compared with the dimensions of the loop,
the magnetic field varies in the same way as the electric field of an electric
dipole (compare Figure 4.8 with Figure 1.22).

There is one obvious feature of the field lines on Figures 4.7 and 4.8 that is
shared by the magnetic field lines arising from *any* current distribution. *The
lines of the field* **B** *are continuous.* Unlike electrostatic field lines, lines of the
field **B** have no beginning and no end. This means that there are no free 'magnetic charges' or 'magnetic poles', a fact which has important consequences for
all of electromagnetism. Let us write this simple experimental fact in the form
of a convenient mathematical equation. Consider a closed surface S around a
volume V. Since the lines of the field **B** are continuous, as many lines enter the
volume V as leave it. This means that the total outward flux of the field **B** over
the surface S is zero,

$$\int_S \mathbf{B} \cdot d\mathbf{S} = 0. \tag{4.18}$$

We may write this equation using the divergence theorem, Equation (1.17),
to give

$$\int_V \operatorname{div} \mathbf{B} \, d\tau = 0.$$

This equation is true whatever the size of the volume V, hence at all points the integrand must be zero, and everywhere

$$\boxed{\operatorname{div} \mathbf{B} = 0.} \tag{4.19}$$

Equation (4.19) is one of the fundamental laws of electromagnetism, on the same footing as Coulomb's law. There have been no observed violations of this law that the divergence of the magnetic field $\mathbf{B}$ is everywhere zero; no free magnetic 'charges' have yet been discovered. The development of the theory of electromagnetism depends on this last negative observation. Since electromagnetic theory always gives the right answer we have very strong indirect evidence for the general validity of Equation (4.19).

4.3 THE MAGNETIC DIPOLE

The whole of this section is devoted to a more detailed discussion of the magnetic properties of small current loops like the one in Figure 4.8. There are two reasons why the small current loop deserves such attention. First it will be seen later that we can evaluate the field generated by any circuit carrying a current in terms of the field due to a small loop. Secondly the circulating electrons in an atom are themselves small current loops and we shall make use of the results of this section when studying the magnetic behaviour of atoms.

4.3.1 Current loops in external fields

We begin by showing that a small loop carrying a current behaves in an external magnetic field rather like an electric dipole in an external electric field.

Consider the small rectangular coil shown in Figure 4.9, which is suspended in a vertical plane, and consists of a single loop of wire. (The input and output leads are wound close together so that we may neglect their magnetic effects.) The coil is placed in a magnetic field $\mathbf{B}$ which is constant over the area of the coil. The field lines of $\mathbf{B}$ are horizontal, at an angle θ to the normal to the plane of the coil, as shown in the diagram. If the coil carries a current I and the wire has a cross-sectional area A, then

$$I = NevA,$$

where N, e and v have the same meanings as in Equation (4.3). In a small length $\mathrm{d}l$ of the coil on the side PQ there are $AN\,\mathrm{d}l$ electrons moving upwards with speed v. The force on these is

$$\begin{aligned} \mathrm{d}\mathbf{F}_1 &= -e\mathbf{v} \wedge \mathbf{B}(AN\,\mathrm{d}l) \\ &= I\,\mathrm{d}\mathbf{l} \wedge \mathbf{B}, \end{aligned} \tag{4.20}$$

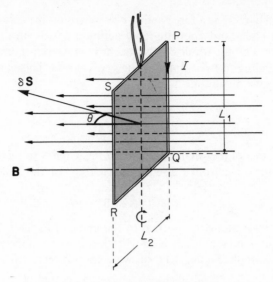

Figure 4.9. A rectangular coil carrying a current, situated in a uniform magnetic field. The coil rotates about an axis perpendicular to the field direction.

where the vector $d\mathbf{l}$ is in the direction of the current along PQ. The total force $\mathbf{F}_1$ on the side PQ thus has magnitude IL_1B, since the side PQ and the field $\mathbf{B}$ are everywhere perpendicular. This force is in the direction shown in Figure 4.10, which is a view looking down on the coil. There is a force $\mathbf{F}_2$ of equal magnitude on the side RS of the coil, but it is in the opposite direction to the

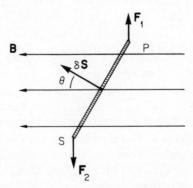

Figure 4.10. A view looking down on the rectangular coil, showing the directions of the forces $\mathbf{F}_1$ and $\mathbf{F}_2$ and the direction of the vector $\delta\mathbf{S}$.

force $\mathbf{F}_1$. Similarly there is a force acting downward on the side QR but this is balanced by an equal and opposite force acting upwards on the top side PS. The resultant force on the whole coil is thus zero. However, there is a net torque $\mathbf{T}$ on the coil tending to twist it about the vertical axis. This can be seen from Figure 4.10 to have magnitude

$$T = L_1 L_2 I B \sin \theta$$
$$= I \, \delta S \, B \sin \theta, \tag{4.21}$$

where δS is the area of the small coil. The torque $\mathbf{T}$ tends to twist the coil anti-clockwise (looking down from above the coil) and so is in the upwards direction, i.e.

$$\mathbf{T} = I \, \delta \mathbf{S} \wedge \mathbf{B}, \tag{4.22}$$

where $\delta \mathbf{S}$ is a vector along the normal to the plane of the coil. Its direction is given by the right-hand screw rule used with the direction of the current around the loop. For example Figure 4.11 shows a loop of current looked at from

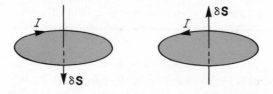

Figure 4.11. The right-hand screw rule. The direction of the vector $\delta \mathbf{S}$ associated with the area of the loop is given by the direction in which a right-hand screw would advance if rotated in the same direction as the current in the loop.

above; if the current is clockwise the vector $\delta \mathbf{S}$ is down; if the current is anti-clockwise the vector $\delta \mathbf{S}$ is up. We have written the torque on the coil in the form of Equation (4.22) because it can be shown that this equation is true for a plane coil of *any* shape when placed in a uniform magnetic field.

In Equation (4.22) the field $\mathbf{B}$ is the external field present at the coil before the current was passed through it. The total field near the coil when it has the current I flowing in it is equal to the sum of the external field and the field arising from the current I. Now the force on a moving charge is related to the total field from all sources (other than the charge itself) at the position of the moving charge, so that it might seem that the derivation above needs to be modified. However, that part of the total field that arises from the current in the coil itself will cause no net force or torque on the coil. In the absence of an external field a coil through which a current is passed does not move or rotate. Hence the field **B** in Equation (4.22) can equally well be taken to be the total magnetic field.

Equation (4.21) enables us to calculate the potential energy U_P of a current loop in an external field. The torque on the coil in Figures 4.9 and 4.10 tends to *decrease* the angle θ. In terms of the potential energy the torque tending to *increase* θ is $-\partial U_\mathrm{P}/\partial\theta$. Hence from Equation (4.21)

$$-\partial U_\mathrm{P}/\partial\theta = -I\,\delta S\,B\sin\theta.$$

Integration of this equation gives

$$U_\mathrm{P} = -I\,\delta S\,B\cos\theta + \text{constant.}$$

If we choose the constant so that the potential energy U_P is zero when θ is equal to $\pi/2$ we have

$$U_\mathrm{P} = -I\,\delta S\,B\cos\theta$$

$$= -I\,\delta\mathbf{S}\cdot\mathbf{B}. \tag{4.23}$$

By considering the torque exerted on the coil by a vertical component of the magnetic field, it can be shown that this equation holds whatever the direction of **B**. Equation (4.23) is similar in form to Equation (2.11),

$$U = -\mathbf{p}\cdot\mathbf{E},$$

which gives the potential energy of an electric dipole, of dipole moment **p**, in an electric field **E**. Forces and torques on a small loop of current in a magnetic field can thus be determined by associating a magnetic dipole moment $I\,\delta\mathbf{S}$ with the loop.

4.3.2   The field of a magnetic dipole

We have already observed in Figure 4.8 that the magnetic field produced by a current loop is similar in form to the electrostatic field of an electric dipole. In the above section we showed that in an external magnetic field a current loop behaves like an electric dipole in an electric field. The equivalence of small current loops and magnetic dipoles was first noted by Ampère during a series of experiments he made before 1820. In these experiments he showed that two small coils carrying currents exerted forces and torques on each other which varied with the distance apart and the relative orientation of the coils in the same way as the forces and torques between two electric dipoles.

We define the *magnetic dipole moment* **m** of a small coil of area δS carrying a current I to be

$$\mathbf{m} = I\,\delta\mathbf{S}, \tag{4.24}$$

where $\delta\mathbf{S}$ is the vector related to the current sense by the right-hand screw rule. The dependence of the field **B** of a very small current loop on position is the same as that of the field of an electric dipole, which can be calculated as outlined in section 1.5.3. For example, on the axis of the loop at large distances r,

$$\mathbf{B} \propto \frac{2\mathbf{m}}{r^3}. \tag{4.25}$$

Let us now examine the constant of proportionality required in this equation. We put

$$\mathbf{B} = \frac{\mu_0}{4\pi} \frac{2\mathbf{m}}{r^3}, \tag{4.26}$$

writing the constant as $\mu_0/4\pi$, where the factor 4π has been introduced in order to simplify the form of some important equations. The constant μ_0 is called the *permeability of free space*. The right-hand side of expression (4.25) has the dimensions [current][length]$^{-1}$, whilst the dimensions of the field $\mathbf{B}$ are, from Equation (4.15), [mass][current]$^{-1}$[time]$^{-2}$. Thus the constant μ_0 has the dimensions [mass][length][charge]$^{-2}$. The numerical value of the constant cannot be chosen arbitrarily, since the units of all the other quantities have already been determined. The numerical value of the constant is exactly $4\pi \times 10^{-7}$ SI units.*

Magnetic dipole moments are measured in A m^2. A flat coil consisting of 2000 turns of thin wire, each of area 50 mm^2 and carrying a current of 100 μA has a dipole moment of $m = 2 \times 10^3 \times 5 \times 10^{-5} \times 10^{-4} = 10^{-5}$ A m^2. Such a coil would typically be found in a multipurpose moving-coil current measuring instrument such as an 'Avometer'. If the coil, with 100 μA flowing in it, were placed in a uniform magnetic field of 1000 G with its magnetic moment per-perpendicular to the field direction, the torque on the coil would be 10^{-6} N m. This is roughly equal to the torque produced by two forces each equal to the weight of 10^{-4} g at the surface of the earth acting over a lever arm of length 1 m. The magnetic field produced by the current of 10^{-4} A in the coil at a point distant 1 m from the centre of the coil and on its axis is equal to 2×10^{-12} T.

4.4   AMPÈRE'S LAW

In this section we proceed from very small current loops to more realistic circuits. We investigate the behaviour of the magnetic field of a current carrying coil both near to and far away from the coil (where it looks like the field of a dipole). We derive an expression relating the line integral of the field $\mathbf{B}$ around a closed path to the current that passes through the area traced out by the path.

---

* As one may guess, the convenient value for the constant μ_0 is no accident. In the SI system of units the permeability of free space is chosen to be exactly equal to $4\pi \times 10^{-7}$. Only two quantities need to have their values fixed by reference to experiment in order to give a consistent account of both electric and magnetic forces. The unit of charge (the coulomb) and the permittivity of free space ε_0 are determined in the SI system from experimental data. The hierarchy of electrical units is discussed in detail in Appendix A. Our somewhat perverse presentation of units so far is the result of proceeding in the most convenient way for the development of the subject as a whole.

This integral relationship is called *Ampère's law*. It is analogous to Gauss' law, Equation (1.12), in electrostatics, and can be used to determine the magnetic fields due to some simple symmetric current distributions. Ampère's law is one of the most important laws of electromagnetism, and we shall see later that it can be expressed in a differential form which enables one to determine the fields of more complicated current distributions.

4.4.1 The field of a large current loop

We will now examine the field of a current-carrying coil in more detail. We will first introduce a method for calculating the field near to the coil. This method considers the coil to be a sheet of magnetic dipoles, and a similar method can be used to determine the electrostatic field of a layer of electric dipoles. By discussing the physical differences between the two situations we derive an expression for the line integral of the field **B** over a closed path threading the coil.

At points near to a coil which carries a current, we cannot treat the whole coil as one dipole, but we can consider it to be made up of very many dipoles by introducing the idea of the current grid. Figure 4.12 shows a loop of wire with a contour which we shall denote by the symobl s. Consider the flat surface S whose rim is the contour s and imagine this surface divided by a fine grid into very many small surfaces δS. If the same current I flows clockwise around the perimeters of all these small surfaces, the currents cancel except at the edge of the grid and the net result is a current I flowing clockwise around the contour s. Thus, for the purpose of calculating its magnetic effects, a current in a flat coil

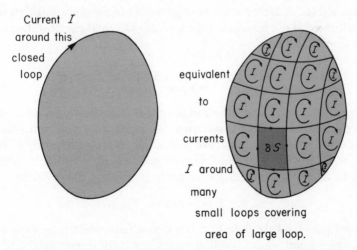

Figure 4.12. The equivalence of a loop of wire carrying a current and a current grid. The current I around the perimeter of an area like δS inside the grid is cancelled by the currents around the perimeters of surrounding areas. The result is a current I around the original loop.

can be replaced by the same current flowing around the perimeters of the very many small surfaces that together make up the surface area of the grid. If we make the grid become more and more fine, the areas δS tend to become vanishingly small elemental areas dS. Each one of these can be considered to give rise to a dipole field at any point, no matter how near the point is to the coil.

Let the direction of the current I around the perimeters of the areas dS define the direction of vectors $d\mathbf{S}$ via the right-hand screw rule. The current grid is equivalent to a sheet of elementary magnetic dipoles, each of moment $d\mathbf{m} = I\,d\mathbf{S}$. The magnetic field at any point can now be calculated from the sum of the fields of these elementary magnetic dipoles.

We may compare our problem, which we have reduced to that of determining the field of a uniform sheet of magnetic dipoles, to the one where we want to know the electrostatic field of a uniform sheet of electric dipoles. The fields outside an electric or a magnetic dipole have the same spatial variation. Hence the magnetic field on the right hand side of the current loop, for example, has the same dependence on position as the electric field on the right-hand side of a sheet of electric dipoles with similar size and shape to the flat surface S.

Let us consider the electrostatic field of a sheet of electric dipoles, and let $\mathbf{p}$ be its electric dipole moment per unit area. Each elemental area dS of the sheet contributes an amount $d\phi$ to the electrostatic potential ϕ at a point P, where

$$d\phi = \frac{1}{4\pi\varepsilon_0}\frac{\mathbf{p}\,d\mathbf{S}\cdot\mathbf{r}}{r^3},$$

from Equation (1.26). Here $\mathbf{r}$ is the position vector of the point P with respect to an origin located at the position of the area dS. The total electrostatic potential from the whole sheet is

$$\phi = \frac{1}{4\pi\varepsilon_0}\int_S dS\frac{\mathbf{p}\cdot\mathbf{r}}{r^3}. \tag{4.27}$$

This expression is simply the dipole moment per unit area times the solid angle subtended at the point P by the surface S. For points very near to the sheet, the solid angle becomes 2π and at these points the electrostatic potential has the values $\pm p/2\varepsilon_0$ according to which side of the sheet the point is situated. The electrostatic field is obtained from the potential ϕ using the equation

$$\mathbf{E} = -\text{grad }\phi.$$

The line integral I_E of the field $\mathbf{E}$, taken over a path which goes from a point at infinity to a point whose position vector is $\mathbf{r}$ (with respect to the centre of the sheet as origin), is thus given by

$$I_E = \int_{\infty}^{\mathbf{r}} \mathbf{E}\cdot d\mathbf{l} = -\phi(\mathbf{r}),$$

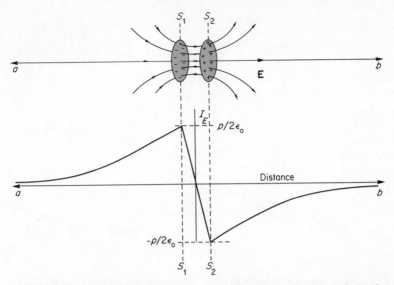

Figure 4.13. The variation of the line integral of the electrostatic field of a sheet of electric dipoles. The integral is taken over a path between two distant points a and b on the axis of the sheet of dipoles and on either side of it. The real positive and negative charges constituting the electric dipole sheet are shown separated on two surfaces S_1 and S_2 in order to make the diagram clearer.

choosing the potential to be zero at infinity. Figure 4.13 shows the variation of the line integral I_E as we traverse a path from point a to point b on the axis of the dipole sheet. Just to the left of the sheet the line integral reaches the value $p/2\varepsilon_0$. As we pass through the sheet there is a region of space, however narrow, over which the direction of the electric field is reversed. This is the space between the layers of positive and negative charge which constitute the dipole sheet. This space is drawn enlarged on Figure 4.13 and the surfaces on which the positive and negative charges are distributed are labelled S_1 and S_2 for clarity. The reversal of the field direction changes the integral I_E to the value $-p/2\varepsilon_0$ just to the right of the charges, and I_E rises back to zero as we proceed to the point b at infinity on the right. If we complete a closed path of integration between a and b by closing the loop at very large distances where the field is zero, over this path

$$\oint \mathbf{E} \cdot d\mathbf{l} = 0,$$

and it can be seen that the reversal of the field between the layers of electric charge is responsible for this result.

Now let us consider the current loop shown in Figure 4.14. We will work out the line integral of the field $\mathbf{B}$ over a path taken from the point a far away from the coil to the point b, and closed at very large distances between a and b.

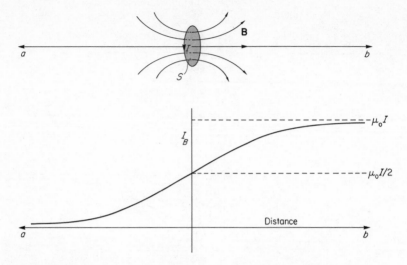

Figure 4.14. The variation of the line integral of the magnetic field **B** of a current in a loop of wire. The integral is taken over a path between two distant points a and b on the axis of the loop and on either side of it.

From the point a to the coil the variation of the line integral I_B has the same shape as the variation of the line integral of the electric field discussed above. To obtain the value of I_B we note that the magnetic field of a dipole is obtained from the formula for the electric field of a dipole by replacing the term $1/4\pi\varepsilon_0$ by the term $\mu_0/4\pi$, and the electric dipole moment by the magnetic dipole moment. Thus the term $d S\,\mathbf{p}$ in Equation (4.27) is replaced by the term $I\,d\mathbf{S}$, and the value of the line integral of the magnetic field just to the left of the coil becomes $\mu_0 I/2$. As the path of integration goes through the current loop the lines of the field **B** are continuous. There are no magnetic 'charges' and the field lines do not reverse direction. The curve showing how the line integral I_B varies with distance is thus continuous through the surface S. On the right-hand side of the loop the curve again has the same shape as the curve for the line integral of the electric field, since the field distributions are the same. The resulting curve of I_B versus distance is shown in Figure 4.14. If we complete a closed path of integration at very large distances between the points a and b, there is no contribution to the integral from this part, and

$$\oint \mathbf{B} \cdot d\mathbf{l} = \mu_0 I.$$

This expression remains true for any closed path that threads the current loop and not just the one on axis that we considered. If we evaluate the line integral of the field **B** over a closed path which does not go through the current loop, the variation of the magnetic field over the whole path is exactly the same as the variation of the electrostatic field of the electric dipole sheet, and

$$\oint \mathbf{B} \cdot d\mathbf{l} = 0.$$

These results can be summarized in *Ampère's law*, which states that the line integral of the field **B** around any closed path is equal to the constant μ_0 times the current flowing through the area enclosed by the path.

$$\oint \mathbf{B} \cdot d\mathbf{l} = \mu_0 I. \tag{4.28}$$

The sense of traversal of the loop when the line integral is performed is determined by the direction of the current together with the right-hand screw rule.

On either side of the current-carrying coil the magnetic field varies in the same way as does the electrostatic field of the electric dipole sheet. This means that to the right and to the left of the coil the field **B** can be obtained by taking the gradient of a scalar function, just as the electrostatic field can. This scalar function is called the *magnetostatic potential*, and we give it the symbol ψ, where

$$\mathbf{B} = -\operatorname{grad}\psi.$$

The magnetostatic potential to be used in the example of a coil carrying a current discussed above is given by analogy with Equation (4.27).

$$\psi = \frac{\mu_0 I}{4\pi} \int_S \frac{d\mathbf{S} \cdot \mathbf{r}}{r^3}$$

$$= \frac{\mu_0 I}{4\pi} \Delta\Omega_P. \tag{4.29}$$

In this equation ψ is the magnetostatic potential at a point P where the coil subtends a solid angle $\Delta\Omega_P$. The potential ψ is not single valued, as the line integral of the field **B** around a closed loop need not be zero. If we traverse a closed path through the loop of Figure 4.14 each time we return to the same point the potential ψ has changed by an amount $\mu_0 I$. This does not prevent magnetic fields from being determined using the magnetostatic potential, as it is the differences in this potential from one point to a nearby point which determine the fields.

We shall see later that the field **B** cannot always be obtained by taking the gradient of a scalar function. If currents are present in a particular region of space the magnetostatic potential cannot be defined in that region. For example it is not possible to use a magnetostatic potential inside a conductor carrying a steady current. Even where the scalar potential is applicable in practical problems it is not always used to calculate magnetic fields because other methods are sometimes simpler.

4.4.2   Examples of the calculation of magnetic fields

Ampère's law enables us to determine the magnetic fields of certain current distributions whose symmetries ensure that the field lines have a simple pattern. The following examples illustrate the use of Ampère's law and of other ideas developed in the previous section.

(i)   The field inside a long solenoid

Consider a long solenoid ($L \gg d$) with N turns of wire per unit length and carrying a current I, as shown in Figure 4.15. Ampère's law requires that the line integral of the field $\mathbf{B}$ over a closed path which runs along the axis of the solenoid and is closed at very large distances be given by

$$\oint \mathbf{B} \cdot d\mathbf{l} = NL\mu_0 I. \tag{4.30}$$

The term on the right-hand side is μ_0 times the total current that flows through the area traced out by the path chosen for the line integral. Here the path is threaded by NL turns of wire each carrying a current I. The path has to be traversed in the direction from P to Q in order that the sign of the term NLI be positive.

The resultant field is the sum of the fields from all the coils of the solenoid, each of which reinforce the field in the interior, but which tend to cancel each other outside. To a first approximation one may consider the field outside the solenoid to be zero and the field inside to be a constant $\mathbf{B}$ with direction from P to Q. Equation (4.30) then becomes

$$BL = NL\mu_0 I$$

or

$$B = \mu_0 NI. \tag{4.31}$$

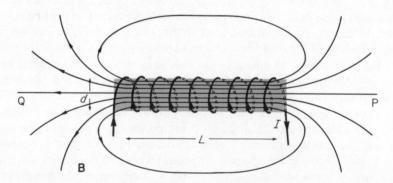

Figure 4.15. The field of a long solenoid carrying a current. The field is very nearly uniform inside the solenoid.

This equation predicts for example that one ampere flowing in a solenoid which has 20 turns per cm produces a field inside equal to about $(20 \times 10^2) \times (4\pi \times 10^{-7}) \times 1.0$; about 25 gauss. The approximations made above become better as the ratio of the length L to the diameter d of the solenoid increases. Equation (4.31) gives the field in the middle of a solenoid to better than 1 % if $L/d > 10$.

The field lines of a typical current-carrying solenoid are shown in Figure 4.15.

(ii)   The field due to a current in a long straight wire

Another simple example is that of the field due to a current I in a long straight wire. In this situation the field has cylindrical symmetry; it must be the same at all points whose perpendicular distances from the wire are equal. Figure 4.16 shows a circle of radius r centred on a point on the wire, and lying in a plane perpendicular to the wire. The field at a point on this circle has constant magnitude B, lies along the tangent to the circle at the point and has a direction given

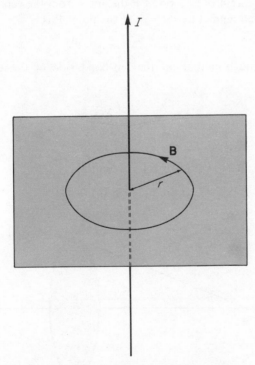

Figure 4.16. The field of a long straight wire carrying a current has cylindrical symmetry. It looks the same at all points on the circle of radius r. Hence it has constant magnitude on the circle and is in a direction tangential to the circle at all points on the circumference, since the line integral of the field around the circle must equal $\mu_0 I$.

by the right-hand screw rule. The line integral of the field **B** around the circle is given by Ampère's law,

$$\oint \mathbf{B} \cdot d\mathbf{l} = 2\pi r B = \mu_0 I.$$

Hence

$$B = \frac{\mu_0 I}{2\pi r}. \tag{4.32}$$

A current of one ampere flowing in a long wire produces a field of 0.2 gauss at a distance of one cm from the wire, a field comparable with the Earth's magnetic field at the surface of the Earth.

(iii)   The field due to a current in a circular loop

The use of the magnetostatic potential and the idea of the current grid are illustrated by this example. Let us calculate the field on the axis of the circular coil shown in Figure 4.17 at a point P distant x from the centre O of the coil. The solid angle subtended by the coil at the point P is

$$\Delta \Omega_P = 2\pi(1 - \cos \theta).$$

The magnetostatic potential on the left-hand side of the coil is given by Equation (4.29)

$$\psi = \frac{\mu_0 I}{4\pi} \Delta \Omega_P,$$

hence

$$\mathbf{B} = -\text{grad}\, \{\mu_0 I(1 - \cos \theta)/2\}$$
$$= -\tfrac{1}{2}\mu_0 I \,\text{grad}\, \{1 - x/(a^2 + x^2)^{1/2}\},$$

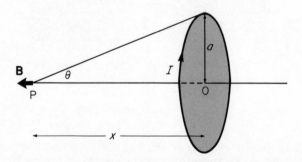

Figure 4.17. The calculation of the magnetic field on the axis of a single circular loop of wire carrying a current. The field lines are shown on Figure 4.8.

where a is the radius of the coil. The gradient of the expression in brackets is along the axis in a direction away from the coil and has magnitude $a^2/(a^2 + x^2)^{3/2}$. The magnetic field is thus given by

$$\mathbf{B} = \mathbf{e}_x \mu_0 I a^2/2(a^2 + x^2)^{3/2}, \tag{4.33}$$

where $\mathbf{e}_x$ is a unit vector in the direction of x increasing along the axis.

The method can be used to calculate fields at points outside thin conductors if the solid angles subtended at the points by the conductors are easy to calculate. If the loop discussed above were a thin coil with N turns of wire, the field given by Equation (4.33) would be multiplied by N. The field on the axis and in the centre of a single loop of radius 1 cm carrying 1 A is given by the equation

$$B = \mu_0 I/2a, \tag{4.34}$$

and is about 0.6 gauss.

(iv)   *The magnetic fields inside atoms*

We will use a very crude model of an atom in order to guess the order of magnitude of microscopic magnetic fields. Consider an electron moving in a circle of radius a around a nucleus with constant speed v. The circular motion of charge is equivalent to a current I around the circle, and gives rise to a magnetic field given by Equation (4.34). The current I is equal to the charge that passes any point on the circle every second, i.e.

$$I = ev/2\pi a .$$

Hence, from Equation (4.34)

$$B = \mu_0 ev/4\pi a^2$$

$$= \frac{\mu_0 e}{4\pi m_e a^3}(m_e va),$$

where m_e is the mass of the electron, and the last equation has been written in terms of the angular momentum $(m_e va)$ of the rotating electron. The reason for doing this is that the angular momentum of an atomic system is restricted to certain definite values by the laws of quantum mechanics. We are already familiar with the quantization of certain physical observables; for example electric charge is always observed in multiples of the electronic charge e. The quantum unit of angular momentum associated with orbital motion of atomic particles is $h/2\pi$ where h is Planck's constant, which has the value 6.6×10^{-34} joule s. The component in a specified direction of the orbital angular momentum of an electron in an atom must be an integral multiple of $h/2\pi$ or zero. Taking the value of $(m_e va)$ to be $h/2\pi$, in order to make a rough estimate of the magnetic field, we obtain

$$B = \frac{\mu_0 eh}{8\pi^2 m_e a^3}. \tag{4.35}$$

Substituting in this expression the value 10^{-10} m for an atomic radius we find that B is about 6 T.

The crude classical model we have used is not a proper description of an atom, and in addition magnetic fields inside atoms and molecules have important contributions from intrinsic magnetic moments possessed by electrons. Nevertheless our calculation does suggest that microscopic magnetic fields are much greater than those typically encountered when currents flow in macroscopic conductors. This is in agreement with experimental observations.

4.5   THE DIFFERENTIAL FORM OF AMPÈRE'S LAW

We have already stated that Ampère's law (4.29) is as basic for the understanding of the effects that arise when steady currents flow in conductors as Gauss' law (1.12) is for the understanding of electrostatics. The next step in the study of magnetostatics is to reduce the integral form of Ampère's law to a differential equation, just as we reduced the integral form of Gauss' law to the differential equation

$$\text{div } \mathbf{E} = \rho/\varepsilon_0.$$

The differential form of Ampère's law is most conveniently expressed in terms of another vector differential operator called *curl*. We will now digress a little in order to introduce this operator.

4.5.1   The operator curl and the vector curl B

The result of performing the curl operation on a vector $\mathbf{B}$ is to produce another vector, curl $\mathbf{B}$. The component of curl $\mathbf{B}$ in a particular direction is defined in terms of the line integral of the vector $\mathbf{B}$ around a closed curve drawn in a plane perpendicular to that direction.

Let us examine what happens to the line integral $\oint \mathbf{B} \cdot d\mathbf{l}$ around the closed path s shown in Figure 4.18, as we make the length of the path s smaller and smaller. This integral will tend to zero as the path s tends to zero. If δS is the area of the surface whose perimeter is the path s the quantity $\oint_s \mathbf{B} \cdot d\mathbf{l}/\delta S$ may tend to a non-zero limit as δS tends to zero. The component of curl $\mathbf{B}$ in the direction of the arrow (given by the direction a right hand screw advances if rotated in the same sense as the circulation of the path s) is defined by the limit

$$(\text{curl } \mathbf{B})_{\text{component}} = \underset{\delta S \to 0}{\text{Limit}} \left\{ \frac{1}{\delta S} \int_s \mathbf{B} \cdot d\mathbf{l} \right\}. \tag{4.36}$$

The operator curl takes different mathematical forms for different systems of coordinates. We will determine its form in a Cartesian coordinate system. The z-component of curl $\mathbf{B}$ can be obtained by taking the limit (4.36) over a closed surface in a plane parallel to the x–y plane. Consider the very small surface shown in Figure 4.19, situated at the point with coordinates x, y and z. The sides of the

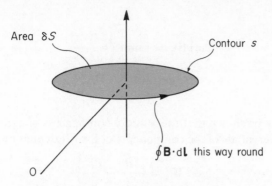

Figure 4.18. In order to obtain the component of curl **B** in the direction of the arrow, the line integral $\oint \mathbf{B} \cdot \mathbf{dl}$ must be calculated as shown, i.e. the loop must be traversed in the direction given by the right hand rule.

surface have lengths δx and δy parallel to the x and y axes respectively. To obtain (curl **B**)$_z$ we have to traverse the path in the direction shown, in which case

$$\oint \mathbf{B} \cdot \mathbf{dl} = B_x(x, y - \tfrac{1}{2}\delta y, z)\,\delta x + B_y(x + \tfrac{1}{2}\delta x, y, z)\,\delta y$$

$$- B_x(x, y + \tfrac{1}{2}\delta y, z)\,\delta x - B_y(x - \tfrac{1}{2}\delta x, y, z)\,\delta y$$

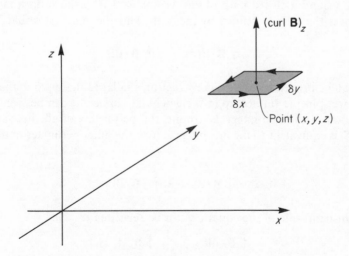

Figure 4.19. The evaluation of the z-component of curl **B** at the point (x, y, z).

Hence

$$(\text{curl } \mathbf{B})_z = \underset{\delta x \delta y \to 0}{\text{Limit}} \left\{ \frac{\oint \mathbf{B} \cdot \mathbf{dl}}{\delta x \, \delta y} \right\}$$

$$= \left(\frac{\partial B_y}{\partial x} - \frac{\partial B_x}{\partial y} \right).$$

Similarly it may be shown that the x and y components of curl $\mathbf{B}$ are given in terms of the differentials of the components of $\mathbf{B}$ by the equations,

$$(\text{curl } \mathbf{B})_x = \left(\frac{\partial B_z}{\partial y} - \frac{\partial B_y}{\partial z} \right)$$

$$(\text{curl } \mathbf{B})_y = \left(\frac{\partial B_x}{\partial z} - \frac{\partial B_z}{\partial x} \right).$$

We can summarize the three equations for the components of curl $\mathbf{B}$ by writing the latter in the form of a determinant.

$$\text{curl } \mathbf{B} = \begin{vmatrix} \mathbf{e}_x & \mathbf{e}_y & \mathbf{e}_z \\ \dfrac{\partial}{\partial x} & \dfrac{\partial}{\partial y} & \dfrac{\partial}{\partial z} \\ B_x & B_y & B_z \end{vmatrix}, \tag{4.37}$$

where $\mathbf{e}_x$, $\mathbf{e}_y$ and $\mathbf{e}_z$ are unit vectors along the x, y and z axes. The forms that the operator curl takes in some other coordinate systems are given in Appendix B.

There is an important theorem called *Stokes' theorem* which relates the line integral around a closed path of any vector field $\mathbf{B}$ to the surface integral of curl $\mathbf{B}$ over the surface defined by the path. This theorem states that

$$\oint_s \mathbf{B} \cdot \mathbf{dl} \equiv \int_S \text{curl } \mathbf{B} \cdot \mathbf{dS}, \tag{4.38}$$

and it can be proved by using a construction similar to that used in the current grid. If we split the surface up into elementary surfaces it can be seen that the summation of the line integrals around the perimeters of all the elementary areas δS_i is equivalent to the line integral over the outer perimeter of the whole surface.

$$\oint_s \mathbf{B} \cdot \mathbf{dl} = \sum_i \oint_{s_i} \mathbf{B} \cdot \mathbf{dl}.$$

The right-hand side of the equation can be rewritten to give

$$\oint_s \mathbf{B} \cdot \mathbf{dl} = \sum_i \frac{1}{\delta S_i} \oint \mathbf{B} \cdot \mathbf{dl} \, \delta S_i$$

$$= \sum_i \text{curl } \mathbf{B} \cdot \mathbf{dS}_i$$

from the definition of curl, Equation (4.36). In the limit of vanishingly small δS_i this leads to Stokes' theorem

$$\oint_s \mathbf{B} \cdot d\mathbf{l} \equiv \int_S \text{curl } \mathbf{B} \cdot d\mathbf{S}.$$

It is not easy to visualize the curl of a vector function. If one is given the form of a magnetic field, only after considerable experience can one quickly answer such questions as 'Is curl $\mathbf{B}$ large, small or zero?', or 'In which direction does curl $\mathbf{B}$ point?'. Since the components of curl $\mathbf{B}$ are all differentials it is obvious that if a vector function is constant everywhere the curl of that function is everywhere zero. Some complicated vector functions however may also have zero curl. For example if we consider the electrostatic field $\mathbf{E}$, since the line integral $\oint \mathbf{E} \cdot d\mathbf{l}$ over any closed path is zero, it follows from Stokes' theorem that curl $\mathbf{E}$ is always zero.

Now that we have introduced the operator curl we can express Ampère's integral relation (4.28) as a differential equation which the field $\mathbf{B}$ must satisfy everywhere. Recalling that the current I through an area S is given by

$$I = \int_S \mathbf{j} \cdot d\mathbf{S},$$

Equation (4.28) can be written

$$\oint_s \mathbf{B} \cdot d\mathbf{l} = \mu_0 \int_S \mathbf{j} \cdot d\mathbf{S}. \tag{4.39}$$

The line integral may be converted into an integral over the surface S by using Stokes' theorem, Equation (4.38),

$$\int_s \mathbf{B} \cdot d\mathbf{l} \equiv \int_S \text{curl } \mathbf{B} \cdot d\mathbf{S}.$$

Substituting in Equation (4.39),

$$\int_S \text{curl } \mathbf{B} \cdot d\mathbf{S} = \mu_0 \int_S \mathbf{j} \cdot d\mathbf{S}.$$

This equation is true whatever the size of the surface S, hence the integrands must be equal everywhere, and

$$\boxed{\text{curl } \mathbf{B} = \mu_0 \mathbf{j}.} \tag{4.40}$$

This is the differential equation we have been seeking which is obeyed at all points by the field $\mathbf{B}$. It relates the field at a point to the current density at that point, and its solution, fitted to the appropriate boundary conditions, will give the magnetic field for a specified current distribution. We discuss the general solution to Equation (4.40) in section 4.5.3. This solution is the *Biot–Savart law*, which can be used to calculate the magnetic field due to currents flowing in thin wires in air or free space. In order to derive the Biot–Savart law from the differential form of Ampère's law it is easiest to introduce a potential from which the field **B** can be derived. We will now do this.

4.5.2   The magnetic vector potential

The Equation (4.40) plays an analogous role in magnetism to the equation

$$\text{div } \mathbf{E} = \rho/\varepsilon_0$$

in electrostatics. The solution of this equation for the field **E** is made easier by using the electrostatic potential ϕ and the equation

$$\mathbf{E} = -\text{grad } \phi.$$

The question arises as to whether there is a potential from which we can always obtain the magnetic field, and whose use would make the solution of Equation (4.40) easier. If a vector function can be obtained everywhere by taking the gradient of a scalar, then the line integral of the vector function around any closed path is zero. It follows that the curl of such a vector function is everywhere zero as we saw in the case of the electrostatic field. This can be proved formally. In Cartesian coordinates the gradient of a scalar function ϕ has components $\partial\phi/\partial x$, $\partial\phi/\partial y$ and $\partial\phi/\partial z$. Hence

$$(\text{curl } [\text{grad } \phi])_z = \frac{\partial}{\partial x}\left(\frac{\partial \phi}{\partial y}\right) - \frac{\partial}{\partial y}\left(\frac{\partial \phi}{\partial x}\right) = 0.$$

Similarly the x and y components of curl grad ϕ are zero, and the scalar satisfies the identity

$$\text{curl grad } \phi \equiv 0.$$

Thus at points where the current density **j** is non-zero we cannot derive the field **B** from a scalar function. The magnetostatic potential ψ, introduced in section 4.4.1, is only meaningful in regions of space where there is no current flowing. We shall now show that the field **B** can be derived quite generally from a vector potential **A**, called the *magnetic vector potential*.

If a vector function of position is obtained from another vector **A** by taking the curl of **A**, the divergence of the vector curl **A** is everywhere zero. Expressed mathematically

$$\text{div curl } \mathbf{A} \equiv 0, \tag{4.41}$$

a result which again can easily be proved by writing down div curl **A** in Cartesian coordinates. (A proof in one particular coordinate system is of course equivalent to a general proof.)

The divergence of the field **B** is everywhere zero, as we saw in section 4.2.2, and so we can always obtain the field **B** by taking the curl of another vector field **A**.

$$\boxed{\mathbf{B} = \text{curl } \mathbf{A}.}$$ (4.42)

This equation is analogous to the equation

$$\mathbf{E} = -\text{grad } \phi$$

for the electrostatic field. Equation (4.42) does not fix the potential **A** uniquely. We may add to **A** any function whose curl is zero and the field **B** obtained by taking the curl of the resultant addition remains the same. In the same way if we add a constant everywhere to the electrostatic potential ϕ, the field **E** remains the same. We are free to make additional assumptions about the potentials **A** and ϕ. The latter is usually defined uniquely by adding the condition that it tends to zero as the distance from a charge distribution tends to infinity. The potential **A** is fixed by adding the condition that, for steady currents,

$$\text{div } \mathbf{A} = 0.$$

The reason for the choice of this particular condition on the vector potential **A** is that it usually makes calculations easier than other choices.

4.5.3 The Biot–Savart law

Equation (4.40) can be solved with the aid of the vector potential **A** to give an expression for the magnetic field that arises when currents flow in thin conductors in free space, or, for all practical purposes, in air. This expression is analogous to Equation (1.8) for the electrostatic field of a distribution of charges, and is called the *Biot–Savart Law*. It gives the magnetic field as the sum of contributions from the infinite number of elementary lengths that comprise the circuit. The derivation of the law is left until Appendix C. Here we will introduce it with reference to a simple example in which the field is already known, viz. a current in a long straight wire. Consider Figure 4.20. The field at the point P (position vector **r** with respect to the origin O) is made up of contributions d**B** from elements of length d*l′* like the one at the point Q. Let the position vector of the point Q be **r′**. The Biot–Savart law tells us that the magnitude of each contribution is proportional to the current *I* in the wire, and inversely proportional to the square of the distance of the element from the point P. The direction of the field d**B** is perpendicular to the vector d**l′** which is in the direction of the current in the wire at the point Q, and perpendicular to the vector $(\mathbf{r} - \mathbf{r'})$ (which is in the direction of the line joining Q to P). The contribution d**B** is given by

$$d\mathbf{B} = \frac{\mu_0 I}{4\pi} \frac{d\mathbf{l'} \wedge (\mathbf{r} - \mathbf{r'})}{|\mathbf{r} - \mathbf{r'}|^3}.$$

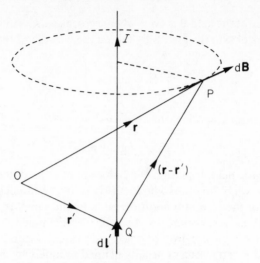

Figure 4.20. The field at the point P is made up
of contributions d**B** from elements like the one
shown at Q.

The total field at the point P is obtained by integrating around the complete
circuit s,

$$\mathbf{B(r)} = \frac{\mu_0 I}{4\pi} \oint_s \frac{d\mathbf{l'} \wedge (\mathbf{r} - \mathbf{r'})}{|\mathbf{r} - \mathbf{r'}|^3}. \tag{4.43}$$

This is the Biot–Savart law.

Let us work out the field at P to verify that we obtain the same result as
obtained using Ampère's law in section 4.4.2. It is simplest to choose the origin
of coordinates where the perpendicular from P hits the wire, as shown on
Figure 4.21. Let the distance OP be r, and the distance OQ be z. The variable $\mathbf{r'}$
in Equation (4.43) now becomes $\mathbf{e}_z z$, where $\mathbf{e}_z$ is a unit vector in the direction of
the current in the wire, and the variable $d\mathbf{l'}$ becomes $\mathbf{e}_z\, dz$. The vector $d\mathbf{l'} \wedge (\mathbf{r} - \mathbf{r'})$
has magnitude QP $dz \sin \theta$, where θ is the angle between the lines OQ and QP,
and no matter where Q lies on the wire always points in the same direction. It is
tangential to the circle and in a direction related to the direction of the current
by the right-hand screw rule. The magnitude of the resultant field is given by

$$B = \frac{\mu_0 I}{4\pi} \int_{-\infty}^{\infty} \frac{r\, dz}{(r^2 + z^2)^{3/2}} \tag{4.44}$$

$$= \frac{\mu_0 I}{2\pi r}, \tag{4.45}$$

in agreement with Equation (4.32).

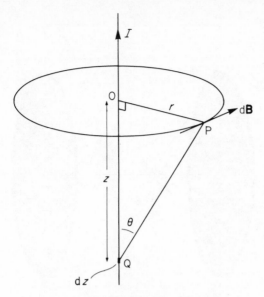

Figure 4.21. The calculation of the field of a
long straight wire carrying a current.

As another example we will calculate the field on the common axis of the two
coils shown in Figure 4.22. Two parallel, coaxial coils arranged like this are often
used to provide a field that cancels out unwanted fields, like the Earth's field,
or fields from iron building supports, etc. Two coils are used because near the
axis they can provide a field which is very nearly uniform over an appreciable
volume, as we shall see. Let us calculate the field at the point P distant x from
the centre O of the left-hand coil. We take the point O to be the origin of coordinates. Consider the contribution $d\mathbf{B}$ to the field of the left-hand coil from
an element $d\mathbf{l}'$ at the top point Q of the coil, as shown. The vector $d\mathbf{B}$ makes an
angle θ with the axis, the same angle as that between the lines OQ and QP.
Contributions from successive elements obtained by proceeding around the coil
all make the same angle θ with the axis and lie in a cone of half angle θ. The
result of the vector integral (4.43) for the resultant field can be seen to be in the
direction from P to O along the axis. Its magnitude is

$$B = \frac{\mu_0 I}{4\pi} \oint \frac{dl' \cos \theta}{|\mathbf{r} - \mathbf{r}'|^2}. \tag{4.46}$$

The path of the integral is around the circular coil and at all points on the path
the terms $\cos \theta$ and $|\mathbf{r} - \mathbf{r}'|^2$ are constant and given by

$$\cos \theta = a/(a^2 + x^2)^{1/2},$$

$$|\mathbf{r} - \mathbf{r}'|^2 = (a^2 + x^2).$$

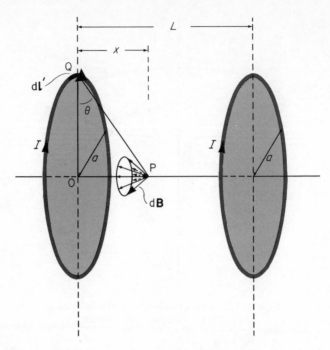

Figure 4.22. The field due to a current in two parallel coaxial circular coils.

Substitution in Equation (4.46) gives

$$B = \frac{\mu_0 I}{4\pi} \frac{a}{(a^2 + x^2)^{3/2}} \oint dl' \qquad (4.47)$$

$$= \mu_0 I a^2 / 2(a^2 + x^2)^{3/2}. \qquad (4.48)$$

The field at P due to the second coil is in the same direction, and its magnitude is simply obtained by replacing the variable x in the above equation by $(L - x)$. Hence the resultant field is

$$B = \frac{\mu_0 I a^2}{2} \left\{ \frac{1}{(a^2 + x^2)^{3/2}} + \frac{1}{(a^2 + (L - x)^2)^{3/2}} \right\}. \qquad (4.49)$$

The field near to any point can be expanded in a Taylor series in terms of the derivatives of the field at the point. If the first few derivatives of the field are zero the field will be very nearly uniform near that point. The first three derivatives with respect to the distance x in expression (4.49) are zero at the point half-way between the coils if the coils are separated by a distance equal to their common radius. Similar, coaxial coils, separated by a distance equal to their common radius, and carrying equal currents in the same direction, are called Helmholtz coils. Suppose that a pair of Helmholtz coils, each of radius 1 m and having 200 turns is used to nullify the effect of the Earth's magnetic field at a place where the magnitude of this field is 0.6 gauss. The current through the coils needed to produce such a field in the middle is given by Equation (4.49) (after putting $L = a = 1$ m, and $x = 0.5$ m) to be about 0.7 A.

4.6   FORCES AND COUPLES ON COILS

We now deal with an important practical topic that arises when currents flow in conductors in air. This is the subject of the forces and couples exerted on current carrying coils when they are placed in a magnetic field generated by other sources.

The forces and couples can be calculated in a straightforward manner using the Lorentz force (4.15). We have already considered a rectangular coil in a uniform field in section 4.3.1. More generally, if a steady current I flows in a very small length dl of a conductor whose cross-sectional area is A, then from Equation (4.3),

$$I = NevA.$$

The number of free electrons in the length dl is $NA\, dl$. If the conductor is situated in a magnetic field that has the value $\mathbf{B}$ at the position of the element dl, each free electron in that element experiences a force $-e\mathbf{v} \wedge \mathbf{B}$. Hence the total force on the element is

$$d\mathbf{F} = -NA\, dl\, e\mathbf{v} \wedge \mathbf{B}$$

$$= I\, d\mathbf{l} \wedge \mathbf{B},$$

where $d\mathbf{l}$ is a vector in the direction of the current in the wire at the position of the element dl. The total force on the whole conductor is obtained by integrating around the closed contour s,

$$\mathbf{F} = I \oint_s d\mathbf{l} \wedge \mathbf{B}. \tag{4.50}$$

If the position vector of the element $d\mathbf{l}$ with respect to an origin of coordinates is $\mathbf{r}$, the total torque on the conductor is

$$\mathbf{T} = I \oint_s \mathbf{r} \wedge (d\mathbf{l} \wedge \mathbf{B}). \tag{4.51}$$

We remind the reader that the field $\mathbf{B}$ in the above equations can be taken to be the field from all sources *other* than the current I in the conductor or can be considered to be the total field. This was discussed in section 4.3.1.

Let us calculate, as a simple example, the force per unit length between two long, thin, parallel wires separated by 4 mm and each carrying a steady current of 50 A. The field of one wire at the position of the other is given by Equation (4.32) to be 2.5×10^{-3} T. The force per unit length is then given by Equation (4.50) to be 0.125 N m^{-1}, i.e. about the weight of 12 grammes at the surface of the Earth. This example shows that the forces between conductors that carry very large currents can give rise to considerable mechanical problems. There may also be large forces between the separate parts of a single conductor carrying a large current. For example, there cannot be a resultant force on an isolated coil that carries a current, but there will be forces on the individual turns of a multi-looped coil from the current in the other turns. A solenoid wound with gaps between the turns will tend to squash up when a current is passed through it. These effects necessitate considerable care in the design of high current coils.

Expressions (4.50) and (4.51) can be used quite generally to calculate forces and torques on current carrying coils, although the calculations are often tedious.

4.6.1   Magnetic flux

In this section we will derive alternative expressions for the forces and torques on coils in terms of the *magnetic flux* through them. These expressions are very useful and often easier to use than Equations (4.50) or (4.51). We will first derive them using the idea of the current grid and its equivalent sheet of magnetic dipoles, and leave the definition of magnetic flux until afterwards.

Consider a closed coil which carries a steady current I. We have seen in section 4.4.1 that the magnetic effects of the current are the same as those of a current grid over the surface S of the coil, and each elemental area $d\mathbf{S}$ of the surface behaves as though it were a magnetic dipole of moment $I\,d\mathbf{S}$. If $\mathbf{B}$ is the magnetic field at the position of an element $d\mathbf{S}$ that arises from sources other than the current I, the dipole $I\,d\mathbf{S}$ has a potential energy

$$dU_P = -I\,d\mathbf{S} \cdot \mathbf{B},$$

from Equation (4.23). The potential energy* of the whole coil in the external field is

$$U_P = -I \int_S d\mathbf{S} \cdot \mathbf{B}. \tag{4.52}$$

If the coil with the current I constant is displaced a distance $d\mathbf{r}$ while a magnetic force $\mathbf{F}$ acts on it, the change of potential energy is given by

$$dU_P = -\mathbf{F} \cdot d\mathbf{r}.$$

The minus sign arises because the coil gains potential energy if it is pushed

---

* This is not the total magnetic energy associated with the entire physical system, which consists of the current I in the coil and the sources that give rise to the external field. The total energy is discussed in Chapter 6.

against the force. From this expression we obtain

$$\mathbf{F} = -\operatorname{grad} U_{\mathrm{P}}$$

$$= I \operatorname{grad} \int_{S} \mathrm{dS} \cdot \mathbf{B},$$

from Equation (4.52). This equation is usually written as

$$\mathbf{F} = I \operatorname{grad} \Phi, \tag{4.53}$$

where

$$\Phi = \int_{S} \mathbf{B} \cdot \mathrm{dS} \tag{4.54}$$

is the magnetic flux through the circuit. Since the lines of the field **B** are continuous the surface integral of the field is the same over any surface whose edge is defined by the coil, i.e. the magnetic flux through any such surface is constant. The sign of the magnetic flux Φ in Equation (4.53) is determined by the directions of the current in the coil and the magnetic field. The direction of the vectors dS in Equation (4.54) is determined by the right-hand screw rule used with the current in the coil giving the screw direction, as illustrated in Figure 4.23. The sign of the magnetic flux is then positive or negative depending on the direction of the field **B**. The dimensions of magnetic flux are [mass] $\times$ [charge]$^{-1}$ $\times$ [time]$^{-1}$ $\times$ [length]^{2}. The SI unit of magnetic flux is called the *weber* (W). The

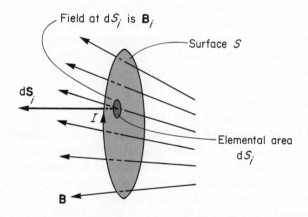

Flux through $S = \sum_{i} \mathbf{B}_{i} \cdot \mathrm{d}\mathbf{S}_{i} = \int_{S} \mathbf{B} \cdot \mathrm{d}\mathbf{S}$

Figure 4.23. The magnetic flux through a surface in a magnetic field.

unit of magnetic field **B** is sometimes called the weber per square metre (W m^{-2}), i.e. 1 weber per square metre $\equiv$ 1 tesla.

The couple on a coil situated in a magnetic field and carrying a current I is given by a formula analogous to Equation (4.53). For example if the coil can rotate about a vertical axis in its plane, and the angle between the plane of the coil and a fixed horizontal line is θ, the torque T on the coil about the vertical axis is given by

$$T = -\frac{\partial u}{\partial \theta} = I \frac{\partial \Phi}{\partial \theta}. \tag{4.55}$$

Equations (4.53) and (4.55) again make it clear that the force and torque on a current-carrying coil can be expressed equally well in terms of the total magnetic field or in terms of the magnetic field due to sources other than the current in the coil itself. If the magnetic flux Φ used in Equations (4.53) and (4.55) is calculated from the *total* magnetic field, the resulting values of the force and torque are unchanged. The total flux Φ_T can be written as the sum of the flux Φ_I due to the current in the coil, and the flux Φ due to the field **B** arising from sources other than the current I. Now

$$\text{grad } \Phi_T = \text{grad } \Phi_I + \text{grad } \Phi$$

$$= \text{grad } \Phi$$

since the flux Φ_I will not change when the coil is displaced slightly at constant current I. Hence the calculated force on the coil (and similarly the torque) remains the same whether Φ_T or Φ is used in the equations.

Moving coil instruments provide an illustration of the above ideas. These instruments are commonly used to measure steady currents over the current range from one microampere to ten amperes. If a current is passed through a coil suspended in a magnetic field there will be a couple on the coil if the magnetic flux through the coil varies with the angle of rotation about the axis of the suspension. The coil will be deflected from its zero position, determined by the suspension, and this deflection can be used to measure the current through the coil. Usually the field in which the coil moves is provided by a permanent magnet with a soft iron core fixed to rotate about the centre line between the poles, as shown in Figure 4.24. The iron core has the effect of shaping the field so that the magnetic flux Φ through the coil, which is attached to the core, is proportional to its angle of deflection θ from its zero position.

$$\Phi = k\theta,$$

where k is a constant for the instrument. If a current I is passed through the coil it will experience a torque kI given by Equation (4.55) and will deflect through

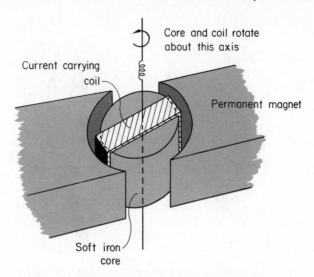

Figure 4.24. The coil of a moving galvanometer suspended
between the poles of a permanent magnet. The coil can
rotate about the axis shown and will experience a deflection when it carries a current. The gap between the pole
faces is typically less than 2 cm in a multipurpose moving
coil meter.

an angle θ given by

$$\alpha\theta = kI \tag{4.56}$$

where α is the restoring couple per radian of twist provided by the suspension.

4.6.2   The forces between two coils

In this section we consider what happens when the external field acting on
one current carrying coil is provided by a current in another coil. The force
between the coils is then proportional to the currents in each coil, but also
depends on their geometry, i.e. their shapes, sizes, numbers of turns, spacing
and orientation with respect to one another. Figure 4.25 shows two arbitrarily
shaped coils, labelled 1 and 2, carrying currents I_1 and I_2. Let $\mathbf{B}_1$ be the field
due to the current I_1 and $\mathbf{B}_2$ be that due to the current I_2. The field $\mathbf{B}_2$ is everywhere proportional to the current I_2, according to the Biot–Savart law, so
that the magnetic flux through coil 1 can be written

$$\Phi_1 = M_{12}I_2 \,,$$

where the constant M_{12} depends on the geometry of the coils and is called
their *mutual inductance*. The mutual inductance is equal to the magnetic flux

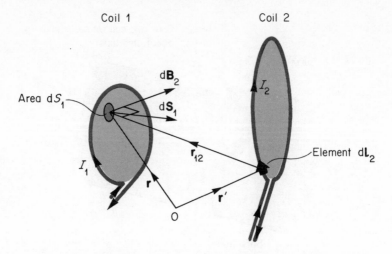

Figure 4.25. The magnetic flux through coil 1 due to a current I_2 in coil 2 is a double integral. All of the elementary lengths dl_2 which comprise coil 2 contribute to the flux through each elementary surface dS_1 of coil 1.

through coil 1 when unit current flows in coil 2. The force on coil 1 is, using Equation (4.53),

$$F_1 = I_1 I_2 \operatorname{grad}_1 M_{12}, \qquad (4.57)$$

where we have written $\operatorname{grad}_1 M_{12}$ in order to show that it is coil 1 undergoing the small displacement when we calculate the gradient. From Equation (4.54), and the definition of mutual inductance

$$M_{12} = \frac{1}{I_2} \int_{S_1} d\mathbf{S}_1 \cdot \mathbf{B}_2.$$

Also, from Equation (4.43)

$$\mathbf{B}_2 = \frac{\mu_0 I_2}{4\pi} \oint_{s_2} \frac{d\mathbf{l}_2 \wedge \mathbf{r}_{12}}{r_{12}^3},$$

where $\mathbf{r}_{12}$ is the vector from the position of the element $d\mathbf{l}_2$ of coil 2 to the position of a surface element $d\mathbf{S}_1$ of coil 1. Hence

$$M_{12} = \frac{\mu_0}{4\pi} \int_{S_1} d\mathbf{S}_1 \cdot \oint_{s_2} \frac{d\mathbf{l}_2 \wedge \mathbf{r}_{12}}{r_{12}^3}.$$

This equation can be shown to reduce to the equation

$$M_{12} = \frac{\mu_0}{4\pi} \oint_{s_1} \oint_{s_2} \frac{d\mathbf{l}_1 \cdot d\mathbf{l}_2}{r_{12}}, \qquad (4.58)$$

which shows, because of the symmetry between the suffixes 1 and 2, that $M_{12} = M_{21} = M$. (Notice that Equation (4.57) now immediately leads to the correct result that the coils exert equal and opposite forces on each other.) Equation (4.58) is known as Neumann's formula for the mutual inductance M. An equally valid form for the force $\mathbf{F}$, on coil 1 is given by Equation (4.50),

$$
\mathbf{F}_1 = I_1 \oint_{s_1} d\mathbf{l}_1 \wedge \mathbf{B}_2
$$

$$
= I_1 \oint_{s_1} d\mathbf{l}_1 \wedge \left\{ \frac{\mu_0 I_2}{4\pi} \oint_{s_2} \frac{d\mathbf{l}_2 \wedge \mathbf{r}_{12}}{r_{12}^3} \right\}. \tag{4.59}
$$

It can be shown with a little vector algebra that the mutual inductance M given by Equation (4.58) leads to a force in agreement with Equation (4.59). From Equations (4.57) and (4.58),

$$
\mathbf{F}_1 = \frac{I_1 I_2 \mu_0}{4\pi} \oint_{s_1} \oint_{s_2} d\mathbf{l}_1 \cdot d\mathbf{l}_2 \, \mathrm{grad} \, \frac{1}{r_{12}}
$$

$$
= -\frac{I_1 I_2 \mu_0}{4\pi} \oint_{s_1} \oint_{s_2} \frac{d\mathbf{l}_1 \cdot d\mathbf{l}_2}{r_{12}^3} \mathbf{r}_{12}. \tag{4.60}
$$

By expanding the triple vector product in the integrand of Equation (4.59) one finds

$$
\mathbf{F}_1 = \frac{\mu_0 I_1 I_2}{4\pi} \oint_{s_1} \oint_{s_2} \left\{ \frac{(d\mathbf{l}_2 \cdot \mathbf{r}_{12})}{r_{12}^3} d\mathbf{l}_1 - \frac{(d\mathbf{l}_1 \cdot d\mathbf{l}_2)}{r_{12}^3} \mathbf{r}_{12} \right\}.
$$

The first term in this integral vanishes, as the integration is carried out over a closed loop, and we are again left with Equation (4.60).

4.7   THE MOTION OF CHARGED PARTICLES IN ELECTRIC AND MAGNETIC FIELDS

The force on a particle of charge q moving with velocity $\mathbf{v}$ in electric and magnetic fields $\mathbf{E}$ and $\mathbf{B}$ is given by Equation (4.16):

$$
\mathbf{F} = q\mathbf{E} + q\mathbf{v} \wedge \mathbf{B}.
$$

If the particle is moving in a vacuum at the surface of the Earth there no nearby molecules with which it can collide or interact, and the only force acting on the particle, other than that due to the electric and magnetic fields, is the force of gravity. When one is considering atomic particles it can easily be shown that the gravitational forces are usually negligible. For example a proton moving with a speed of 2×10^5 cm s^{-1} (which is roughly the speed of hydrogen molecules at room temperature) perpendicular to the Earth's magnetic field experiences a force about 10^6 times greater than the gravitational force. The electric force on a proton in a field of 10^{-7} V m^{-1} is about equal to the gravitational force. If we neglect the gravitational force, Newton's law for the particle in vacuum becomes

$$\frac{d\mathbf{p}}{dt} = q\mathbf{E} + q\mathbf{v} \wedge \mathbf{B},\tag{4.61}$$

where $\mathbf{p}$ is the momentum of the particle. For given fields $\mathbf{E}$ and $\mathbf{B}$ one must solve Equation (4.61) to find the position of the particle as a function of time.

4.7.1   The motion of a charged particle in a uniform magnetic field

A simple example which is often encountered is that of the motion of a charged particle of mass m in a uniform magnetic field. Since the magnetic force always acts at right angles to the motion, the force cannot increase the energy of the particle, and its speed will remain constant. If the particle is moving in a plane perpendicular to the direction of the uniform magnetic field $\mathbf{B}$, the force $q\mathbf{v} \wedge \mathbf{B}$ on the particle is constant in magnitude and always at right angles to the direction of motion. The particle therefore moves in a circle of radius R, where

$$\frac{mv^2}{R} = qvB,\tag{4.62}$$

or

$$R = \frac{mv}{qB}.\tag{4.63}$$

This is illustrated in Figure 4.26.

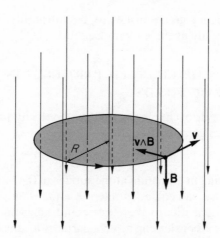

Figure 4.26. The path of a positively charged particle moving perpendicular to a uniform magnetic field. The magnetic force is perpendicular to both the field and the velocity of the particle, and the path is a circle.

It can be seen from Equation (4.63) that particles of the same ratio of momentum to charge move in circles of the same radius in a uniform magnetic field. Particles of a given value of (mv/q) are often selected by accepting only those that traverse a path of fixed radius when moving in a uniform field. Figure 4.27 shows a beam of charged particles, all with the same mass and charge but having

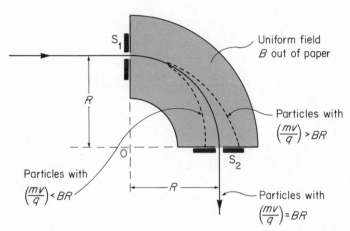

Figure 4.27. The trajectory of a positively charged particle entering a uniform magnetic field at slit S_1 and leaving at slit S_2. There is a uniform magnetic field out of the paper over the shaded region. Particles with different speeds are bent in circles of different radii.

a spread of velocities. We consider a somewhat unrealistic beam in which all the particles are moving in the same direction without any component of velocity perpendicular to this direction. The beam enters through a narrow slit S_1 into a region where there is a uniform magnetic field in a direction perpendicular to the velocity of the particles. The defining slit S_2 allows particles to leave the field only if they have moved in an arc of a circle centre O, where $OS_1 = OS_2 = R$. The selected particles then have the value of (mv/q) given by Equation (4.63)*. An equivalent statement is that the slits and magnetic field together select those particles which have a constant value for the quantity (mE/q^2), where E is the energy of the particle. This arrangement is used in mass spectrographs, which select particles of the same q and E, but differing mass m, by varying the magnetic field. The same technique can be used to measure the energies of particles from accelerators. From Equation (4.63)

$$\left(\frac{2mE}{q^2}\right)^{1/2} = BR.$$

* The Equation (4.63) has been derived non-relativistically. It is valid for any speed of the moving particle however, if one uses the relativistic mass given by $m = m_0(1 - v^2/c^2)^{-1/2}$, where m_0 is the rest mass of the particle.

The quantity BR for protons is plotted against energy in Figure 4.28. From this figure it can be seen that a magnetic field of ≈ 0.14 T is required to bend 1 MeV protons in a circle of radius 1 m.

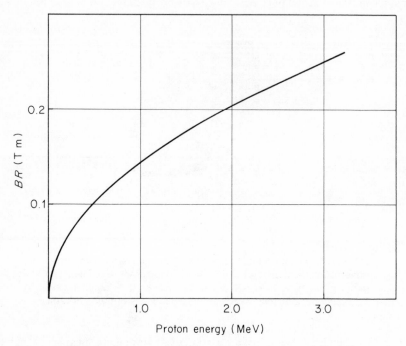

Figure 4.28. The parameter (BR) for protons moving in a uniform field **B**, plotted against proton energy.

★   **4.7.2   Magnetic quadrupole lenses**

As another example of the motion of charged particles in magnetic fields we will discuss the focusing action of a magnetic quadrupole lens on a beam of moving charged particles.

Consider the idealized situation shown in Figure 4.29, in which all the particles in the beam are moving with the same speed parallel to the z-axis, but are spread out over a limited area in the x–y plane. They enter a region of length l over which there is a magnetic field which is designed to focus the beam at the point F, distant f from the field region. If the magnetic field had a perfect focusing action the lens would bring the beam to a point focus at the point F, in a manner similar to that in which a thin convergent glass lens focuses a plane wave at its focal point.

Let us first consider a beam whose profile is a narrow slit in the x direction. A particle moving at a distance x_i off the axis of the lens must be deflected

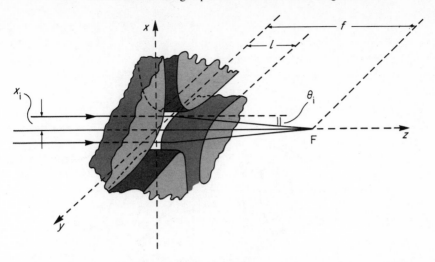

Figure 4.29. The focusing action of a quadrupole lens. Particles moving parallel to the z-axis with non-zero values of the coordinate x are deflected in the magnetic field and are all focused at the point F.

through an angle θ_i which is proportional to the distance x_i in order that all particles pass through the focus F. A *magnetic quadrupole* field within the region between the planes $z = 0$ and $z = l$ will have this effect on the charged particles. If the particles are positively charged the field lines are similar to those shown in Figure 4.30, which is a view looking from the focus F. In this magnetic quadrupole field the field is given by

$$\mathbf{B}(x, y, z) = \mathbf{e}_x gy + \mathbf{e}_y gx, \qquad (4.64)$$

where g is a constant. The x-component of the field is proportional to the coordinate y, and the y-component to the coordinate x. The field is usually provided by current carrying coils wound around iron pole pieces, and the constant g depends on the current and the distances apart of the pole pieces and their sizes. The length l in Figure 4.29 corresponds to the length of the pole pieces. In practice the field will depend on the coordinate z because of end effects, but we will neglect these, and assume that the field is given by Equation (4.64) for $0 < z < l$, and is equal to zero elsewhere.

The equation of motion of a particle in the space between the pole pieces is

$$m\ddot{\mathbf{r}} = q(\mathbf{v} \wedge \mathbf{B}),$$

where m and q are the mass and charge of the particle. If the length l is small compared to the distance f, the velocity is unchanged to a good approximation, and the vector $\mathbf{v}$ is constant, with magnitude equal to v, the original speed along the z direction. With this approximation the components of the above equation

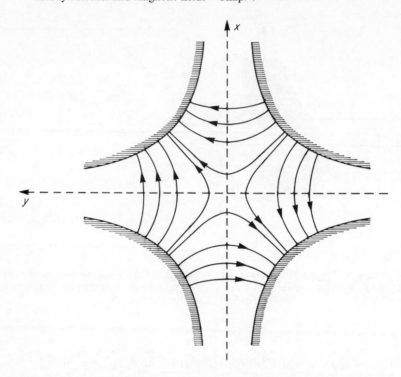

Figure 4.30. The field lines of the magnetic quadrupole field which will focus a beam of positively charged particles moving in the z-direction if the initial beam profile is a slit in the x-direction.

may be written:

$$m\ddot{x} = -qrB_y = -qgx \qquad (4.65)$$

$$m\ddot{y} = qB_x = qgy \qquad (4.66)$$

$$m\ddot{z} = 0. \qquad (4.67)$$

These equations must be solved with the initial conditions that at time $t = 0$ (the time the particle enters the field region), $z = y = 0$; $dx/dt = dy/dt = 0$; and $x = x_i$. Equation (4.65) describes simple harmonic motion in the x direction and its solution is

$$x = \alpha \cos \omega_1 t + \beta \sin \omega_1 t, \qquad (4.68)$$

where α and β are constants and $\omega_1^2 = gqv/m$. The solution to Equation (4.67) is $z = vt$, and Equation (4.68) may be written, now regarding z as the independent variable, as

$$x = \alpha \cos \omega z + \beta \sin \omega z,$$

where $\omega^2 = gq/mv$. The constants α and β in this equation are determined by the initial conditions to be $\alpha = x_i$, and $\beta = 0$, and we have finally

$$x = x_i \cos \omega z.$$

At $z = l$, when the particle leaves the field,

$$\frac{dx}{dz} = -\omega x_i \sin \omega l.$$

If $\omega l \ll 1$, which is usually the case and corresponds to the condition used above that $f \gg l$, dx/dz may be rewritten

$$\frac{dx}{dz} = (-\omega^2 l)x_i. \tag{4.70}$$

The solution to Equation (4.66) shows that the y coordinate is unchanged throughout. Equation (4.70) then tells us that particles initially moving parallel to the z-axis at a distance x_i are bent through angles proportional to the distance x_i. To a first approximation all the particles in the hypothetical beam we started with pass through the point F on Figure 4.29.

The focal length f is given by

$$f = \frac{x_i}{(\omega^2 l)x_i} = \frac{1}{\omega^2 l} = \frac{mv}{gql}. \tag{4.71}$$

It can be seen that the focusing action of a magnetic quadrupole lens (with fixed dimensions and current) is the same for all particles that have the same value of the quantity (mE/q^2), where m, E and q are mass, energy, and charge of the particles respectively.

If an analysis similar to that above is made for a beam whose profile is a narrow slit in the y direction, it can be shown that the action of the quadrupole field given by Equation (4.64) is defocusing. Particles initially off the axis of the lens in the y direction are bent farther off. Thus for a realistic beam profile the lens described above would make the profile at the focus F into a long slit along the y direction. To give a better focusing action for realistic beams two magnetic quadrupole lenses are used, one placed very close to the other, the pair constituting a *quadrupole doublet*. The field in the first lens focuses in one direction, and the field in the second is arranged to focus in a direction at right angles. The combined action is focusing in both directions, rather like the way in which a pair of thin glass converging and diverging lenses can be made to give a converging combination.

The field required in a magnetic quadrupole lens to give it a specific focal length for moving charged particles can be calculated from Equation (4.71). For example if a lens of length 10 cm is required to have a focal length of 3 m for 50 MeV protons, the field gradient g, given by

$$g = \frac{mv}{qf\,l},$$

has to be about $3.3\,\mathrm{T\,m}^{-1}$.

PROBLEMS 4

4.1 Estimate the mean velocity at which oxygen molecules carrying a net charge $(-e)$ would move between two parallel plate electrodes 10 cm apart and having a potential difference of 100 V between them. Estimate the negative current if each electrode is of area $10\,\mathrm{cm}^2$. The oxygen between the plates is at N.T.P.; at N.T.P. the r.m.s. speed of the molecules is $3 \times 10^4\,\mathrm{cm\,s}^{-1}$ and the mean free path between collisions is 10^{-5} cm. One part in 10^{11} of the oxygen molecules is ionized.

4.2 Two spherical buoys float half submerged in a calm deep sea. Their radius a is very much smaller than their separation b. Calculate the resistance between them if the resistivity of the water is ρ.

4.3 A small sphere is uniformly charged throughout its volume, and is rotating with constant angular velocity. Determine its magnetic moment in terms of the total charge Q on the sphere, the angular momentum L of the sphere, and its mass M.

4.4 Two very long thin wires carrying equal and opposite currents of 10^3 A are placed parallel to the x-axis at $y = 0$ and $z = \pm 0.5$ cm. Calculate the magnetic field **B** in the x–y plane and determine its maximum gradient.

4.5 A long sheet of conductor of thickness 1 cm and height 20 cm carries a current of 10^4 A distributed uniformly within it. Calculate the magnetic field along a line ab perpendicular to the surface cutting the sheet half way up as shown in Figure 4.31.

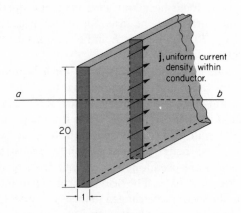

Figure 4.31.

Consider distances from the sheet small compared with 20 cm.

What is the field inside the sheet? What is the force on an electron travelling near the speed of light parallel to the sheet?

4.6 A long straight wire of circular cross-section carries a current I uniformly distributed. What is the magnetic field inside the wire?

4.7 A cable of circular cross-section and diameter 2 cm has a long cylindrical hole of diameter 1 mm drilled in it parallel to the cable axis. The distance between the axis of the hole and cable axis is 5 mm. If the cable has a uniform steady current density of 10^5 A m^{-2} flowing in it, calculate the magnetic field
  (i) at the centre of the cable
  (ii) at the centre of the hole.

4.8 Using Equation (4.32) for the field **B** outside a long wire carrying a current verify that curl **B** = 0 outside the wire. Determine curl **B** inside the wire.

4.9 A flat coil having 50 turns of radius 30 cm is in series with another flat coil of 100 turns each of radius 2 cm, the coils being coaxial and 10 cm apart. Estimate the force between the coils when a current of 2 A is passing.

4.10 The magnetic field of the Earth is found to fall off according to an inverse cube law at several Earth radii. The field can thus be roughly described in terms of a magnetic dipole at the centre of the Earth. If the field strength is 10^{-4} G at 15 Earth radii in the equatorial plane what is the magnetic dipole moment of the Earth? (Radius of Earth $\simeq$ 6000 km).

4.11 A particle of mass m, charge q, moving with velocity **v** enters a region where there is a uniform magnetic field **B**. The vectors **v** and **B** make an angle θ with each other. Determine the motion of the particle.

4.12 An electron moving in outer space has a component of velocity $0.01c$ in the direction of a magnetic field of 10^{-5} gauss. How many revolutions does it make whilst travelling between two points one light year apart along a line of magnetic field? (Treat the problem non-relativistically.)

CHAPTER

5

Magnetic materials

Up to now we have been discussing magnetic fields in free space, or in air. In air the distances apart of the molecules are relatively large, and the influence of the air molecules on the magnetic fields is very small. We have seen how a steady current gives rise to a steady magnetic field, and have obtained equations describing the field's behaviour. Suppose that we put a sample of material into a magnetic field. How is the field changed and what are the new field equations? How do we describe what happens to the material? What new phenomena arise that can be measured and used to learn something about the material's microscopic structure? These questions are discussed in this chapter.

5.1 MAGNETIZATION

If a specimen is placed in a magnetic field, the field in the vicinity of the sample is changed. How much it is changed depends on the material and on the shape of the sample, but for most specimens it is not altered very much—typically only by about one part in 10^5. For certain materials, however, the field at some points nearby can be increased by a factor of one hundred or so over its value in the absence of the specimen. Such materials are called *ferromagnetic*. Non-ferromagnetic materials, which produce only a small change in field, can be either *diamagnetic* or *paramagnetic*. If a diamagnetic rod is inserted into a solenoid that carries a current, the field near the end of the solenoid is decreased very slightly. If a paramagnetic rod is inserted the field is increased slightly. Glass and copper are examples of diamagnetic materials; oxygen and titanium are paramagnetic; iron, cobalt and nickel, and certain alloys of these metals, are ferromagnetic.

The most easily observed characteristics of the different materials are their behaviours in strong non-uniform fields. A ferromagnetic material is strongly pulled into the field. A diamagnetic material is repelled towards field-free regions, but so weakly that sensitive apparatus is required to measure the repulsive force. A paramagnetic material is pulled into the magnet. Although the attraction is usually weak, it can sometimes be strong enough to observe in a simple way. If a flask full of liquid oxygen is placed near the gap in the yoke of a strong electromagnet the liquid may flow out of the flask and into the gap between the pole pieces.

The explanation of these effects lies in the interaction of microscopic magnetic dipole moments with the fields in which the materials are placed. Let us restrict our attention to liquids and gases, in which the atoms and molecules move independently. These atoms or molecules may have *permanent* magnetic dipole moments. If they do have permanent moments, the material is paramagnetic; otherwise it is diamagnetic. If paramagnetic molecules are placed in a magnetic field there is a net alignment of the dipoles in the field direction. This is exactly analogous to the way in which polar molecules on average acquire an effective electric dipole moment in the direction of an applied electric field, as discussed in section 2.2.2. In a non-uniform magnetic field a paramagnetic material will thus experience a force in the direction of increasing field.

All atoms and molecules acquire an induced magnetic dipole moment when placed in a magnetic field. Induced magnetic dipole moments, unlike induced electric dipole moments, are in a direction *opposite* to the field in the specimen. Thus, if the molecules of a substance have no permanent moments, i.e. the substance is diamagnetic, a sample inserted into a non-uniform magnetic field will experience a force in the direction of decreasing field strength.

On the atomic scale atoms or molecules in a magnetic field are said to be *magnetized*. The microscopic magnetization can vary from atom to atom but can be averaged over volumes which contain very many atoms to give a function which is smoothly and slowly varying with position on a macroscopic scale. This function is called the *magnetization* of the medium, and is given the symbol $\mathbf{M}$. It is defined as the magnetic moment per unit volume, and is given by

$$\mathbf{M} = N\langle\mathbf{m}\rangle, \tag{5.1}$$

where N is the number of molecules per unit volume, and $\langle\mathbf{m}\rangle$ is their average dipole moment in the direction of the field. As in electrostatics we consider only isotropic materials, for which the magnetization $\mathbf{M}$ can be simply related to the field. The dimensions of magnetization are [current] $\times$ [length]$^{-1}$, and it is measured in amperes per metre. As in the electrical case, magnetized atoms produce magnetic fields that fall off more rapidly with distance than the dipole field. These higher order fields need to be described by more complex atomic properties than the average effective dipole moment $\langle\mathbf{m}\rangle$. However, they make essentially no contribution to the macroscopic magnetic field, and when discussing this field it is safe to ignore the higher order microscopic fields completely. Thus, the macroscopic effects of atomic magnetization can be adequately represented by the magnetization **M**.

Before we discuss the consequences of magnetization on a macroscopic scale we will use simple atomic models which approximate the behaviour of electrons in atoms in order to gain insight into the origin of diamagnetism and paramagnetism. There are two chief contributions to the magnetic dipole moment of an atom or molecule. The first arises from the orbital motions of the electrons, which correspond to current loops. The second arises because electrons themselves have permanent (intrinsic) magnetic dipole moments. These are roughly the same size as those due to the orbital motions. The magnetic dipole moments are closely related to the angular momenta of the electrons; the orbital moment to the orbital angular momentum, and the intrinsic moment to an intrinsic angular momentum possessed by electrons. This latter angular momentum, the electron spin, can for many purposes be regarded as due to the electron spinning around an axis through its centre, like a top. The proper description of angular momentum in atomic systems requires the use of quantum mechanics and is outside our scope. We need only mention one result pertinent to the present discussion. This is that the component of the spin angular momentum of an electron, measured in any specified direction, can only have the values $\pm \hbar/2$, where $\hbar$ is Planck's constant divided by 2π. Compare this with the assertion in section 4.4.2 that the component of the orbital angular momentum of an electron in a specified direction can take only the value zero or an integral multiple of $\hbar$.

If the electrons in an atom or molecule have zero resultant angular momentum that atom or molecule has no permanent magnetic dipole moment*. A liquid or gas consisting of such atoms or molecules is diamagnetic. If there is a resultant electronic angular momentum there is a permanent moment, and the liquid or gas is paramagnetic. In the solid state, where the atoms or molecules do not move independently, and where a particular electron is not necessarily attached to a single molecule, the situation is more complicated. In a metal for example there are contributions to the magnetization from the conduction electrons and there may be contributions from the ions fixed in the lattice. These contributions vary with temperature and with field strength, and usually result in a small effect which can be either paramagnetic or diamagnetic. Ferromagnetic materials are a special case of paramagnetism. In certain conductors there is an interaction between electrons on the lattice ions and the conduction electrons tending to align all of the latter in the same direction. Complete alignment can occur in large magnetic fields resulting in a very large enhancement of the total field.

---

* There may be a permanent magnetic dipole moment arising from the contribution of the nucleus or nuclei of the atom or molecule. This will be very small, and we will ignore it.

5.1.1 Diamagnetism

In this section we consider the origin of diamagnetism. We will describe how atoms acquire induced magnetic dipole moments in a direction opposite to an applied magnetic field.

We first use the simple atomic model outlined in section 4.4.2 to relate the magnetic moment of the orbital motion of the electrons in an atom to their total orbital angular momentum. Consider an electron (i) moving in a circle of radius r_i about a central nucleus. If the electron has speed v_i, the equivalent current around the circle is

$$I = \frac{ev_i}{2\pi r_i},$$

and the magnetic moment of the current loop has magnitude

$$m_i = \frac{ev_i}{2\pi r_i}\pi r_i^2$$

$$= \frac{1}{2}\frac{e}{m_e}L_i,$$

where $L_i = m_e\omega_i r_i^2$ is the angular momentum of the ith electron, and ω_i is its angular speed.

The direction of the vector $\mathbf{m}_i$ is in the opposite direction to that of the angular momentum $\mathbf{L}_i$, as shown in Figure 5.1, hence

$$\mathbf{m}_i = -\frac{e}{2m_e}\mathbf{L}_i. \tag{5.2}$$

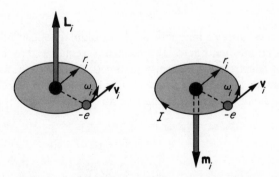

Figure 5.1. The angular momentum and magnetic moment of an electron moving in a circular orbit. The current is in the opposite direction to the motion of the negatively charged electron, and the orbital magnetic moment $\mathbf{m}_i$ is opposite to the orbital angular momentum $\mathbf{L}_i$.

The total magnetic dipole moment resulting from the orbital motion of all the electrons in the atom is the vector sum

$$\mathbf{m} = -\frac{e}{2m_e}\sum_i \mathbf{L}_i. \tag{5.3}$$

If the resultant orbital angular momentum $\sum_i \mathbf{L}_i$ is zero, and if the resultant intrinsic angular momentum of the electrons is zero, the atom has no permanent dipole moment and is diamagnetic. The magnetic effects of a material which consists only of diamagnetic atomic particles are entirely due to induced magnetic moments.

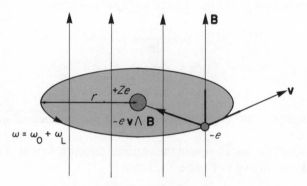

Figure 5.2. The effect of a magnetic field on the circular orbit of an electron whose orbital angular momentum is in the same direction as the field. The angular velocity ω_0 is increased by the amount ω_L.

Now suppose we apply a magnetic field to a diamagnetic atom. The resultant intrinsic angular momentum of the electrons remains zero but the orbital motions of the electrons are slightly changed. Consider the orbit shown in Figure 5.2 in which an electron of mass m_e is moving with angular velocity ω_0 in a circle of radius r. The central acceleration is $\omega_0^2 r$, where

$$m_e\omega_0^2 r = Ze^2/4\pi\varepsilon_0 r^2,$$

and hence

$$\omega_0 = \left(\frac{Ze^2}{4\pi\varepsilon_0 m_e r^3}\right)^{1/2}. \tag{5.4}$$

When a field $\mathbf{B}$ is switched on, perpendicular to the plane of the orbit, and in the direction shown, the central force on the electron is increased by an amount $-e\mathbf{v} \wedge \mathbf{B}$, where $\mathbf{v}$ is the velocity of the electron. This extra force is very small compared with the atomic forces holding the electron in the atom, and it is a reasonable assumption that the electron stays in its original orbit after the field is switched on. The angular velocity of the electron in the orbit changes to the value ω, where

$$m_e\omega^2 r = \frac{Ze^2}{4\pi\varepsilon_0 r^2} + e\omega rB.$$

Solving this quadratic equation for the angular velocity ω gives

$$\omega = \left(\frac{Ze^2}{4\pi\varepsilon_0 m_e r^3}\right)^{1/2} + \frac{eB}{2m_e}, \tag{5.5}$$

if the condition

$$B^2 \ll \frac{m_e Z}{\pi\varepsilon_0 r^3}$$

is satisfied. For all normal fields this condition is met. The result (5.5) tells us that in the presence of the field the angular velocity of the electron is increased

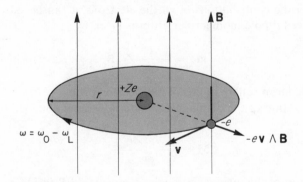

Figure 5.3. The effect of a magnetic field on the circular orbit of an electron whose orbital angular momentum is opposite to the field. The angular velocity ω_0 is decreased by the amount ω_L.

by an amount $(eB/2m_e)$, a quantity known as the *Larmor angular frequency* and often written ω_L. The increase in angular frequency gives the electron shown in Figure 5.2 an increase in angular momentum equal to $m_e\omega_L r^2$.

Now consider another orbit perpendicular to the applied field, but with the electron going around in the opposite direction, as in Figure 5.3. The angular velocity of the electron in the absence of the field is still given by Equation (5.4). When the field is present, the magnetic force acting on the electron is now outwards. Following the same argument as in the first case we obtain for the new angular velocity ω' of the electron

$$\omega' = \omega_0 - \frac{eB}{2m_e}.$$

Hence in both cases the electrons acquire an additional orbital angular momentum $m_e \omega_L r^2$ in the direction of the applied field. This gives them both an induced magnetic moment of magnitude

$$m_{\text{induced}} = \frac{e}{2m_e} \cdot m_e \omega_L r^2$$

from Equation (5.2). The induced moment is opposite to the increase in angular momentum, i.e. opposite to the applied field, and

$$\mathbf{m}_{\text{induced}} = -\frac{e^2 r^2}{4m_e}\mathbf{B}, \tag{5.6}$$

on substituting for the Larmor angular frequency ω_L.

An atom with atomic number Z has Z electrons. These electrons are in orbits with different radii and with different inclinations to the applied field. Taking the average of the contributions of all the electrons, it can be shown that the effective induced dipole moment per atom is given by

$$\langle \mathbf{m} \rangle = -\frac{e^2}{6m_e}Zr_0^2\mathbf{B}, \tag{5.7}$$

where r_0^2 is the mean square radius of the electron orbits.

The magnetization of the diamagnetic material is now obtained from Equation (5.1).

$$\mathbf{M} = -\frac{Ne^2 Zr_0^2}{6m_e}\mathbf{B}. \tag{5.8}$$

A determination of the magnetization for a given field $\mathbf{B}$ inside a sample can thus provide an estimate of the radius of the atoms.

The field $\mathbf{B}$ in Equations (5.7) and (5.8) is strictly speaking the local magnetic field at the position of the atom, i.e. the total field less the contribution of the atom itself. However, for non-ferromagnetic materials the atoms have only a very small magnetic effect—for most practical purposes the local field can be taken to be the same as the macroscopic field $\mathbf{B}$ in the absence of the sample.

We have not explained the mechanism whereby the electrons in the orbits shown in Figures 5.2 and 5.3 acquired the extra angular momentum $m_e \omega_L r^2$. This is discussed in the next chapter.

5.1.2 Paramagnetism

Some atoms and molecules have a non-zero electronic angular momentum, and thus have a permanent magnetic dipole moment. The orientation of the dipole moments of individual molecules in a liquid or a gas is random in the absence of a magnetic field. When a field is applied there is a tendency for them to line up in the field direction and provide a paramagnetic effect. This process is analogous to the alignment of polar molecules in an electric field as discussed in section 2.2.2. The molecules will also acquire induced magnetic dipole moments, but this diamagnetic effect is usually smaller than the paramagnetism due to the permanent moments.

The derivation of the magnetization of a paramagnetic gas or liquid in terms of the permanent molecular magnetic dipole moments follows exactly the arguments presented in section 2.2.2. The theory given there was a good approximation because the potential energy of an electric dipole in an electric field was small compared with the thermal energy ($\sim kT$, where k is Boltzmann's constant, and T the absolute temperature) of a molecule. We can easily show that the same is true for the potential energy of a microscopic magnetic dipole in a magnetic field. Permanent atomic magnetic dipole moments are of the same order of magnitude as the resultant orbital moment of the atomic electrons. This moment is given by Equation (5.3). The resultant orbital angular momentum $\sum_i \mathbf{L}_i$ is typically no more than a few times $\hbar$ ($\hbar \equiv$ Planck's constant divided by 2π), and so a typical atomic dipole moment has a magnitude about equal to the quantity ($e\hbar/2m_e$). Atomic magnetic dipole moments are often expressed in units of this quantity, which is called the *Bohr magneton*, and given the symbol m_B,

$$m_B = \frac{e\hbar}{2m_e}. \tag{5.9}$$

The numerical value of m_B is 9.273×10^{-24} A m^2. The potential energy of a magnetic dipole moment equal to one Bohr magneton, aligned with a field of 10 T (a very large field) is given by Equation (4.23)

$$U_P = -m_B B,$$

and is equal to $\sim 10^{-22}$ joules, or $\sim 6 \times 10^{-4}$ eV. At room temperature the thermal energy kT is about 2.5×10^{-2} eV, and so the potential energy of atomic magnetic dipoles in a magnetic field is almost always much less than kT.

The derivation of the magnetization can now be pursued in a manner identical to that followed in section 2.2.2. The result is similar to Equation (2.13),

$$\mathbf{M} = \frac{Nm^2}{3kT}\mathbf{B}. \tag{5.10}$$

In this formula m is the permanent dipole moment of the molecules and N is the number of molecules per unit volume. This result corresponds to an average dipole moment per molecule in the direction of the field

$$\langle\mathbf{m}\rangle = \frac{m^2}{3kT}\mathbf{B}. \tag{5.11}$$

Again the field **B** should strictly speaking be the local field $\mathbf{B}_{\text{local}}$. However, as for diamagnetic materials, the two fields are nearly the same and for most practical purposes they need not be distinguished.

The total magnetization of a paramagnetic material must include the induced moments, and adding their contribution to Equation (5.10) we obtain

$$\mathbf{M} = N\left\{\frac{m^2}{3kT} - \frac{e^2}{6m_e}Zr_0^2\right\}\mathbf{B}. \tag{5.12}$$

Let us substitute typical numbers into this equation in order to estimate the relative magnitudes of the paramagnetic and diamagnetic terms. For a material with Z equal to 50, and atoms which have a mean radius of 10^{-10} m and permanent moments of one Bohr magneton, the ratio of the diamagnetic to the paramagnetic term in Equation (5.12) is about one half at room temperature. For lighter atoms the ratio is correspondingly smaller, and it is a safe general rule that for liquids and gases made up of atoms with a permanent magnetic moment, the diamagnetism is weaker than the paramagnetism.

5.1.3 Ferromagnetism

Some substances, like iron, exhibit very large paramagnetic effects, and are called ferromagnetic. We will not discuss ferromagnetism at the atomic level in any detail. Below a certain temperature, called the *Curie temperature*, the spins of the conduction electrons in macroscopic sized portions called *domains* of certain conductors, are all aligned parallel. The cause of this is an interaction between the free electrons and those on the lattice ions. The total alignment within domains occurs only when the lattice ions have electrons in certain atomic orbitals. This happens in iron, cobalt and nickel and certain alloys of these metals. In a piece of unmagnetized iron the domains, which typically have volumes in the range 10^{-10} to 10^{-12} m^3, are oriented at random. In a weak external field, there is a net alignment of the domains in the field direction, giving a very large paramagnetic effect. As the field is increased those domains which are already aligned become enlarged at the expense of the others*. This is illustrated in Figure 5.4.

5.2 THE MACROSCOPIC MAGNETIC FIELD INSIDE MEDIA

If a substance is placed in a magnetic field arising from currents in nearby conductors, there is a field inside the material and the atoms or molecules are magnetized. The atomic magnetic field, $\mathbf{B}_{\text{atomic}}$, varies rapidly with position, even inside a single atom, and has contributions from both the atomic constituents of the material and from the nearby currents. The *macroscopic field* **B** in the medium is the average of the field $\mathbf{B}_{\text{atomic}}$ taken over volumes very small on

---

* For a discussion of ferromagnetism and all magnetism at a more advanced level, see e.g. the volume on Solid State Physics, by H. E. Hall, in the Manchester Physics Series.

1 mm

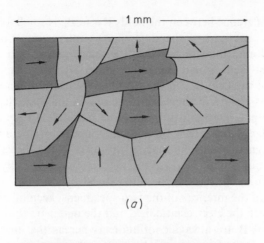

(a)

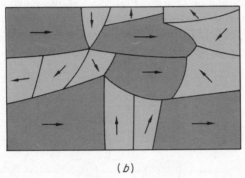

(b)

Figure 5.4. The arrangement of the domains
within a ferromagnetic material in (a) weak and
(b) strong magnetic fields.

a macroscopic scale but containing very many atoms. Changes of field on the
atomic scale are smoothed out and the macroscopic field is a relatively slowly
and smoothly varying function of position. The magnetizing field acting on a
molecule is the field $\mathbf{B}_{\text{local}}$. This is equal to the field $\mathbf{B}_{\text{atomic}}$ less the contribution
from the magnetization of the molecule itself, analogously to the field $\mathbf{E}_{\text{local}}$ in
electrostatics.

The field $\mathbf{B}$ is made up of two contributions. One is the original field $\mathbf{B}_0$
present when the specimen was absent. (This assumes that the current distributions which give rise to the field $\mathbf{B}_0$ are unaffected by the magnetization of the
sample. This is almost always true.) The other contribution is the field $\mathbf{B}_M$ due
to the magnetization $\mathbf{M}$ of the sample. The total field $\mathbf{B}$ is the sum of the fields
from the two sources,

$$\mathbf{B} = \mathbf{B}_0 + \mathbf{B}_M. \tag{5.13}$$

Although the difference between the fields **B** and $\mathbf{B}_0$ is small for non-ferromagnetic materials, the sign of the difference is interesting. For paramagnetic materials it is found that the value of the macroscopic field inside is greater than the original field. For diamagnetic materials the value of the field **B** is less than that of the field $\mathbf{B}_0$. In the case of a dielectric in an electric field the polarization **P** is always in the same direction as the total electric field **E**, yet the effect of the polarization is to reduce the initial field $\mathbf{E}_0$ present before the dielectric was placed in position. In the analogous case of a paramagnetic sample placed in a magnetic field, the magnetization is in the direction of the total magnetic field, but acts to increase the field present originally. The difference between the two situations can be understood by drawing simplified microscopic pictures of the interiors of the two specimens, keeping in mind the basic difference between the electrostatic field and the magnetic field. The lines of the fields $\mathbf{B}_{\mathrm{atomic}}$ and **B** are always continuous, whereas the lines of the electrostatic field begin or end on charges. Figure 5.5 shows schematically the atomic field due to atoms polarized within a dielectric. The average atomic field **E** is less than the original field $\mathbf{E}_0$ because of the reversal of the field direction within

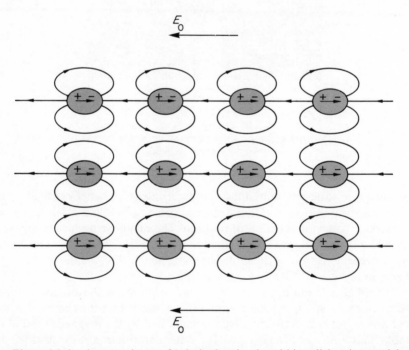

Figure 5.5. A *schematic* picture of polarized molecules within a dielectric material placed in an electric field. The electric field within the material is less than the original electric field $\mathbf{E}_0$ because the average field produced by the polarized molecules is in the opposite direction to $\mathbf{E}_0$.

the atomic dipoles. Figure 5.6 shows the circulating current loops that represent the effective magnetic dipoles of the atoms of a magnetized paramagnetic material. In this case there are no changes of direction of the atomic magnetic field within the atoms, and the averaged atomic field **B** is greater than the original field $\mathbf{B}_0$.

The field $\mathbf{B}_M$ in ferromagnetic materials can be several orders of magnitude greater than those in other substances. The total field **B** inside a ferromagnetic sample is usually very much larger than the original field.

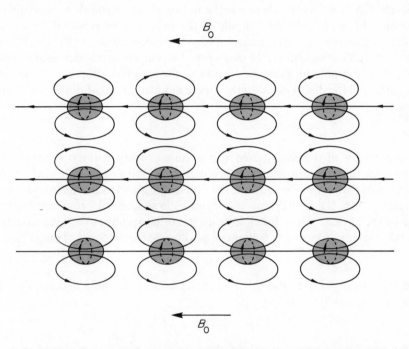

Figure 5.6. A *schematic* picture of magnetized molecules within a paramagnetic material placed in a magnetic field. The magnetic field within the material is greater than the field $\mathbf{B}_0$ originally present because the average field produced by the magnetized molecules is in the same direction as $\mathbf{B}_0$.

5.2.1 The currents equivalent to a magnetized body

In this section we discuss how a magnetized body can be replaced by currents in order to determine the magnetic field arising from the magnetization. This is is necessary step in the chain of reasoning which leads to equations satisfied by the magnetic field in the presence of magnetic materials.

Let us first estimate the field $\mathbf{B}_M$ in a simple example before proceeding to the general case. Consider a long cylindrical solenoid, having N turns per unit length of thin wire carrying a current I. The field $\mathbf{B}_0$ inside the solenoid is uniform, in a direction parallel to the axis, and has magnitude

$$B_0 = \mu_0 NI, \tag{5.14}$$

as given by Equation (4.31). If the solenoid is completely filled with a paramagnetic rod the magnetic field inside will remain uniform. The rod will acquire a uniform magnetization $\mathbf{M}$ in a direction along its axis. If the rod is divided into very thin slices of thickness δx perpendicular to the axis, each slice is equivalent to a uniform sheet of magnetic dipoles. The total magnetic dipole moment of each slice is $\pi r^2 \, \delta x M$, where r is the radius of the solenoid. As discussed in section 4.4.1, each individual dipole of the sheet can be replaced by a small circulating current I_M. All the currents inside the disc cancel each other, leaving a current I_M flowing around the disc's rim. This current is called a *magnetization current*. The dipole moment $\pi r^2 I_M$ due to the magnetization current around the surface of the disc is equal to the dipole moment $\pi r^2 \, \delta x M$ of the whole disc, hence

$$I_M = M \, \delta x.$$

A unit length of the magnetized rod is thus equivalent (for the purpose of calculating the magnetic field to which it gives rise) to a current flowing around the outside of the rod of magnitude per unit length equal to the magnitude of the magnetization. Using the same arguments as in section 4.4.2 to derive the field due to the current in the solenoid coils, it can be seen that the magnetization surface current gives rise to a uniform field inside the solenoid of magnitude

$$B_M = \mu_0 M. \tag{5.15}$$

The total field B inside the solenoid is the sum of this field and the field B_0 given by Equation (5.14),

$$B = \mu_0(NI + M).$$

We can use the results obtained above to show that the local field in paramagnetic or diamagnetic materials is usually very nearly the same as the macroscopic field. To do this we have to show that the magnetization of such materials produces only a very small field $\mathbf{B}_M$ compared with the original field $\mathbf{B}_0$ in which the material was placed. If this is so Equation (5.13) becomes $\mathbf{B} \simeq \mathbf{B}_0$, and the local field to the same approximation is also equal to the original field. In the simple example considered above let us first assume that the field $\mathbf{B}_M$ produced by the magnetized rod is indeed very much smaller than the field $\mathbf{B}_0$. In this case the magnetization can be obtained from Equation (5.12) by replacing the field $\mathbf{B}$ by $\mathbf{B}_0$, and the field arising from the magnetization is given by Equations (5.12) and (5.15) to be

$$B_M = \mu_0 N \left\{ \frac{m^2}{3kT} - \frac{e^2}{6m_e} Z r_0^2 \right\} B_0.$$

The diamagnetic term in this equation is usually less than the paramagnetic term and we will neglect it. With this approximation

$$\frac{B_M}{B_0} = \frac{\mu_0 N m^2}{3kT}.$$

Taking the permanent magnetic dipole moment m equal to one Bohr magneton, and the number of molecules per unit volume N equal to 5×10^{28} m^{-3} (a typical value for condensed matter), the ratio B_M/B_0 is given by the above equation to be 3×10^{-4} at room temperature. If the diamagnetic term had been included the ratio would have been even smaller. Our conclusions are thus consistent with the original assumption, and we have shown that to a good approximation the field B_M can be neglected compared with the field B_0. The magnetization of non-ferromagnetic materials makes very little difference to the magnetic field.

We will now deal with the more general case. The magnetization current flowing around the surface of a uniformly magnetized material is analogous to the polarization charges on the surface of a uniformly polarized dielectric. A non-uniformly polarized dielectric (for the purpose of calculating the electrostatic field to which it gives rise) is equivalent to a volume distribution of charge, in addition to charges over its surface. Similarly a non-uniformly magnetized material gives rise to magnetic fields as though it carried a volume distribution of current, in addition to currents over its surface. In order to see how such volume distributions of current arise consider the slab of magnetized material shown in Figure 5.7. The magnetization is everywhere in the y-direction. Its magnitude is independent of the coordinates y and z, but decreases as the coordinate x increases. The effective microscopic moments that constitute the magnetization may again be regarded as circulating atomic currents. A suitably sized current circulating in a loop of atomic radius will give the same magnetic effects on a macroscopic scale as the average effective magnetic dipole moment of the atoms. Figure 5.8 shows the atomic current loops over part of the plane ABCD cut through the magnetized slab. The circulating currents in the loops decrease as the coordinate x increases, corresponding to the decreasing effective dipole moments of the atoms. The result is a net downward current over any line parallel to the z-axis, since the current up just to the right of the line is less than the current down just to the left of it. These magnetization currents, specified by a magnetization current density $\mathbf{j}_M$, will give rise to magnetic fields just like the currents discussed in Chapter 4. It can be seen that volume distributions of currents only appear if a specimen is non-uniformly magnetized.

We will now relate the volume and surface currents at each point to the magnetization at that point. Let us look at things on a macroscopic scale. Consider elementary volumes, with sides δx, δy and δz, within a magnetized body, as shown in Figure 5.9. In a volume element A, centred on the point

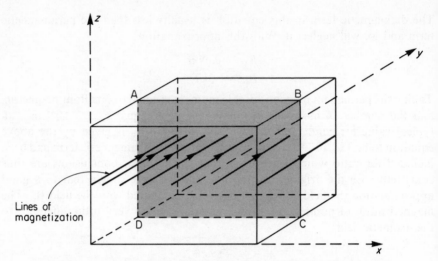

Figure 5.7. A volume within a magnetized magnetic material. The magnetization is in the y-direction and varies only with the coordinate x, decreasing as
x increases.

(x, y, z), the average circulating atomic current can be resolved into three
components I_1, I_2 and I_3. These correspond to the x, y and z-components of the
magnetic moment $\mathbf{M} \, \delta\tau$ of the volume element, where $\delta\tau$ is the volume of the
element and $\mathbf{M}$ the magnetization at its position. The z component of the current
density at any point has contributions from currents like I_1 and I_2, but not I_3.

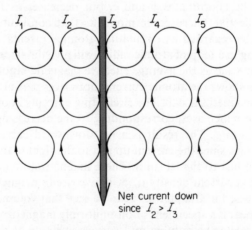

Figure 5.8. A schematic representation of the
atomic current loops which give rise to the
magnetization over part of the plane ABCD
of the magnetized slab shown in Figure 5.7.

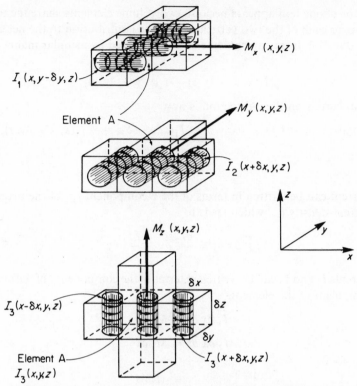

Figure 5.9. Very small macroscopic volume elements within a magnetized material. The element A is centred on the point (x, y, z). The currents I_1, I_2 and I_3 are the macroscopic circulating currents equivalent to the magnetization of the element A.

Let us calculate the net current in the z direction over the area $\delta x \, \delta y$ of the element A. If the current I_2 varies with the distance x (it would decrease as x increases if we were considering the magnetized slab shown in Figure 5.7) the upward current over the left-hand face of the element is different to the current up the right-hand face. The current up the left-hand face is equal to

$$I_2(x, y, z) - I_2(x - \delta x, y, z),$$

and the current up the right-hand face is equal to

$$I_2(x + \delta x, y, z) - I_2(x, y, z).$$

The net upward current over the area $\delta x \, \delta y$ of element A contributed by the circulating currents I_2 is thus equal to

$$\tfrac{1}{2}\{I_2(x, y, z) - I_2(x - \delta x, y, z) + I_2(x + \delta x, y, z) - I_2(x, y, z)\}$$
$$= \tfrac{1}{2}\{I_2(x + \delta x, y, z) - I_2(x - \delta x, y, z)\}.$$

The factor of one half appears because two volume elements share the upward current over each of the two vertical faces. The contribution to the net current in the z direction from currents like I_1 is determined in a similar manner to be given by

$$\tfrac{1}{2}\{I_1(x, y - \delta y, z) - I_1(x, y + \delta y, z)\}.$$

The total current in the z-direction is now

$$\tfrac{1}{2}\{I_2(x + \delta x, y, z) - I_2(x - \delta x, y, z) + I_1(x, y - \delta y, z) - I_1(x, y + \delta y, z)\}$$

$$= \frac{\partial I_2}{\partial x}\,\delta x - \frac{\partial I_1}{\partial y}\,\delta y.$$

This current can be written in terms of the z-component j_{M_z} of the magnetization current-density $\mathbf{j}_M$, which leads to

$$j_{M_z}\,\delta x\,\delta y = \frac{\partial I_2}{\partial x}\,\delta x - \frac{\partial I_1}{\partial y}\,\delta y.$$

The currents I_1 and I_2 can be written in terms of the components of the magnetic dipole moment of the element,

$$I_1\,\delta y\,\delta z = M_x\,\delta\tau$$

$$I_2\,\delta x\,\delta z = M_y\,\delta\tau$$

Hence

$$I_1 = M_x\,\delta x$$

$$I_2 = M_y\,\delta y.$$

Substitution of these expressions into the previous equation gives

$$j_{M_z}\,\delta x\,\delta y = \frac{\partial M_y}{\partial x}\,\delta x\,\delta y - \frac{\partial M_x}{\partial y}\,\delta x\,\delta y,$$

from which we obtain for the z-component of the magnetization current density

$$j_{M_z} = \frac{\partial M_y}{\partial x} - \frac{\partial M_x}{\partial y}.$$

The right-hand side of this equation is the z-component of the curl of the magnetization $\mathbf{M}$. (See Equation (4.37).) So we may write

$$j_{M_z} = (\operatorname{curl}\mathbf{M})_z. \tag{5.16}$$

The x- and y-components of the vector curl $\mathbf{M}$ can be shown in a similar fashion to be equal to the x- and y-components of the magnetization current density $\mathbf{j}_M$.

$$j_{M_x} = \frac{\partial M_z}{\partial y} - \frac{\partial M_y}{\partial z} = (\operatorname{curl}\mathbf{M})_x \tag{5.17}$$

$$j_{M_y} = \frac{\partial M_x}{\partial z} - \frac{\partial M_z}{\partial x} = (\text{curl } \mathbf{M})_y.$$

(5.18)

Equations (5.16), (5.17) and (5.18) can be summarized in the form

$$\mathbf{j}_M = \text{curl } \mathbf{M}.$$

(5.19)

This equation is the one we have been seeking in this section and tells us how to replace a volume distribution of magnetization by an equivalent current distribution throughout the volume.

Over the surface of a magnetized body there may be surface currents, as we have seen. We will now determine these surface currents in terms of the magnetization of the body just at the surface. If the magnetization is perpendicular to the surface, the surface current is zero, as demonstrated in Figure 5.10. The surface current $\mathbf{j}_{M_s}$ is a current per unit length that flows in a vanishingly thin

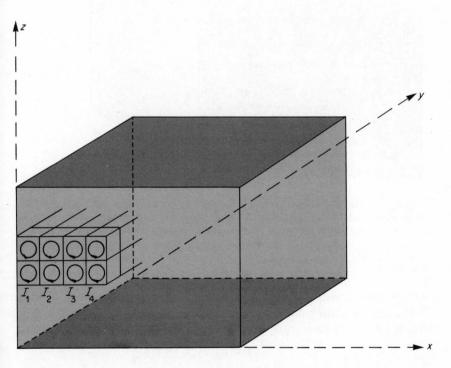

Figure 5.10. The circulating currents in the volume elements over the front face of the magnetized block shown in Figure 5.7 give zero surface current over the front surface as the thickness of the volume elements tends to zero.

layer (on a macroscopic scale) in the surface. If the magnetization is parallel to the surface, the surface current is a maximum. This suggests that the surface current is proportional to the vector product of the magnetization **M** and the unit vector **n** in the direction of the outward normal to the surface.

$$\mathbf{j}_{M_s} \propto \mathbf{M} \wedge \mathbf{n}. \tag{5.20}$$

The constant of proportionality can be determined by a procedure similar to the one used to obtain the equivalent currents inside the specimen. Figure 5.11

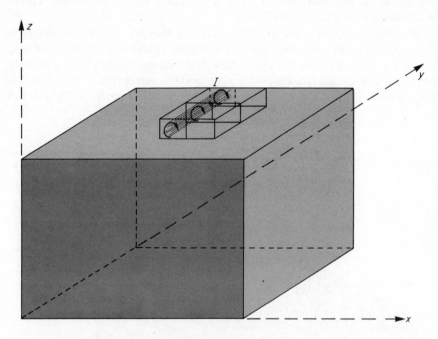

Figure 5.11. The circulating currents in the volume elements over the top face of the magnetized slab shown in Figure 5.7 give rise to surface currents over the top surface as the height of the volume elements tends to zero.

shows volume elements at the top surface of the magnetized block of Figure 5.7. The net circulating atomic current I gives an element a magnetic dipole moment $I \, \delta x \, \delta z$, which is equal to $M \, \delta x \, \delta y \, \delta z$. The current I is given by

$$I = j_{M_s} \, \delta y,$$

and equating the two expressions for the dipole moment

$$M \, \delta x \, \delta y \, \delta z = j_{M_s} \, \delta x \, \delta y \, \delta z.$$

Hence

$$j_{M_s} = M. \tag{5.21}$$

The magnitude of the surface current per unit length is equal to the magnitude of the magnetization, as we found earlier, and Equation (5.20) becomes

$$\mathbf{j}_{M_s} = \mathbf{M} \wedge \mathbf{n}. \tag{5.22}$$

5.2.2   Magnetic susceptibility and atomic structure

We have established that the magnetization of a non-ferromagnetic material is proportional to the magnetic field **B** for isotropic, homogeneous non-ferromagnetic materials. We may thus write

$$\mathbf{M} = \chi_B \frac{\mathbf{B}}{\mu_0}, \tag{5.23}$$

where χ_B is a dimensionless constant, called the *magnetic susceptibility* of the material. Its value may vary with temperature, but at constant temperature the relationship between the magnetization and the field is linear.

For ferromagnetic materials the dependence of the magnetization on the field is non-linear. Also for most ferromagnetic materials there is no unique value of the magnetization corresponding to a given value of the field **B**. If the external field in which a piece of iron is situated changes, the resulting magnetization of the iron depends on the state of magnetization before the change. This phenomenon is called hysteresis and is discussed in section 5.4. For the remainder of this section we discuss paramagnetic and diamagnetic materials only.

One way in principle of measuring magnetic susceptibilities is to determine the field near a sample magnetized by an external field. However, this method is not sensitive enough, since the change in the field arising from the presence of the sample is so small. Magnetic susceptibilities are usually measured by determining the force on a magnetized sample in an inhomogeneous magnetic field. This is discussed in the next chapter. Measurements of the magnetic susceptibilities of materials give information on their microscopic structures. Combining Equations (5.12) and (5.23) we obtain the relation

$$\chi_B = \mu_0 N \left\{ \frac{m^2}{3kT} - \frac{e^2}{6m_e} Z r_0^2 \right\} \tag{5.24}$$

for the susceptibility of a liquid or gas. The two contributions to the susceptibility can be separated by performing experiments at different temperatures, as discussed in section 2.2.2 for dielectric materials. Interpretation of the results gives the permanent moment m of the molecules, and their mean square radius.

Table 5.1 gives the magnetic susceptibilities of several materials at a temperature of 293 K. Consider the noble gases as an example. Their atoms have no permanent magnetic dipole moments, since the total angular momentum of the electrons is zero for all the noble gas atoms. They all have the same number of

Table 5.1. The magnetic susceptibilities of some common
substances at 293 K.

Material	Susceptibility χ_B
Gold	-3.6×10^{-5}
Copper	-0.98×10^{-5}
Germanium	-1.5×10^{-5}
Tungsten	$+6.8 \times 10^{-5}$
Aluminium	$+2.3 \times 10^{-5}$
Magnesium	$+1.2 \times 10^{-5}$
Glass	-1.1×10^{-4}
Fused Quartz	-6.2×10^{-5}
Nitrogen (76 cm Hg pressure)	-6.7×10^{-9}
Oxygen (76 cm Hg pressure)	$+1920.0 \times 10^{-9}$
Helium (76 cm Hg pressure)	-1.05×10^{-9}
Neon (76 cm Hg pressure)	-3.76×10^{-9}
Argon (76 cm Hg pressure)	-11.0×10^{-9}
Krypton (76 cm Hg pressure)	-11.1×10^{-9}
Xenon (76 cm Hg pressure)	-24.6×10^{-9}
Water	-9.1×10^{-6}
Sodium Chloride	-1.38×10^{-5}

atoms per unit volume at normal temperature and pressure, equal to 2.69×10^{25} m^{-3}, and Equation (5.24) can be used to calculate their root mean square radii. Table 5.2 shows the calculated values. The sizes of the atoms are roughly equal as predicted by quantum mechanics. The sizes are grouped around a radius of about 0.55 Å. This is a surprisingly good estimate of the size of a noble gas atom in view of the simplicity of theory we have used.

As another example of the microscopic information which can be obtained from susceptibilities let us estimate the magnetic dipole moment of the oxygen molecule. At 293 K and atmospheric pressure oxygen is strongly paramagnetic. This behaviour can be contrasted with the weak diamagnetism of nitrogen. We will thus neglect the induced magnetism and attribute the susceptibility of

Table 5.2. The root mean square radii of the noble gas atoms obtained from
their magnetic susceptibilities at normal temperatures and pressure.

	Z	$\chi_B \times 10^{-9}$	$r_0 \times 10^{-10}$ m
He	2	-1.0	0.575
Ne	10	-3.8	0.486
Ar	18	-11.0	0.620
Kr	36	-16.1	0.530
Xe	54	-24.6	0.536

oxygen as given in Table 5.1 to the permanent dipole moments alone. With this approximation Equation (5.24) becomes

$$\chi_B = \frac{\mu_0 N m^2}{3kT}.$$ (5.25)

Substitution of the appropriate numbers into this equation gives a value for the permanent moment m equal to $\sim 2.6 \times 10^{-23}$ A m²; i.e. about two and one half times a Bohr magneton.

5.3   THE FIELD VECTOR H

Suppose we need to calculate very precisely the magnetic induction field near the end of a short solenoid which has a rod of paramagnetic material inside it. How do we do such a calculation? The field is the addition of two parts, one part $\mathbf{B}_0$ due to the current in the solenoid, the other $\mathbf{B}_M$ due to the magnetization of the rod. The part $\mathbf{B}_0$ can be calculated from the Biot–Savart law, but the contribution $\mathbf{B}_M$ depends on the magnetization $\mathbf{M}$ of the material of the rod, and we do not know the distribution of magnetization, although we may know a relation between the magnetization $\mathbf{M}$ and the total magnetic field $\mathbf{B}$ at each point in the rod. Thus $\mathbf{M}$ depends on $\mathbf{B}$, which is the quantity we want to calculate. The problem is most easily treated by introducing a third magnetic field vector, called the *magnetic intensity*, and given the symbol $\mathbf{H}$. The field vector $\mathbf{H}$ obeys simple boundary conditions and has the important property that its line integral around any closed path depends only on free currents caused by the flow of electrons in wires, for example, and is independent of magnetization currents. The use of the field $\mathbf{H}$ is analogous to the use of the electric displacement in electrostatics. There the field vector $\mathbf{D}$ was introduced in order to avoid direct reference to polarization charges. In electrostatics we worked with the vectors $\mathbf{E}$ and $\mathbf{D}$, instead of $\mathbf{E}$ and the polarization $\mathbf{P}$. Similarly with magnetic media it turns out to be easier to work with the vectors $\mathbf{B}$ and $\mathbf{H}$ than to work with the vectors $\mathbf{B}$ and $\mathbf{M}$.

5.3.1   Ampère's law for the field H

The line integral of the field $\mathbf{B}$ around a closed path is equal to the current I through the area defined by the path, times the permittivity of free space μ_0, as discussed in section 4.4.1. If the path encloses magnetized material there will in general be a magnetization current through the area defined by the path. This magnetization current gives rise to a magnetic field in the same way as currents due to the motion of free charges. Hence it must be included in the relationship involving the line integral of the field $\mathbf{B}$.

We first consider a simple example before proceeding to the general case. Figure 5.12 shows part of a long solenoid filled with a paramagnetic rod. We will work out the line integral of the field $\mathbf{B}$ over the path ABCD. The line

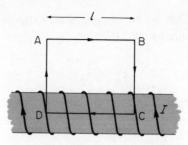

Figure 5.12. Part of a long solenoid completely filled with a paramagnetic rod. The rod is uniformly magnetized. The magnetic field within the rod is everywhere parallel to the axis, and is zero outside. The line integral of the field **B** over the closed loop ABCD has a contribution only from the part CD, of length l.

integral is

$$\oint \mathbf{B} \cdot d\mathbf{l} = \oint (\mathbf{B}_0 + \mathbf{B}_M) \cdot d\mathbf{l},$$

where $\mathbf{B}_0$ and $\mathbf{B}_M$ have their usual meanings. From Equation (5.15)

$$\mathbf{B}_M = \mu_0 \mathbf{M},$$

where **M** is the magnetization, which is zero outside the rod. The line integral is thus equal to

$$\oint \mathbf{B} \cdot d\mathbf{l} = \oint \mathbf{B}_0 \cdot d\mathbf{l} + \mu_0 \oint \mathbf{M} \cdot d\mathbf{l},$$

and rearranging

$$\oint (\mathbf{B} - \mu_0 \mathbf{M}) \cdot d\mathbf{l} = \oint \mathbf{B}_0 \cdot d\mathbf{l}.$$

From Equation (5.14)

$$B_0 = \mu_0 N I,$$

where N is the number of turns per unit length of the solenoid and I the current. Hence

$$\oint (\mathbf{B} - \mu_0 \mathbf{M}) \cdot d\mathbf{l} = \mu_0 I_f, \tag{5.26}$$

where $I_f = NIl$ is the total 'free' current through the area ABCD. We now define the new auxiliary field vector, the magnetic intensity **H**, to be given at each point by the equation

$$\mu_0 \mathbf{H} = \mathbf{B} - \mu_0 \mathbf{M},$$

or

$$\mathbf{H} = \frac{1}{\mu_0}\mathbf{B} - \mathbf{M}, \tag{5.27}$$

where **B** and **M** are the magnetic induction field and the magnetization at the point. In terms of the magnetic intensity, Equation (5.26) becomes

$$\oint \mathbf{H} \cdot d\mathbf{l} = I_f,$$

and we see that the line integral of the field **H** over the closed path depends only on the total current due to the motion of free charges which pass through the area defined by the path. The dimensions of magnetic intensity are current divided by length, the same as the dimensions of magnetization. In SI units magnetic intensity is measured in amperes per metre.

The magnetic field **H** may be introduced in a more general way, beginning with Ampère's law (4.28). The total current through a closed loop may include contributions from magnetization currents I_M as well as from currents I_f due to the motion of free charges, and we have

$$\oint \mathbf{B} \cdot d\mathbf{l} = \mu_0 I_f + \mu_0 I_M.$$

Writing this equation in terms of the corresponding current densities $\mathbf{j}_f$ and $\mathbf{j}_M$

$$\oint \mathbf{B} \cdot d\mathbf{l} = \mu_0 \int_S \mathbf{j}_f \cdot d\mathbf{S} + \mu_0 \int_S \mathbf{j}_M \cdot d\mathbf{S}, \tag{5.28}$$

where S is any surface whose perimeter is the path of the line integral. From Equation (5.19)

$$\mathbf{j}_M = \text{curl } \mathbf{M},$$

where **M** is the magnetization, and Equation (5.28) can be written

$$\oint \mathbf{B} \cdot d\mathbf{l} = \mu_0 \int_S \mathbf{j}_f \cdot d\mathbf{S} + \mu_0 \int_S \text{curl } \mathbf{M} \cdot d\mathbf{S}.$$

The second term on the right-hand side of this equation can be transformed into a line integral using Stokes' theorem (4.38), and after rearranging we have

$$\oint \left(\frac{\mathbf{B}}{\mu_0} - \mathbf{M} \right) \cdot d\mathbf{l} = \int_S \mathbf{j}_f \cdot d\mathbf{S}.$$

In terms of the magnetic field $\mathbf{H}$ this equation becomes

$$\oint \mathbf{H} \cdot d\mathbf{l} = \int_S \mathbf{j}_f \cdot d\mathbf{S}. \tag{5.29}$$

This expression relating the line integral of the field $\mathbf{H}$ over a closed path to the 'free' current through the loop is the integral expression of Ampère's law for the field $\mathbf{H}$. The differential form of the law may be obtained by rewriting Equation (5.29), using Stokes' theorem, in the form

$$\int_S \text{curl } \mathbf{H} \cdot d\mathbf{S} = \int_S \mathbf{j}_f \cdot d\mathbf{S}.$$

This equation is true for any surface S, hence at every point

$$\text{curl } \mathbf{H} = \mathbf{j}_f. \tag{5.30}$$

The curl of the magnetic intensity is everywhere equal to the current density due to the motion of free charges.

In the following sections of this book we shall frequently omit the subscript f on currents and current densities when it is clear from the context that the currents referred to arise from the motion of free charges. We will usually use the subscript f only when we wish to emphasize the difference between 'free' currents and magnetization currents.

For diamagnetic or paramagnetic materials which are isotropic and homogenous the magnetization is proportional to the magnetic field $\mathbf{B}$, as expressed in Equation (5.23),

$$\mathbf{M} = \chi_B \frac{\mathbf{B}}{\mu_0}. \tag{5.31}$$

This relationship is also approximately valid for ferromagnetic materials over small ranges of the field $\mathbf{B}$. For these materials the susceptibilities are much higher than for non-ferromagnetic materials, and vary with magnetic field. Equation (5.31) enables us to write the magnetic intensity in terms of the field $\mathbf{B}$ and the susceptibility. Substitution of the expression (5.31) for the magnetization into Equation (5.27) gives

$$\mathbf{H} = \frac{1}{\mu_0}\mathbf{B} - \frac{\chi_B}{\mu_0}\mathbf{B},$$

from which we obtain the equation

$$\boxed{\mathbf{B} = \mu\mu_0\mathbf{H},}$$ (5.32)

where

$$\mu = (1 - \chi_B)^{-1}.$$ (5.33)

The constant μ is dimensionless and is called the *relative permeability* of the material. For non-ferromagnetic materials the relative permeability is very nearly unity, but for ferromagnetic materials it may have large values, typically around 1000, and like the susceptibility varies with the magnetic field.

If we insert expression (5.32) for the field **H** into Equation (5.30) we obtain

$$\text{curl}\left(\frac{\mathbf{B}}{\mu\mu_0}\right) = \mathbf{j}_f.$$

If μ is constant *everywhere*, this reduces to the equation

$$\text{curl } \mathbf{B} = \mu\mu_0\mathbf{j}_f,$$ (5.34)

which is similar in form to Equation (4.40) except for the factor μ. Thus for a given distribution of currents flowing in an infinite medium of relative permeability μ, the magnetic field **B** everywhere is *increased* by the factor μ over the field produced by the same distribution of currents in free space. The situation can be compared with the analogous one in electrostatics. The electric field due to a system of charges at rest in an infinite medium of relative permittivity ε is *reduced* everywhere by the factor ε compared with the field due to the same system of charges in free space.

Unlike the magnetic field **B**, the field **H** has sources, and the lines of the field **H** are not continuous. Taking the divergence of each side of Equation (5.27)

$$\text{div } \mathbf{H} = -\text{div } \mathbf{M},$$ (5.35)

and div **H** is non-zero wherever the magnetization **M** is discontinuous. For example Figure 5.13 shows the lines of the fields **B** and **H** in and around a magnetized paramagnetic rod. At the ends of the rod, where the magnetization changes discontinuously, more lines of the field **H** occur outside than inside. Over a small volume enclosing part of the left end of the rod more lines of the magnetization flow in than out, and more lines of magnetic intensity flow out than in.

If the free charge current distribution is specified, and the geometry of magnetic media present is known, Equation (5.30) can in principle be solved to give the magnetic intensity everywhere. Such a situation occurs in the example with which we began this section in which we wanted to know the field near the end of a rod inserted in a solenoid. At the boundaries between different media there will be difficulties, due to the fact that the field **H** has sources there. However,

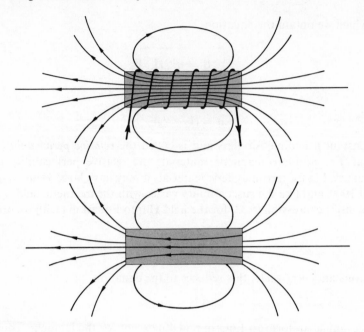

Figure 5.13. The lines of the field **B** (upper diagram) and **H** (lower diagram) for a short solenoid with a paramagnetic rod inside. The relative permeability of the hypothetical paramagnetic material has been taken to be about three in order to clarify the difference between the fields.

we will find that at these boundaries the fields **H** and **B**, related by the equation **B** = $\mu\mu_0$**H** must obey certain conditions. These enable one to match the solutions to Equation (5.30) in different regions of space containing materials of constant permeabilities. If we can find fields **H** that satisfy Equation (5.30) in each of the regions of space, and which obey the boundary conditions, these fields make up the solution we are looking for, since the uniqueness theorem is valid for magnetostatics as well as for electrostatics.

5.3.2   The boundary conditions on the fields B and H

In this section we will derive the conditions that the field vectors **B** and **H** must obey at the boundaries between different media.

We can derive a condition on the field **B** based on the equation

$$\text{div } \mathbf{B} = 0.$$

This equation means that the flux of the field **B** out of any closed surface is zero. The field lines are continuous and therefore as many lines enter the volume enclosed by the surface as leave it. Consider the closed surface shown in Figure 5.14. This is the surface of a small disc of height Δh that sits astride the boundary

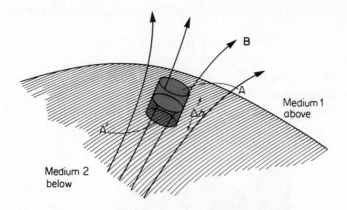

Figure 5.14. The boundary condition on the magnetic field **B**.

between two magnetic media. As we make the height Δh tend to zero, the total flux of magnetic induction over the surface of the disc has contributions from the top and bottom ends only, and

$$\int_A \mathbf{B} \cdot d\mathbf{S} + \int_{A'} \mathbf{B} \cdot d\mathbf{S} = 0. \tag{5.36}$$

Now $\mathbf{B} \cdot d\mathbf{S} = B_\perp \, dS$, where $B_\perp$ is the component of the field **B** normal to the plane of the surface dS. Equation (5.36) becomes

$$\int_A B_{1\perp} \, dS = \int_{A'} B_{2\perp} \, dS,$$

where $B_{1\perp}$ is the component of the field **B** in medium 1 normal to the interface, and $B_{2\perp}$ is the component in medium 2 normal to the interface. Note that the signs are correct because $d\mathbf{S}$ is a vector perpendicular to the plane of dS and in a direction *out* of the volume enclosed by the surface. The above equation is true whatever the size of the (equal) areas A and A'. The only way this can be so is for the condition

$$\boxed{B_\perp \text{ continuous}} \tag{5.37}$$

to hold at every point on the surface separating the two media. The normal component of the magnetic field **B** at every point on the interface between two media are equal. Note that this is not true for the field **H** because at the boundary the magnetization **M** is discontinuous, and the field **H** has sources.

We can derive a condition on the field **H** based on Equation (5.29).

$$\oint \mathbf{H} \cdot d\mathbf{l} = I, \tag{5.38}$$

where I is the free charge current that passes in a direction given by the right-hand rule through the area defined by the path over which the line integral is taken. Consider a path as shown in Figure 5.15, where the lengths PQ and RS follow

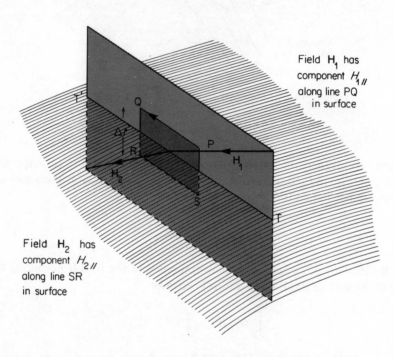

Figure 5.15. The boundary condition on the magnetic intensity **H**.

the contour of the surface on either side of the line TT′ drawn in the interface, and are in the direction of the surface component of the magnetic intensity. The lengths QR and PS lie along normals to the surface and are of length Δt. We will apply Equation (5.38) to this path as the length Δt becomes smaller and smaller, tending to zero, in which case

$$\int_P^Q \mathbf{H}_1 \cdot d\mathbf{l} + \int_R^S \mathbf{H}_2 \cdot d\mathbf{l} = I.$$

As the area of the surface tends to zero the current I tends to zero, otherwise infinite amounts of charge would flow through the vanishingly small area

PQRS. Infinite current density can occur if one of the media has infinite electrical conductivity, but for the moment we will ignore this special case. The current I in the above equation then reduces to zero, and we have

$$H_{1\parallel}PQ - H_{2\parallel}RS = 0.$$

Here $H_{1\parallel}$ and $H_{2\parallel}$ are the maximum components of the fields $\mathbf{H}_1$ and $\mathbf{H}_2$ in the surface either side of the interface between the media. (The sides PQ and RS of the rectangle were drawn parallel to the projections of the vectors $\mathbf{H}_1$ and $\mathbf{H}_2$ on the surface.) The boundary condition on the field $\mathbf{H}$ is thus finally that

$$\boxed{H_{\parallel\,\text{continuous}}} \tag{5.39}$$

at all points across the boundary between two media.

Equations (5.37) and (5.39) are the boundary conditions on the fields $\mathbf{B}$ and $\mathbf{H}$ we set out to derive at the beginning of this section. They are of great help in the solution of problems in magnetism, as the remainder of this chapter illustrates.

5.4 MAGNETS

We have shown how the magnetic fields $\mathbf{B}$ and $\mathbf{H}$ obey certain relations. At all points in a steady magnetic field

$$\text{div } \mathbf{B} = 0,$$

$$\text{curl } \mathbf{H} = \mathbf{j}_f,$$

and

$$\mathbf{B} = \mu_0(\mathbf{H} + \mathbf{M}),$$

where the symbols have their usual meanings. The surface integral $\int_S \mathbf{B} \cdot d\mathbf{S}$ over any *complete* surface is zero, and the line integral $\oint \mathbf{H} \cdot d\mathbf{l}$ around any closed path is equal to the current due to free charges that passes through the area defined by the path in a direction given by the right-hand rule. At the boundary between different media there are discontinuities in the fields, but the latter have to obey certain conditions at these boundaries, viz

$$B_{1\perp} = B_{2\perp},$$

and

$$H_{1\parallel} = H_{2\parallel},$$

where $B_\perp$ is the component of the magnetic field $\mathbf{B}$ normal to the interface and $H_\parallel$ is the component of magnetic intensity tangential to the interface. We can now solve some simple problems in magnetism.

5.4.1 Electromagnets

Consider a toroidal sample of magnetic material overwound uniformly with coils of wire that carry a current I, as shown in Figure 5.16. Take a circular path of radius r inside the toroid. The line integral of the magnetic intensity

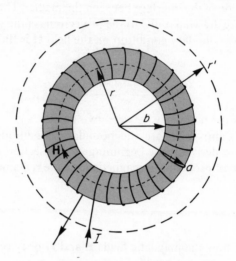

Figure 5.16. A toroidal shaped magnetic material, uniformly overwound with coils carrying a current.

around this path in a clockwise direction is equal to $N_t I$ where N_t is the total number of turns on the toroid,

$$\oint \mathbf{H} \cdot d\mathbf{l} = N_t I \qquad (5.40)$$

But by symmetry the magnetic intensity must have the same magnitude at all points on this circle and be tangential to the circle, hence

$$2\pi r H = N_t I,$$

and

$$H = \frac{N_t I}{2\pi r}.$$

The magnetic field $\mathbf{B}$ inside the toroid is thus given by

$$B = \frac{\mu \mu_0 N_t I}{2\pi r},$$

where μ is the relative permeability of the material. Consider a different circular path of radius r' outside the coils as shown in Figure 5.16. If there were a field **H** outside it would again be tangential and have the same magnitude at all points on the circle. But now the total current which passes through the circle is zero and so from Equation (5.40)

$$\oint \mathbf{H} \cdot \mathbf{dl} = 0.$$

Hence

$$2\pi r' \cdot H = 0,$$

and the field **H** outside is zero. The same argument can be applied to any point outside the coils, so that outside, the fields are everywhere zero.

If the sample were removed from within the coils Equation (5.40) would still be true. If the number of turns on the coil were large, so that the coil was close packed, to a good approximation the field **H** would again be tangential to circles centred on the middle and the centre. The magnetic intensity would thus be unchanged, but the magnetic field **B** would be reduced by the factor μ.

Suppose now that there is a small gap of length l cut in the toroidal sample, as shown in Figure 5.17. An arrangement similar to this is often used to make an

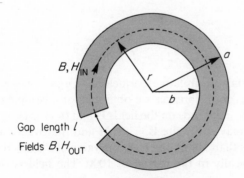

Figure 5.17. A toroidal shaped magnetic material with a wedge-shaped gap cut in it. The toroid is uniformly overwound with coils carrying a current, as in Figure 5.16.

electromagnet. If the core is made of a ferromagnetic material the field **B** in the gap is greatly enhanced over its value in the absence of the core. How do we determine the field in the gap? As a first approximation we will assume that there are still no fields outside that toroidal region of space occupied by the core and gap, i.e. that all the magnetic flux produced by the current in the coils and the magnetization of the material remains inside the coils. If this is so, then again the fields are tangential to a circle of radius r, as drawn, and at all points on this circle within the material the magnetic intensity has the same magnitude. At the interfaces between the material and the air in the gap the magnetic fields are normal to the boundary, and so the magnetic field **B** has the same magnitude B in the medium and in the gap. If the relative permeability of the material of the sample is μ, the magnetic intensity inside the sample is

$$H_{IN} = B/\mu\mu_0,$$

and the magnetic intensity in the gap is

$$H_{OUT} = B/\mu_0,$$

if we put the relative permeability of air equal to unity. Taking the complete path around the circle

$$\oint \mathbf{H} \cdot \mathbf{dl} = N_t I,$$

where N_t is the number of turns on the winding and I is the current. Evaluating this line integral gives

$$(2\pi r - l)\frac{B}{\mu\mu_0} + \frac{lB}{\mu_0} = N_t I, \tag{5.41}$$

or

$$B = \frac{\mu\mu_0 N_t I}{(2\pi r - l + l\mu)}. \tag{5.42}$$

For ordinary materials the relative permeability μ is very near unity and Equation (5.42) reduces to the answer we obtained for a complete toroid. For ferromagnetic materials μ depends on the fields **H** and **B**. However, even though μ is not constant, we may still write $\mathbf{B} = \mu\mu_0\mathbf{H}$ and the equations above still apply. For fields **B** less than about one tesla the value of μ defined by the equation $\mathbf{B} = \mu\mu_0\mathbf{H}$ is typically in the region of 1000. The field is then approximately given by

$$B = \frac{\mu\mu_0 N_t I}{(2\pi r + l\mu)}.$$

Unless the gap is very small, $l\mu \gg 2\pi r$ and

$$B = \frac{\mu_0 N_t I}{l}.$$

As we expect, a ferromagnetic material, due to its large magnetization, increases the field **B** in the gap by a large amount in comparison with the value $\mu_0 N_t I/2\pi r$ in an empty coil.

In order to calculate more precisely the field in the gap of a particular electro-magnet it is necessary to know the relation between the fields **B** and **H** within the material, i.e. to know the value of the permeability μ at each value of the field **H**. The *B–H curve* for soft iron, a material often used in electromagnets, is shown in Figure 5.18. This curve can be obtained approximately by measuring the field **B** near the end of a rod of soft iron inserted into a solenoid. The field at a point on axis near the end of the solenoid is given by (see example {5.8})

$$B = \frac{\mu\mu_0 N_t I}{2L}(\cos \alpha - \cos \beta),$$

where the angles α and β specify the position of the point, N_t is the total number of turns and I the current. L is the length of the solenoid. The magnetic intensity inside the rod can be determined to be roughly

$$H = \frac{N_t I}{L},$$

by using Equation (5.38) and neglecting the intensity outside the solenoid. It can be seen that if the field B is measured as a function of the current through the solenoid, I, the permeability of the rod can be determined at each value of the magnetic intensity. Since

$$B = \mu\mu_0 H$$

the *B–H* curve for the material can be plotted.

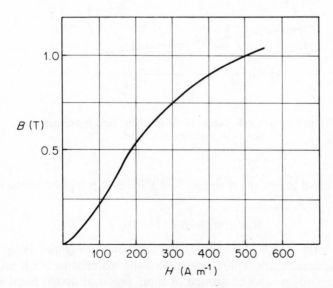

Figure 5.18. The relation between the magnetic field **B** and the magnetic intensity for soft iron.

If the electromagnetic of Figure 5.17 were made with soft iron, the field in the gap could be calculated with the aid of the *B–H* curve for soft iron shown in Figure 5.18. Rewriting Equation (5.41) in the form

$$(2\pi r - l)H + \frac{lB}{\mu_0} = N_t I, \tag{5.43}$$

we have a linear equation which the intensity *H* and the magnetic field *B* within the material have to satisfy. Figure 5.18 shows another relation which the fields have to satisfy, and their actual values inside the material can be obtained graphically by finding the intersection of the straight line (5.43) with the *B–H* curve. For example if the product of the number of turns on the solenoid, N_t, and the current, *I*, is 2000 ampere turns; the gap length, *l*, is 1 cm, and the overall

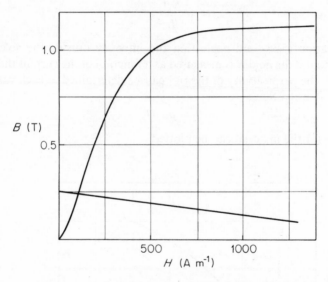

Figure 5.19. The calculation of the magnetic field in the gap
of a soft iron electromagnet.

length of the toroid is 1 m, the magnetic field *B* in the gap is given by the intersection of the line

$$H + (10^5/4\pi)B = 2 \times 10^3$$

with the *B–H* curve. The field is about 0.24 T, as shown in Figure 5.19.

Real electromagnets are seldom made with the material in the form of a toroid, or with coils wound all around. A more practical design is shown in Figure 5.20. The field in the gap can be estimated roughly by making the same assumption as before, that all of the magnetic flux lies inside the iron. The total magnetic flux Φ through any area S obtained by slicing through the magnetic material, e.g. the shaded area in Figure 5.20, is constant, and given by

$$\Phi = \int_S \mathbf{B} \cdot d\mathbf{S}.$$

We now divide the magnet up into n segments of lengths Δl_n, which have either different cross-sectional areas S_n, and/or permeabilities μ_n, and write

$$\Phi = B_n S_n, \tag{5.44}$$

where B_n is the average induction field in the nth segment. B_n is normal to a cross

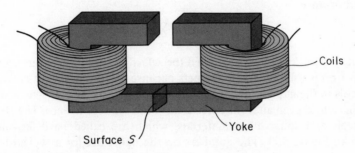

Figure 5.20. An electromagnet. The field in the gap is less than that given by Equation (5.45) due to flux leakage around the pole faces.

section through the nth segment. For a complete path drawn through the magnet material and the gap,

$$\oint \mathbf{H} \cdot d\mathbf{l} = N_t I,$$

where the symbols have the usual meanings. Rewriting the line integral in terms of the field $\mathbf{B}$ we have

$$\oint \frac{\mathbf{B} \cdot d\mathbf{l}}{\mu \mu_0} = N_t I,$$

or

$$\sum_n \frac{\Phi \, \Delta l_n}{\mu_n \mu_0 S_n} = N_t I,$$

from Equation (5.44). Hence,

$$\Phi = \frac{N_t I}{\displaystyle\sum_n \frac{\Delta l_n}{\mu_n \mu_0 S_n}}. \tag{5.45}$$

This equation can be used to determine the flux Φ and hence the field in the gap from the relation

$$B_{\text{gap}} = \frac{\Phi}{S_{\text{gap}}}.$$

In Equation (5.45) the quantity $(N_t I)$ is often called the *magnetomotive force* (m.m.f.) in the *magnetic circuit*, by analogy with the e.m.f. V provided by a battery in an electrical circuit. Equation (5.45) is very similar to Ohm's law

$$I = V/R,$$

which gives the current I through a series of resistors R_n in series, which have a total resistance

$$R = \sum_n R_n.$$

The quantities $\Delta l_n / \mu_n \mu_0 S_n$ are called the *reluctances* of the various segments.

Soft iron is used in electromagnets because when the current is switched off the fields in the gap and the iron become very nearly zero. If steel is used there remains a field even after the current has been reduced to zero. The B–H curves for steels and some other materials, which are called hard ferromagnetics, look like Figure 5.21. The numbers on the figure are for a typical hard steel. Suppose such a B–H curve is measured by putting a steel rod inside a solenoid, as discussed earlier. As the current is increased from zero, the curve follows the line oa. It levels off at a when the steel becomes saturated. At that point the magnetization M of the steel has reached its maximum value, its *saturation magnetization* M_S, when all the individual moments are aligned. The magnetic field $\mathbf{B}$ increases after the point a only by virtue of the term $\mu_0 \mathbf{H}$ in the equation

$$\mathbf{B} = \mu_0 (\mathbf{H} + \mathbf{M}).$$

If the current is reduced below the saturation level and back to zero, the B–H curve follows the line ab. For currents in the reverse direction the magnetic intensity $\mathbf{H}$ needs to be quite large for the field $\mathbf{B}$ to be reduced to zero at the point c. At the point d we have reached saturation in the reverse direction. The whole curve $abcdefa$ is called the *hysteresis curve* of the material. The value of the field $\mathbf{B}$ at the point b on the curve is called the *remanence* of the material: if this is large the material can be used in a permanent magnet, which produces a flux of magnetic induction even in the absence of external currents. For soft iron the remanence is very nearly zero.

The hysteresis curve shown in Figure 5.21 is the *major hysteresis curve* for the steel, in which the steel is taken up to saturation for each direction of the field $\mathbf{H}$. If the sample is not taken up to saturation along the initial portion oa it will return from the point a' of Figure 5.22 via a path which in general has a different shape to the shape of the major hysteresis loop. There are an infinite number of minor hysteresis curves like $a'b'c'd'e'f'a'$, and so the magnetic field $\mathbf{B}$ in a hard

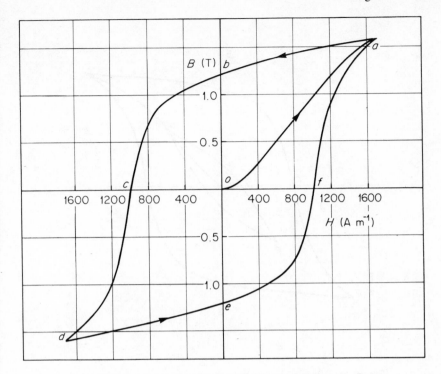

Figure 5.21. The *B–H* curve for a typical hard steel.

ferromagnetic material can have many different values for the same value of the intensity **H**. The values of **B** and **H** that exist in a specimen depend on the particular magnetic treatment the specimen has received in the past, i.e. on its magnetic history.

The hysteresis effect persists to some extent in the soft materials used in electromagnets and this has practical consequences. Suppose for example we wanted to measure the susceptibility of a paramagnetic material. This could be done by measuring the force on a sample placed in the gap of an electromagnet. If accurate measurements are required it is not sufficient to monitor only the current in the magnet. Identical currents will not always produce exactly the same magnetic field because of uncertainties due to hysteresis. The actual magnetic field in the gap must be measured.

The use of the approximations of the magnetic circuit for the calculations of fields due to electromagnets can produce errors of up to about 15 %, or more if the iron is near saturation, when there is appreciable flux leakage out of the iron. The field lines in the gap bulge outwards, corresponding to a smaller field than given by the assumption that the field lines pass straight across the gap.

Saturation magnetization is approached for most ferromagnetic materials at fields near to 1.5 T. In order to obtain such fields, magnetic intensities approach-

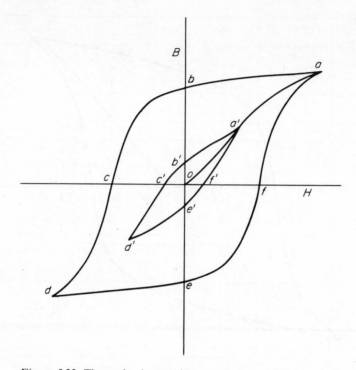

Figure 5.22. The major hysteresis curve and one minor curve for
hard steel.

ing 10^4 amperes/m are required. These can be obtained for example by passing 20 A through a multilayered solenoid which has 10^3 turns/m, or 2 A through 10^4 turns/m. To do this involves a large power loss in the resistance of the coils and large cooling problems. Nowadays fields in excess of one tesla may be obtained with *superconducting magnets*. Some metals, like tin or niobium, or metallic compounds, have the property that when they are cooled to very low temperatures, typically below 10 K, their electrical conductivity becomes infinite; they have become superconductors. If a current of 20 A is injected into a superconducting solenoid which has 10^5 turns/m a field on axis of about 2.5 T is obtained without power losses. There is no iron of course within the superconducting coils and the space within the solenoid is available for example for experiments on the behaviour of materials in strong magnetic fields.

★   5.4.2   **Permanent magnets**

As discussed above, if a piece of steel is magnetized in an external field and then removed from the field it remains magnetized. Its magnetism can only be removed by an external demagnetizing field or by heat treatment. We will not go into the microscopic reasons why the domains in the permanent magnet arrange themselves so as to keep the magnetization. The description of ferromagnetism and the different behaviour of different ferromagnetic materials is left to courses on the solid state of matter. The occurrence of permanent magnetism is of great practical importance as it enables us to provide magnetic fields without power consumption.

The chief point to grasp when considering fields due to permanent magnets is that the magnetization $\mathbf{M}$ of the material does not arise by virtue of an external field. If there are no external fields the material is still magnetized, and the magnetization itself acts as the source of the fields $\mathbf{B}$ and $\mathbf{H}$. These fields obey the usual boundary conditions, and the relation

$$\mathbf{B} = \mu_0(\mathbf{H} + \mathbf{M})$$

still holds at every point. The boundary conditions that the magnetic fields obey impose certain limitations on the distribution of magnetism within a permanent magnet. For example, a piece of material which has straight cut edges with sharp corners cannot be uniformly magnetized. The boundary conditions would imply a discontinuity in the field outside the sharp corners, and this is not allowed. However, the edge effects are rather small, and permanent magnets are uniformly magnetized throughout almost all of their volume.

What is the field around a permanent magnet consisting of a rod of (nearly) uniformly magnetized material? There are no free currents flowing, and if we take the line integral $\oint \mathbf{H} \cdot d\mathbf{l}$ around a closed loop passing through the magnet, then $\oint \mathbf{H} \cdot d\mathbf{l} = 0$. This implies that the magnetic intensity inside the rod must be in a direction opposite to that outside, and opposite to the direction of the field $\mathbf{B}$, since the lines of the field $\mathbf{B}$ are continuous. The lines of the fields $\mathbf{B}$ and $\mathbf{H}$ of the magnetized rod are shown in Figure 5.23. Compare this drawing with Figure 5.13. At distances from the rod much greater than the length of the rod the field looks like the field of a dipole. The field far away from a magnetized specimen of any shape looks like a dipole field, since the more complex parts of the total field fall off more rapidly with distance than does the dipole part. The magnitude of the dipole field is determined by the magnetic dipole moment of the rod, which is simply its volume times the magnetization $\mathbf{M}$.

The magnitude of the field generated by any permanent magnet depends on the magnetization of the magnetic material, and the magnetization is affected by the shape of the hysteresis curve of the magnetic material and the geometry of the magnet. Consider a C-shaped permanent magnet. As we did when working out the field of an electromagnet we shall assume that the lines of the field $\mathbf{B}$ pass straight across the gap between the pole faces. The field $\mathbf{B}$ is normal to the pole faces and is continuous across their surfaces. Its magnitude B in the gap is thus the same as its magnitude in the yoke of the magnet. The magnetic intensity H_g in the gap is thus given by

$$H_g = \frac{B}{\mu_0}.$$

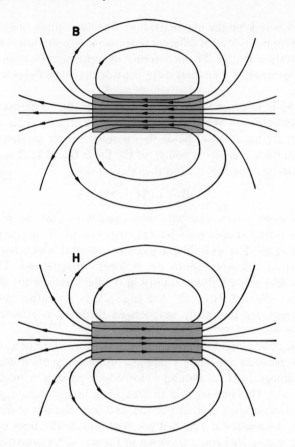

Figure 5.23. The magnetic fields of a nearly uniformly
magnetized bar magnet.

Since there are no free currents flowing the line integral of the field **H** over any complete path is zero, and

$$H_g l + H_i d = 0,$$

where l and d are the lengths of the gap and yoke respectively. Hence the field H_i in the yoke is

$$H_i = -H_g \frac{l}{d} = -\frac{Bl}{\mu_0 d}. \tag{5.46}$$

The values of B and H in the yoke must lie on the line defined by this equation.

The highest magnetic field **B** in the material of the magnet occurs when the material is in a state corresponding to some point on its major hysteresis curve. Figure 5.24 shows the relevant part of this curve for a commercial permanent magnet material. In our example the actual fields in the magnet are obtained by finding the intersection of the line given by Equation (5.46) with the B–H curve. For a gap length of 2 cm and length of yoke 30 cm, the field in the gap obtained in this way is equal to about 1 T. The actual field in the gap will be somewhat less than this due to flux leakage.

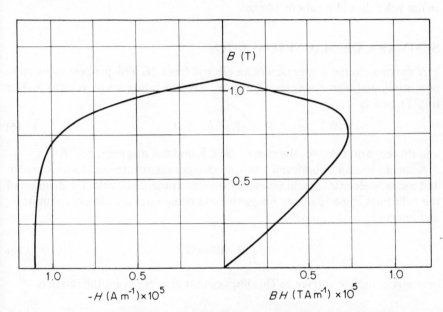

Figure 5.24. The hysteresis curve for a typical commercially available permanent magnetic material. The appropriate quadrant of the B–H curve is shown together with the curve showing how the BH product varies with the field B.

As can be seen from Figure 5.24 the maximum field obtainable, whatever the lengths of gap and yoke, is about 1 T. Normally one is interested not in obtaining the highest field but in obtaining a reasonably high field using the smallest volume of material. This is done when the ratio of the field energy stored in the gap to the volume of material used is a maximum. If the cross-sectional area of the yoke is A the energy stored in the gap is

$$U = \tfrac{1}{2}BH_g Al$$
$$= \tfrac{1}{2}BH_i Ad,$$

from Equation (5.46). The volume of the yoke is Ad, hence using the material most efficiently corresponds to maximizing the BH product in the material. The maximum BH product of a substance is thus a measure of its ability to produce high fields with small volumes. The permanent magnet should be designed so that it operates at the point on the B–H curve corresponding to the maximum value of BH. On the right-hand side of Figure 5.24 the quantity BH is plotted against B for the material whose B–H curve is shown on the left-hand side. It can be seen that for optimum use of the material the field in the gap and yoke should be about 0.8 T, when the BH product has its maximum value of about 7×10^4 T A m^{-1}. Thus if the gap length of a magnet is to be 2 cm, the length of the yoke should be about 14 cm.

SUMMARY OF MAGNETOSTATICS

A moving charge q experiences an electric force $q\mathbf{E}$, independent of its state of motion, and also a magnetic force $q\mathbf{v} \wedge \mathbf{B}$ depending on its velocity $\mathbf{v}$. The total force $\mathbf{F}$ is

$$\mathbf{F} = q\mathbf{E} + q\mathbf{v} \wedge \mathbf{B}, \tag{4.16}$$

and this equation defines the electric field $\mathbf{E}$ and the magnetic field $\mathbf{B}$.

A steady stream of uniformly moving charges constitutes an electric current, and a steady electric current gives rise to a steady magnetic field. If a diagram of the field lines is made of any magnetic field $\mathbf{B}$ the lines are always continuous, and hence

$$\int_S \mathbf{B} \cdot d\mathbf{S} = 0, \tag{4.18}$$

over any complete surface S. This implies that at every point the relation

$$\text{div } \mathbf{B} = 0 \tag{4.19}$$

holds. The field $\mathbf{B}$ obeys Ampère's law

$$\oint \mathbf{B} \cdot d\mathbf{l} = \mu_0 I, \tag{4.28}$$

where I is the current in a right-hand sense through the closed loop and μ_0 is the permeability of free space. This integral relation reduces to the differential relation

$$\text{curl } \mathbf{B} = \mu_0 \mathbf{j} \tag{4.40}$$

in terms of the current density $\mathbf{j}$ and the vector operator curl.

If matter is present in a magnetic field the atoms acquire an effective magnetic dipole moment $\langle \mathbf{m} \rangle$. We can then define a macroscopic function $\mathbf{M}$, called the magnetization of the medium,

$$\mathbf{M} = N\langle \mathbf{m} \rangle, \tag{5.1}$$

where N is the number of atoms or molecules per unit volume. For isotropic paramagnetic ($\chi_B > 0$), or diamagnetic ($\chi_B < 0$), substances

$$M = \chi_B \frac{B}{\mu_0}, \tag{5.23}$$

where χ_B is the magnetic susceptibility, and B is the total magnetic field to which the magnetization of the material now contributes.

The magnetization of a sample gives rise to the same magnetic effects as a volume distribution of current within the sample given by

$$j_M = \text{curl } M, \tag{5.19}$$

and surface currents over the sample. Over a closed loop Ampère's law becomes, including the magnetization currents I_M,

$$\oint B \cdot dl = \mu_0 I_f + \mu_0 I_M,$$

which reduces to the differential form,

$$\text{curl } B = \mu_0 j_f + \mu_0 \text{ curl } M.$$

This is an equation the fields B and M have to obey at all points.

Instead of working with the vectors B and M, the magnetic intensity H is introduced as an auxiliary vector.

$$H = \frac{B}{\mu_0} - M, \tag{5.27}$$

whence

$$\text{curl } H = j_f. \tag{5.30}$$

Magnetic fields obey this equation everywhere, *and* obey the boundary conditions

$$B_{\perp \text{ continuous}} \tag{5.37}$$

$$H_{\parallel \text{ continuous}}. \tag{5.39}$$

For non-ferromagnetic materials

$$B = \mu\mu_0 H \tag{5.32}$$

where the constant μ is given by

$$\mu = (1 - \chi_B)^{-1}, \tag{5.33}$$

and is called the relative permeability of the material. For ferromagnetic materials, the relationship between B and H is non-linear, and for hard ferromagnetics it is multiple-valued as shown in hysteresis curves. Hard ferromagnetics may retain some magnetization in the absence of external fields and are used to make permanent magnets.

PROBLEMS 5

5.1 The magnetic susceptibility of a compound whose molecular weight is 400 and whose density is 2×10^3 kg m^{-3} is given by the formula

$$\chi = \frac{7.3 \times 10^{-2}}{T},$$

where T is the absolute temperature. Calculate the permanent dipole moment associated with each molecule. Make a rough estimate of the diamagnetic correction to the above equation at room temperature.

5.2 Nickel is ferromagnetic at ordinary temperatures. At very large magnetic fields it becomes saturated, and corresponding values of B and H are 1.275 T and 5×10^5 A m^{-1}. Calculate the magnetic moment per Ni atom in Bohr magnetons. The density of Ni is 9×10^3 kg m^{-3}.

5.3 A small transverse gap is cut in a soft-iron anchor-ring (toroid) overwound with a coil carrying a current. The ring has a mean radius of 10 cm, the gap is 0.1 cm long and there are 200 turns of wire carrying 5 A. If the B–H relation for the iron is given by the table below, what is the field in the gap?

H	(Am^{-1})	40	80	160	240	320	480	800	1600
B	(T)	0.1	0.2	0.6	0.85	1.0	1.2	1.4	1.5

Compare the power required to maintain a field of 0.6 T in the gap with that required to maintain a field of 1.2 T.

5.4 A magnet core has the shape shown in Figure 5.25. The cross-section of the iron has the same shape and size everywhere. Two identical coils carrying the same current (in directions such that there is a large field in the gap) are wound on the arms of the yoke. If the current through one coil is stopped, but maintained at its original value through the other one, and the permeability of the iron is assumed to be constant, how does the field in the gap change?

5.5 Sketch the lines of the fields **B** and **H** in the following situations:
   (i) a sphere of material of large constant permeability placed in an originally uniform magnetic field,
   (ii) an isolated permanent magnet in the form of a uniformly magnetized sphere.
Use an appropriate change of scale for the density of the lines of the two different fields.

5.6 A cylindrical specimen of iron of length 10 cm and diameter 5 cm is very nearly uniformly magnetized parallel to its axis. It has a dipole moment of 75 A m^2. Estimate the magnetic field B inside the iron.

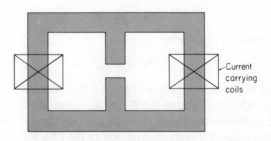

Figure 5.25.

5.7 The magnetic field **B** inside a superconducting material is 'frozen' at the value it had when the material became superconducting. A bar magnet is brought near to the horizontal surface of a rigid superconductor in which the magnetic field **B** is zero. Sketch the lines of the field **B** around the magnet and show that outside the superconductor they are the same as those due to the real magnet together with a fictitious image magnet located inside the superconductor. If the bar magnet is released make a rough prediction of its subsequent behaviour.

5.8 Estimate the magnetic field **B** at a point P on the axis of a long solenoid of length L, filled with a rod whose relative permeability may be taken to have the constant value μ. The total number of turns on the solenoid is N_t and the current I. The point P is outside the solenoid, and the diameter of the nearest and of the solenoid subtends an angle 2β at P; the diameter of the farthest end subtends an angle 2α at P.

5.9 It is sometimes necessary to know the field of a uniformly magnetized permanent magnetic material of a given shape, for example a uniformly magnetized sphere. Show that the method of determining the fields in these situations is very similar to the solution of problems in electrostatics. Determine the field of a uniformly magnetized sphere by following the treatment of a dielectric sphere in a uniform electric field discussed in section 3.6.

CHAPTER

6

Electromagnetic induction and magnetic energy

The last two chapters dealt with magnetostatics—the magnetic effects which arise when steady currents flow. In this chapter we discuss what happens when currents are changing with time. Non-steady currents give rise to non-steady magnetic fields. It is an experimental observation that they also give rise to electric fields. In the following sections we will relate the electric field to the changing magnetic field and examine some of the consequences of the relationship.

6.1 ELECTROMAGNETIC INDUCTION

6.1.1 Electromotive force

In section 4.1.1 we introduced the concept of electromotive force (e.m.f.). A battery has an e.m.f. of V volts if it maintains a constant potential difference V between its terminals. If the terminals are connected by a conductor of resistance R a current equal to V/R flows around the circuit completed by the battery and the external resistance. The electric field is everywhere static and the line integral of the field $\mathbf{E}$ over any closed loop is zero. The line integral of the field $\mathbf{E}$ over any path between two points A and B is constant, and is equal to the potential difference between the two points. If the two points A and B are the positive and negative terminals of the battery, the e.m.f. of the battery is

$$V = \phi(\mathbf{r}_A) - \phi(\mathbf{r}_B) = \int_A^B \mathbf{E} \cdot d\mathbf{l}. \tag{6.1}$$

When a current flows in a wire joining the terminals, the battery must do work to keep the potential difference between the terminals constant. If a charge q moves from terminal A to terminal B around the external conductor, the work done by the battery is Vq.

$$Vq = q \int_A^B \mathbf{E} \cdot d\mathbf{l}$$

$$= \int_A^B \mathbf{F} \cdot d\mathbf{l}$$

where $\mathbf{F}$ is the force on the charge q. Hence

$$V = \frac{1}{q} \int_A^B \mathbf{F} \cdot d\mathbf{l}. \tag{6.2}$$

Equation (6.1) or (6.2) may be used to define the e.m.f. of a battery whose terminals are the points A and B.

An e.m.f. can exist in a closed circuit as well as across the terminals of a battery. Whenever the magnet flux through a closed circuit is changing, the line integral of the force on a charge, integrated around the whole circuit is not zero. In these circumstances work must be done to take a charge around the circuit, and the work done may be used to define an e.m.f., in the same way as the e.m.f. of a battery is defined in Equation (6.2). The e.m.f. V around the circuit is

$$V = \frac{1}{q} \oint \mathbf{F} \cdot d\mathbf{l}. \tag{6.3}$$

If the closed circuit consists of a conductor of resistance R, and contains no source of e.m.f. other than the one due to changing magnetic fields, then a current $I = V/R$ will flow around the circuit.

6.1.2 Motional electromotive force

In this section we give an example of a situation in which an e.m.f. exists in a simple closed circuit, and deduce its value.

Consider first a metal rod moving with constant velocity $\mathbf{v}$ in a direction perpendicular to a uniform magnetic field $\mathbf{B}$, as shown in Figure 6.1. There is a magnetic force given by Equation (4.15) on each electron and positive charge in the rod. The free electrons in the rod will move towards the end a and collect there, setting up a charge distribution which gives rise to an electric field which halts further migration of free electrons. The electric field set up, $\mathbf{E}$, gives rise to a force $-e\mathbf{E}$ on the electrons exactly equal and opposite to the force $-e\mathbf{v} \wedge \mathbf{B}$ which acts on them because of their motion in the magnetic field, and hence

$$-e\mathbf{E} = e\mathbf{v} \wedge \mathbf{B}.$$

The electric field in the rod is thus given by

$$\mathbf{E} = -\mathbf{v} \wedge \mathbf{B}. \tag{6.4}$$

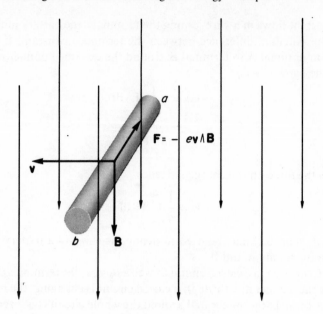

Figure 6.1. A metal rod moving with velocity **v** perpendicular to a uniform field **B**.

The potential difference V_{ba} between the ends a and b of the rod is given by

$$V_{ba} = \int_b^a \mathbf{E} \cdot d\mathbf{l},$$

and hence

$$V_{ba} = vBL, \tag{6.5}$$

where L is the length of the rod.

As an example consider an aeroplane with a wing span of 50 m flying hori-
zontally at 800 km per hour at a place where the vertical component of the
earth's magnetic field is 2×10^{-5} T. There is a voltage difference between the
wing tips, given by Equation (6.5), equal to 0.22 volts.

The potential difference between the ends of the rod shown in Figure 6.1
produces no current flow. If a metal rod moves on stationary conducting rails,
perpendicularly to a uniform magnetic field, as shown in Figure 6.2, a current
will flow. In this situation charges do not build up on the ends a and b of the
rod; electrons traverse the complete circuit from b to a and around to b again,
and continue to circulate. A stationary observer sees a magnetic force acting on
the free electrons in the rod and this force pushes them over to end a. At the end a
they push more electrons around the complete circuit, and a continuous current

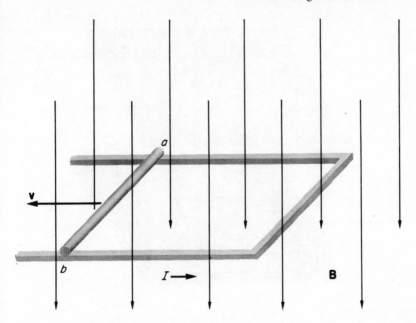

Figure 6.2. A metal rod moving with velocity **v** on stationary conducting rails.
A uniform field **B** acts as shown.

flows. The line integral of the force on a charge q evaluated around the complete
circuit is given by

$$\oint \mathbf{F} \cdot d\mathbf{l} = qvBL,$$

where L is the length of the rod. The sole contribution to the line integral comes
from the portion ab of the loop. The e.m.f. in the closed circuit is given by

$$\frac{1}{q}\oint \mathbf{F} \cdot d\mathbf{l} = vBL. \qquad (6.6)$$

This we may call a *motional e.m.f.* because it arises from the motion of a conductor in a magnetic field. The motional e.m.f. causes a current I to flow, given by

$$I = \frac{vBL}{R}, \qquad (6.7)$$

where R is the resistance of the circuit. The instantaneous power expended is
$(vBL)^2/R$ and the work is derived from the mechanical force needed to move the
rod along the rails.

Consider now the idealized situation shown in Figure 6.3, in which a uniform
magnetic field **B** exists over a sharply defined region of space. A rectangular coil

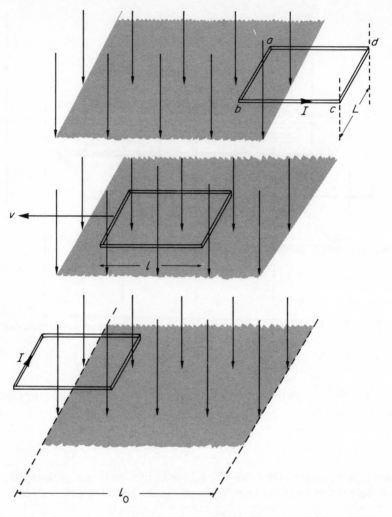

Figure 6.3. A rectangular circuit moving with constant speed v through a
region of uniform magnetic field.

abcd moves in a direction perpendicular to the field **B**, with a speed v. While the
coil is entering the field there is an e.m.f. of magnitude vBL in the circuit, and
this e.m.f. gives rise to a current in the direction shown. When the whole coil is
moving within the uniform field, the e.m.f. and current are zero, since the magnetic
forces along the sides *ab* and *cd* cancel each other out. When the coil leaves the
field there is again an e.m.f. equal to vBL but it now acts in the opposite direction.
If we arbitrarily let the current be positive as the coil enters the field, the curve
of current against time is as shown in Figure 6.4, where R is the resistance of the

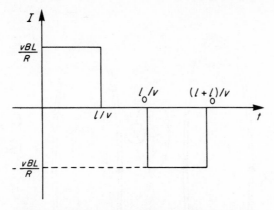

Figure 6.4. The variation of current with time for the moving circuit shown in Figure 6.3. The current is not exactly as shown; the induced currents do not change suddenly, but take some time to build up and die away.

coil. The source of energy that keeps the currents flowing is the force that is moving the coil.

The e.m.f. in the coil as it enters or leaves the field can be expressed in terms of the total magnetic flux Φ through the coil. This flux is the sum of the part BLx due to the external field, where x is the length of the coil in the field, and the part due to the induced current in the coil itself. Since the induced current and the area of the coil are constant, this latter contribution to the magnetic flux is constant, and

$$\frac{d\Phi}{dt} = BL\frac{dx}{dt}$$

$$= BLv.$$

The magnitude of the e.m.f. is thus given by

$$|\,\text{e.m.f.}\,| = \frac{d\Phi}{dt}. \tag{6.8}$$

From the point of view of an observer on the coil, the coil is stationary, and he observes a magnetic field sweep past him for a certain period of time. He measures a current flow in the coil during those times when the magnetic field does not cover the whole coil. Since the coil is stationary there is no net magnetic force on the free electrons, and he attributes the current flow to an electric field. Thus a static magnetic field for one observer looks like a magnetic field plus an electric field to an observer in uniform motion relative to the first*. Hence if a Helmholtz

---

* See Chapter 14 for a discussion of how the electric and magnetic fields transform from one frame of reference to another frame in uniform relative motion.

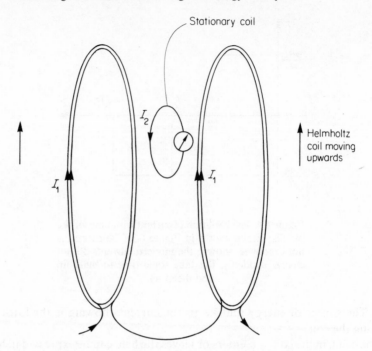

Figure 6.5. A Helmholtz coil carrying a current I_1 moves past a stationary loop of wire. A current I_2 is induced in the loop.

coil carrying a current I_1 is moved past a loop of wire in series with a current
meter, as shown in Figure 6.5, a current I_2 will flow one way in the loop as the
coil approaches, and then in the other way as the coil recedes. Again the energy
dissipated in the small coil as the currents I_2 flow is supplied by the mechanical
force needed to move the Helmholtz coil.

6.1.3 Faraday's law

We have seen that there is a motional e.m.f. in a circuit moving in a magnetic
field in such a way that the magnetic flux through the circuit changes. An e.m.f.
is also produced in a stationary circuit when the magnetic flux through the
circuit changes because the sources of the magnetic field are moving. *It is an
experimental observation that an e.m.f. appears in a circuit when the magnetic
flux through the circuit changes from any cause.* Such an e.m.f. is called an
induced e.m.f., and is related to the *induced electric field* caused by the changing
magnetic field.

Consider two stationary coils, one carrying a current which gives rise to a
magnetic flux through the second. If the current in the first coil changes there

is an induced e.m.f. in the second coil which causes a current to flow. The induced e.m.f. V is equal to the line integral of the induced electric field $\mathbf{E}$ around the coil,

$$V = \oint \mathbf{E} \cdot d\mathbf{l}.$$

If the resistance of the coil is R the current is

$$I = \frac{V}{R} = \frac{1}{R} \oint \mathbf{E} \cdot d\mathbf{l}. \tag{6.9}$$

Experiments show that the current is proportional to the rate of change with time of the total magnetic flux Φ through the circuit. Hence the size of the induced e.m.f. is proportional to the differential $d\Phi/dt$.

$$\left| \oint \mathbf{E} \cdot d\mathbf{l} \right| \propto \frac{d\Phi}{dt}. \tag{6.10}$$

The direction of the current and e.m.f. is given by *Lenz's law*. This law says that the current induced is in a direction such that it produces a magnetic flux tending to oppose the original change of flux, i.e. tending to keep the total flux through the circuit constant. Lenz's law can be expressed mathematically. Figure 6.6 shows a coil in a magnetic field $\mathbf{B}$. The magnetic flux through the coil is equal to

$$\Phi = \int_S \mathbf{B} \cdot d\mathbf{S},$$

where for simplicity we choose S to be the plane surface whose rim is the coil, and the vectors $d\mathbf{S}$ to be in the direction which makes the flux positive. If the field is reduced the flux through the coil falls, and a current is induced which tends to stop the flux falling, i.e. it produces a field in the same direction as the field $\mathbf{B}$. The induced current is thus in the direction shown in the lower drawing of Figure 6.6, and the line integral of the induced electric field is positive if we traverse the coil clockwise, looking from the right, in evaluating the integral. This sense of traversal of the coil is the one we must use since it is the one required by the right-hand screw rule together with the direction already chosen for the vectors $d\mathbf{S}$. In our example the rate of change of flux $d\Phi/dt$ is negative, and so in order to obtain a positive e.m.f., clockwise around the coil, Equation (6.10) must be written

$$\oint \mathbf{E} \cdot d\mathbf{l} \propto -\frac{d\Phi}{dt}. \tag{6.11}$$

If the quantities in Equation (6.11) are in SI units the constant of proportionality is found to be unity, as suggested by Equation (6.8), which applies to motional e.m.f.'s.

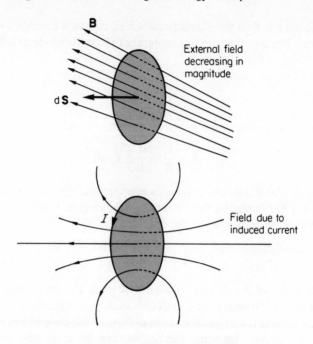

Figure 6.6. An illustration of Lenz's law. If the field **B** over the closed loop of wire (viewed from the left) decreases, a current I is induced in the loop. The direction of the current is such that the field to which it gives rise tends to stop the flux through the loop from falling, and so is in the direction shown in the lower figure.

All of the results of observations on induced e.m.f.'s are summarized in *Faraday's law*, which says that the e.m.f. induced in a circuit is equal to the negative rate of change with time of the magnetic flux through the circuit.

$$\oint \mathbf{E} \cdot d\mathbf{l} = -\frac{d\Phi}{dt}, \tag{6.12}$$

or equivalently

$$\oint \mathbf{E} \cdot d\mathbf{l} = -\frac{d}{dt} \int_S \mathbf{B} \cdot d\mathbf{S}, \tag{6.13}$$

where S is any surface whose rim is the loop under consideration. Faraday's law applies equally to any loop drawn in free space or in media, and not solely to a loop which is an actual conducting circuit, moving or otherwise. It is one of the fundamental laws of electromagnetism, on a par with Ampère's law and Gauss' law. The observation that changing magnetic fields produce electric fields was first made by Faraday around 1830, and independently by Henry and by Lenz near the same time.

6.1.4 Examples of induction

In this section we consider some examples of electromagnetic induction, and their analysis using Faraday's law, in order to become further acquainted with the use of the law and the ideas behind it.

(i) A flat circular coil has 100 turns of wire each of radius 10 cm. A uniform magnetic field B exists in a direction perpendicular to the plane of the coil. The field is decreasing in strength at the rate of $10^{-2}\,T\,s^{-1}$. The back e.m.f. induced in the coil is equal to (number of turns) $\times$ (area) $\times$ $(dB/dt) \approx 0.03$ V. If the resistance of the coil is $0.5\,\Omega$, an induced current ≈ 0.06 A will flow. In this calculation we have neglected the flux through the coil due to its own field. This field is constant in this example, and so will have no effect. At the beginning and end of the change in the external field the current in the coil is not constant, and the resultant flux changes induce further e.m.f.'s. This effect, which is called self-inductance, is discussed in section 6.2.

(ii) Faraday's law enables us to understand the origin of the changed energies of the electrons in atomic orbits when an external field is switched on. The electrons (e.g. in the orbits of Figures 5.2 and 5.3 used in the discussion of diamagnetism) are either speeded up or slowed down by the electric field induced when the external field is switched on.

Consider the orbit shown in Figure 5.2. The area of the orbit is πr^2. The flux of magnetic induction Φ through this area increases from zero* to the value Φ_f during the time it takes to establish the external field **B**.

$$\Phi_f = B\pi r^2. \tag{6.14}$$

While the field is changing there is an e.m.f. in the orbit given by

$$\oint \mathbf{E} \cdot d\mathbf{l} = -\frac{d\Phi}{dt}.$$

The e.m.f. acts in a direction opposite to the electron's motion and so gives a force on the electron which speeds it up. Its angular frequency increases from ω_0 to the final value ω_f. At an intermediate time t, when the angular frequency is ω the electron makes a fraction $\omega\,dt/2\pi$ of a complete turn in a time dt. The

---

* The flux through the orbit due to the motions of the other charges in the atom is to a good approximation constant and we neglect its effect in this analysis.

work done on the electron during this small time by the induced electric field is

$$dW = \frac{\omega}{2\pi} dt \cdot e \frac{d\Phi}{dt}$$

$$= \frac{e\omega}{2\pi} d\Phi.$$

This work is equal to the increase in kinetic energy of the electron, assuming that the electron's orbit remains unchanged, and hence

$$\frac{e\omega}{2\pi} d\Phi = d(\tfrac{1}{2} m_e r^2 \omega^2)$$

$$= m_e r^2 \omega \, d\omega,$$

from which

$$\frac{d\omega}{d\Phi} = \frac{e}{2\pi r^2 m_e}.$$

By integrating this equation we determine the change in angular frequency,

$$\omega_f - \omega_0 = \frac{e\Phi_f}{2\pi r^2 m_e}$$

$$= \frac{eB}{2m_e},$$

from Equation (6.14). The change in angular frequency is equal to the Larmor angular frequency ω_L. This result is the same as obtained in section 5.1.1.

(iii) The betatron, which is a machine designed to accelerate electrons to energies of about 50 MeV, gives a practical demonstration that Faraday's law holds for the electric field around a closed loop in free space, and not only for loops within conducting materials. In the betatron a changing magnetic field is used to accelerate electrons in a circular orbit in an evacuated chamber. The field is designed so that the electrons, as they are accelerated, remain in the same orbit.

Consider an electron moving in an orbit of radius r in a vacuum chamber situated between the poles of an electromagnet, as shown in Figure 6.7. If the magnetic field increases an e.m.f. is induced in the orbit. Since there is symmetry about the axis of the magnet, the electric field induced in the magnet gap depends only on the perpendicular distance from the axis. For the same reason the field at any point is in a direction tangential to a circle drawn around the axis and passing through the point. The electric field induced in the electron's orbit therefore has the same magnitude E at all points in the orbit.

$$E = \frac{1}{2\pi r} \frac{d\Phi}{dt}, \tag{6.15}$$

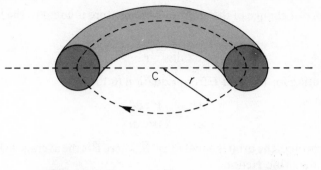

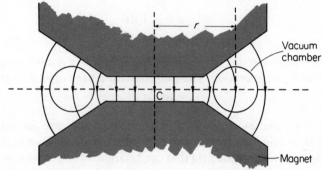

Figure 6.7. A schematic representation of a betatron. The top drawing shows the vacuum chamber situated in a magnetic field which decreases with distance from the centre C. If the field is in the direction shown and increases in magnitude the electrons are accelerated in a clockwise direction, looking downwards. The bottom drawing shows a section through the electromagnet and vacuum chamber.

where Φ is the magnetic flux through the orbit. The direction of the induced field in Figure 6.7 is anticlockwise viewed from above. This field produces a force on the electron accelerating it in a clockwise direction, as indicated by the arrow on Figure 6.7.

If the electron is to stay in its orbit the field B over the circle of radius r must at all times satisfy Equation (4.63),

$$B = \frac{p}{er},\tag{6.16}$$

where p is the momentum of the electron. Differentiating this equation with respect to time

$$\frac{\mathrm{d}B}{\mathrm{d}t} = \frac{1}{er}\frac{\mathrm{d}p}{\mathrm{d}t}.$$

Since the rate of change of the electron's momentum is equal to the force eE,

$$\frac{dB}{dt} = \frac{E}{r},$$

and substituting for the field E from Equation (6.15)

$$\frac{dB}{dt} = \frac{1}{2\pi r^2} \frac{d\Phi}{dt}.$$

The flux Φ through the orbit is equal to $\pi r^2 \bar{B}$, where $\bar{B}$ is the average field over the area πr^2 of the orbit. Hence

$$\frac{dB}{dt} = \frac{1}{2} \frac{d\bar{B}}{dt},$$

and for the electron to stay in the same orbit this equation must be satisfied at all times during the acceleration. The condition for a fixed orbit is thus that

$$B = \tfrac{1}{2}\bar{B},$$

and this is arranged by making the magnetic field decrease outwards from the middle by increasing the gap between the pole faces, as shown in the figure.

If the maximum magnetic field that can be attained at the orbit of the electron is B_{max} Equation (6.16) enables us to calculate the maximum momentum acquired by the electrons,

$$p_{max} = er B_{max}.$$

The electrons in the betatron are moving at speeds comparable with the speed of light in free space, c, and the calculation of their energy has to be done relativistically. The formulae used so far are valid in relativistic dynamics. The relation between the total energy E of the electrons, their rest mass m_e and momentum p is

$$E^2 = c^2 p^2 + m_e^2 c^4.$$

In this example $c^2 p^2 \gg m_e^2 c^4$ and hence $E \approx cp$; $E_{max} \approx c p_{max}$. The maximum energy acquired by the electrons E_{max} is thus given by

$$E_{max} = cer B_{max}.$$

For $B_{max} = 0.25$ T, and $r = 0.5$ m the electrons will be accelerated to energies of about 37 MeV.

(iv) A memory element in the core of a computer consists of a small anchor ring of ferrite material. A *ferrite* is somewhat similar to ferromagnetic materials except that its saturation magnetization does not correspond to full alignment of the atomic dipoles and that it has a high resistivity. This latter property makes ferrites very useful for applications when it is important to reduce heating losses from induced currents. Different ferrites exhibit a wide range of magnetic properties, but the *B–H* curve for a typical ferrite used in a computer memory is as shown in Figure 6.8 and has a characteristic rectangular shape. The memory element normally exists in one of the two states of magnetization corresponding to the points *a* and *b* on the figure. Its state of magnetization is changed by

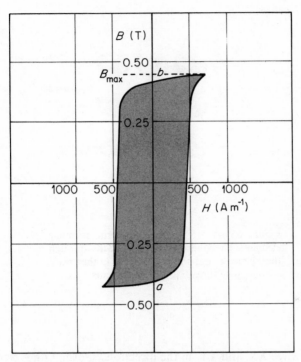

Figure 6.8. The *B–H* curve for a typical ferrite used for memory elements in a computer core. The element switches from the state of magnetization represented by the point *a* to that represented by the point *b*.

passing simultaneous current pulses in the same direction through two wires (the 'write' wires) passing through the centre of the anchor ring. The mean radius of the anchor ring of a typical element is about 0.3 mm. As can be seen from Figure 6.8 a magnetic intensity of about 500 A m^{-1} is needed to change the direction of magnetization of the ferrite, and to provide this the total current in the two write wires (divided equally) has to be nearly 1 A. This figure is obtained by using the integral form of Ampère's law for the field **H**, Equation (5.29), and following the same arguments as in section 4.4.2 (ii).

Two write wires pass through each element because two wires are needed to define the position of a memory element within a two dimensional array. Figure 6.9 shows a 3 × 3 array of elements. Write wires run up and down through each element and others run from left to right through each element. If the element labelled A is the one being written on, current pulses of about 0.5 A pass

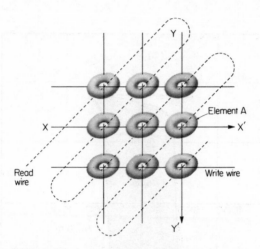

Figure 6.9. A 3 × 3 array of ferrite memory elements. The solid lines are 'write' wires through the elements and the dotted line is the 'read' wire threading all the elements.

through the wires XX' and YY' in the directions shown. Other elements in the same row or column as element A have current pulses through only one of their write wires. This current will not significantly affect the magnetization of the ferrite ring if the B–H curve is rectangular like the one in Figure 6.8. The content of a particular memory element, i.e. its state of magnetization, is 'read' by detecting the e.m.f. induced in a circuit made by passing a third wire through all the elements in the array. Simultaneous current pulses, each of about 0.5 A in our example, are passed through the write wires of the element. Depending on the initial state of magnetization, the field $\mathbf{B}$ in the element will either change direction or not. If it changes there is an induced e.m.f. in the 'read' circuit; if it does not change there is no e.m.f. The cross-sectional area A of a typical memory element is about 4×10^{-8} m^2, and the flux change in the read circuit if the element switches is $2B_{max}A$, i.e. about $2 \times 0.4 \times 4 \times 10^{-8}$ T m^2, as obtained from Figure 6.8. The switching time is around 5×10^{-7} s and hence the e.m.f. induced is about 6×10^{-2} V.

There are many other practical applications and results of electromagnetic induction. One of the most important applications is the generation of electrical power in power stations. Some kinds of electric motor depend on induction for their operation. Induced e.m.f.'s are also used to couple together different parts of a circuit as in power transformers. Induced currents are not always welcome, however, and in the iron of transformers and electromagnetic machinery they are the cause of unwanted energy losses.

6.1.5   The differential form of Faraday's law

The integral form of Faraday's law can be reduced to a differential equation which the fields have to satisfy at every point. The procedure is very similar to that used in reducing the integral form of Ampère's law to a differential equation.

The line integral of an induced electric field around a stationary loop is related by Equation (6.13) to the rate of change with time of the magnetic flux through a surface S whose edge is on the loop

$$\oint \mathbf{E} \cdot d\mathbf{l} = -\frac{d}{dt} \int_S \mathbf{B} \cdot d\mathbf{S}.$$

Using Stokes' theorem (4.38), the left-hand side of this equation can be identified with the integral over the surface S of the curl of the vector $\mathbf{E}$,

$$\oint \mathbf{E} \cdot d\mathbf{l} = \int_S \operatorname{curl} \mathbf{E} \cdot d\mathbf{S}.$$

We thus have

$$\int_S \operatorname{curl} \mathbf{E} \cdot d\mathbf{S} = -\frac{d}{dt} \int_S \mathbf{B} \cdot d\mathbf{S}$$

$$= \int_S -\frac{\partial \mathbf{B}}{\partial t} \cdot d\mathbf{S}.$$

We note that it is permissible to change the order of integration and differentiation because we are considering a surface S which does not change shape with time and is in a fixed position. The total time differential outside the integral becomes a partial derivative under the integral sign, as we are only concerned with the changes in field $\mathbf{B}$ with time at the positions of the elemental areas $d\mathbf{S}$. This equation is true for any surface $\mathbf{S}$, and hence at all points the fields must obey the relation

$$\boxed{\operatorname{curl} \mathbf{E} = -\frac{\partial \mathbf{B}}{\partial t}.} \qquad (6.17)$$

This equation holds for the induced electric field, but it applies equally to a sum of induced and electrostatic fields. The latter field has the property that its line integral around any closed loop is zero, and hence its curl is everywhere zero. The *total electric field from all sources* is related to the magnetic field **B** at all points by Equation (6.17), which is the differential form of Faraday's law.

We note that since the line integral of the induced electric field around a closed loop is not zero, we cannot define a scalar potential from which the field can be derived. The induced electric field, or an electric field in general, has to be derived from a more complicated potential. In section 4.5.2 we saw that since at all points the divergence of the field **B** was zero, we could obtain the field **B** by taking the curl of another vector **A**, the magnetic vector potential. From Equation (4.42)

$$\mathbf{B} = \text{curl } \mathbf{A}.$$

Substitution of this expression for the field **B** into Equation (6.17) gives

$$\text{curl } \mathbf{E} = -\frac{\partial}{\partial t}(\text{curl } \mathbf{A})$$

or

$$\text{curl }\left(\mathbf{E} + \frac{\partial \mathbf{A}}{\partial t}\right) = 0.$$

A solution to this equation is

$$\mathbf{E} = -\frac{\partial \mathbf{A}}{\partial t},$$

but the general solution is obtained by adding to the right-hand side of the above equation any function whose curl is zero. The gradient of any scaler function ϕ has zero curl everywhere, and we may write

$$\mathbf{E} = -\frac{\partial \mathbf{A}}{\partial t} - \text{grad } \phi. \tag{6.18}$$

The function ϕ reduces to the electrostatic potential when the fields are independent of time. Equation (6.18) is a general expression for the electric field in terms of the vector and scalar potentials **A** and ϕ. We will not make use of it now, but it is needed in some problems where charge or current distributions change rapidly with time. These problems are discussed in Chapter 13.

Like the lines of the field **B**, the lines of the induced electric field are continuous. The divergence of the induced field is therefore everywhere zero. Equations (1.16) and (2.19),

$$\text{div } \mathbf{E} = \frac{\rho}{\varepsilon_0}$$

and

$$\text{div } \mathbf{D} = \rho_f,$$

obtained in the earlier chapters on electrostatic fields thus need no modification when one is discussing the total electric field from all sources, or the total electric displacement.

6.2 SELF-INDUCTANCE AND MUTUAL INDUCTANCE

6.2.1 Self-inductance

An important application of Faraday's law is when the flux through a circuit arises solely from the current in the circuit itself. Any change in this current then induces an e.m.f. in the coil.

Consider a long solenoid, of length l, which has N turns per unit length and carries a current I. The magnetic field B at all points within the solenoid is given to a good approximation by Equation (4.31)

$$B = \mu_0 NI.$$

The magnetic flux through each turn of the coil is $\mu_0 NI \pi r^2$, where r is the radius of the coils. The total flux linkage with the Nl turns on the complete solenoid is

$$\Phi = \mu_0 NI \cdot \pi r^2 \cdot Nl.$$

As the current I in the solenoid changes there will be an e.m.f. V induced in the coils, which opposes the change of current.

$$V = -\frac{d\Phi}{dt} = -\mu_0 N^2 \pi r^2 l \frac{dI}{dt}.$$

This e.m.f. may be written as

$$V = -L\frac{dI}{dt},$$

where

$$L = \mu_0 N^2 \pi r^2 l. \tag{6.19}$$

L is called the *self-inductance* of the coil and depends only on the number of turns and the size of the coil.

For a coil of any description the back e.m.f. V induced when the total flux linkage Φ changes, is

$$V = -\frac{d\Phi}{dt}$$

$$= -\frac{d\Phi}{dI} \cdot \frac{dI}{dt}$$

$$= -L\frac{dI}{dt},$$

where

$$L = \frac{d\Phi}{dI}. \tag{6.20}$$

If the flux linkage Φ is proportional to the current I through the coil, Equation (6.20) may be written

$$L = \frac{\Phi}{I}. \tag{6.21}$$

Equation (6.21) defines the self-inductance of a coil in air, where the field produced is always proportional to the current. Equation (6.20) must be used when the relation between flux and current is non-linear, as it is if the coil contains an iron core. In this case the self-inductance varies with the current, and Equation (6.20) defines an incremental self-inductance, valid over a limited range of current only.

The dimensions of inductance can be determined from Equation (6.21). They are those of mass times length squared divided by charge squared*. The SI unit of inductance is called the *henry*, symbol H. A self-inductance of one henry will produce a back e.m.f. of one volt when the current through it changes at the rate of one ampere per second. The self-inductance of a solenoid which is 50 cm long and has 5000 turns each of radius 5 cm is given by Equation (6.19) to be about 0.5 H. The henry is a very large unit, and coils used in circuits typically have inductances of the order of mH. The self-inductance of a wedding ring is about 10^{-8} H.

The self-inductance of a coil is important when the coil forms part of an alternating current circuit, as discussed in Chapter 7. At very high frequencies the current distribution over the cross-sectional area of the wire of the coils changes as a result of the 'skin effect' discussed in Chapter 11. This causes a small change of the inductance with frequency, which is normally not very important.

6.2.2 Mutual inductance

Figure 6.10 shows two long solenoids, one wound on top of the other. The length of each solenoid is l, and the common radius is r. If the bottom coil has N_1 turns per unit length and carries a current I_1, the magnetic flux through each turn of the top coil is $\mu_0 N_1 I_1 \cdot \pi r^2$, and the total flux linking the top coil is $\mu_0 N_1 N_2 l \pi r^2 I_1$, where N_2 is the number of turns per unit length on the top coil. If there is a change in the current I_1, there will be an e.m.f. V_2 induced in the top coil, where

$$V_2 = -\mu_0 N_1 N_2 l \pi r^2 \frac{dI_1}{dt}.$$

---

*The permittivity of free space μ_0 can be seen to have the dimensions of inductance divided by length. Its value is often quoted in units of henries per metre ($\mu_0 = 4\pi \times 10^{-7}$ H m^{-1}). The SI unit of magnetic flux is sometimes called the weber. One weber is equal to 1 T m^2. The unit of magnetic induction field **B** is sometimes quoted as webers m^{-2}.

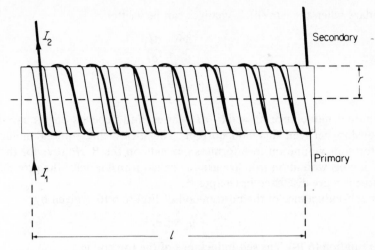

Figure 6.10. A mutual inductance made by winding one long solenoid on top of another.

This equation can be written

$$V_2 = -M_{12}\frac{dI_1}{dt},$$ (6.22)

where

$$M_{12} = \mu_0 N_1 N_2 l\pi r^2.$$

If we reverse the roles of the top and bottom coils by letting a changing current I_2 in the top coil induce an e.m.f. V_1 in the bottom coil it can be seen that

$$V_1 = -M_{21}\frac{dI_2}{dt},$$ (6.23)

where

$$M_{21} = \mu_0 N_1 N_2 l\pi r^2.$$

The quantities M_{12} and M_{21} thus have the same value and can be given the symbol M. M is called the *mutual inductance* of the two coils, and depends only on their construction and geometry.

The result that the induced e.m.f.'s in the two coils are given by Equations (6.22) and (6.23) with $M_{12} = M_{21} = M$ is always true. The mutual inductance between two coils is equal to the flux through one due to unit current in the other if the coils are in air, where the field produced is proportional to current.

If the flux Φ_2 through the secondary coil of a mutual inductance is not proportional to the current I_1 in the primary coil, the voltage V induced in the secondary when the current I_1 changes can be written

$$V = -\frac{d\Phi_2}{dI_1}\cdot\frac{dI_1}{dt}$$

$$= -M\frac{dI_1}{dt}.$$

The mutual inductance $M = d\Phi_2/dI_1$ varies, and depends on the non-linear relationship between flux and current. The incremental mutual inductance used for coils wound on iron formers depends on the B–H curve for the iron. Pairs of coils wound on iron are usually called transformers; the properties of transformers are discussed in Chapter 8.

The self-inductance of the bottom coil of Figure 6.10 is given by

$$L_1 = \mu_0 N_1^2 l\pi r^2$$

from Equation (6.19). The self-inductance of the top coil is

$$L_2 = \mu_0 N_2^2 l\pi r^2.$$

Hence the mutual inductance of the two coils can be written

$$M = (L_1 L_2)^{1/2}. \tag{6.24}$$

This result depends on the assumption that all of the flux produced by one coil passes through the other coil. In a real situation this is not so, and the mutual inductance is less than that given by Equation (6.24). We may write

$$M = k(L_1 L_2)^{1/2} \tag{6.25}$$

where the dimensionless number k, called the coefficient of coupling, is less than or equal to unity.

A car ignition coil is a familiar example of a mutual inductance. In a typical example two insulated coils, one of 16,000 turns the other of 400 turns, are wound over each other. The length of each coil is 10 cm and the turns all have roughly the same radius of 3 cm. A current of 3 A is passed through the primary coil and broken in about 10^{-4} seconds, so that $dI/dt \sim 3 \times 10^4 \, A\,s^{-1}$. The voltage induced in the secondary circuit is given by Equation (6.22), and is about 6000 V. This causes a spark to jump across the gap in a spark plug and ignite a gasoline–air mixture.

The mutual inductance of two coils in which the currents are considered to flow in very thin filaments can be calculated with the aid of the Biot–Savart law (4.43). This has already been discussed in section 4.6.2, and Equation (4.58) gives an expression for the mutual inductance. It is interesting to note that the same technique cannot be directly applied to the calculation of self-inductance. If a current flows in a vanishingly thin wire coil (the situation to which the Biot–Savart law applies) the flux through the coil is infinite. In order to calculate the self-inductance of a real coil an integration must be performed over the current filaments constituting the current in the wire. Each filament is vanishingly thin but carries a vanishingly small current, and if the integration is done carefully no singularities arise. For accurate calculations of mutual inductances the same procedure has to be adopted.

6.3 MAGNETIC ENERGY

In this section we consider magnetic energy. In electrostatics the work done in bringing a system of charges to their final configuration is stored as energy in the electrostatic field. In magnetostatics some of the work done in setting up a current distribution is stored as energy in the magnetic field. We will derive expressions for the total magnetic energy associated with a system of currents and apply the results to some practical situations.

6.3.1 Energy and forces in magnetic fields

The magnetic energy stored in an inductor

Let us suppose that at some time $t = 0$ an inductive coil is placed across the terminals of a battery of e.m.f. V. We will work out the current I_T in the coil at a time T. The coil is equivalent to a self-inductance L which has negligible resistance, in series with a resistor of resistance R. At any time the voltage drop across the coil is equal to $L \, dI/dt$ where I is the current, and the voltage drop across the resistor is RI, hence

$$V = L\frac{dI}{dt} + RI.$$

The total work W_B done by the battery in raising the current to the value I_T is given by

$$W_B = \int_0^T VI \, dt$$

$$= L \int_0^T I\frac{dI}{dt} \, dt + R \int_0^T I^2 \, dt$$

$$= \tfrac{1}{2}LI_T^2 + R \int_0^T I^2 \, dt.$$

The second term on the right-hand side of this equation represents the irreversible conversion of electrical energy into heat. The first term is an amount of energy stored in the inductance after time T, and can be recovered when the magnetic field is switched off. If at time T the battery is removed and replaced by a short circuit, at all times later than T the following equation is obeyed:

$$L\frac{dI}{dt} + RI = 0.$$

Solving this equation for the current I with the initial condition that at time

$t = T$, the current $I = I_T$, we have

$$I = I_T \exp\left\{-\frac{R}{L}(t - T)\right\}.$$

The current decays away exponentially and dissipates energy as heat in the resistor. The total energy appearing as heat after the removal of the battery is given by the integral

$$\int_T^\infty I^2 R \, dt = \tfrac{1}{2} L I_T^2,$$

and the heat energy appearing is equal to the energy stored in the inductance. The energy may be considered to be stored reversibly in the magnetic field caused by the current. An inductance is thus a device for storing magnetic energy, analogously to the manner in which a capacitance stores electrostatic energy.

In the same way as the total energy stored in an electrostatic field can be expressed in terms of the electric fields $\mathbf{E}$ and $\mathbf{D}$, the total energy stored in a magnetic field can be related to the magnetic fields $\mathbf{B}$ and $\mathbf{H}$. In order to determine this relation we will first calculate the total magnetic energy stored in a system of circuits through which steady currents pass.

The total magnetic energy of a system of currents

Suppose there are n stationary circuits and that at an instant of time t the current through the ith circuit is I_i and the magnetic flux through the circuit is Φ_i. The instantaneous induced e.m.f. in the ith circuit is equal to $-d\Phi_i/dt$ and the rate at which work is being done against this induced e.m.f. by the battery supplying the current I_i is given by

$$\frac{dq_i}{dt} \cdot \frac{d\Phi_i}{dt} = I_i \frac{d\Phi_i}{dt}.$$

The total work δW_b done by all the batteries against the induced e.m.f.'s in time δt is given by

$$\delta W_b = \sum_i^n I_i \frac{d\Phi_i}{dt} \delta t. \tag{6.26}$$

If the circuits remain fixed in position during the flux changes the value of the work δW_b given by this equation is equal to the increase in the total magnetic energy stored in the system δU^*. To determine the total magnetic energy U stored in a system of currents I_{0i} flowing in circuits which have magnetic fluxes

---

* The total work done by the batteries in fact includes another term due to Joule heating, which is an irreversible energy loss. This term does not concern us in this section, nor does it influence any of the conclusions reached.

Φ_{0i} through them, Equation (6.26) has to be integrated over the time T it takes to establish the currents. The energy U is independent of the way in which the currents are built up to their final values I_{0i}, and so, for the purposes of calculating the energy, let us assume that they are all built up together linearly with time. In this case the current I_i at time t is given by

$$I_i = \frac{I_{0i}}{T}t.$$

If the circuits are situated in linear materials the fluxes Φ_i are proportional to the currents and

$$\Phi_i = \frac{\Phi_{0i}}{T}t.$$

The restriction on the validity of this equation is important; the equation would not apply for example if a ferromagnetic material were in the space between two of the circuits. The total energy U is now determined by integrating Equation (6.26) over the time T, and using the above equations for I_i and Φ_i.

$$U = \sum_i^n \int_0^T I_i \frac{\partial \Phi_i}{\partial t}dt$$

$$= \sum_i^n \frac{I_{0i}\Phi_{0i}}{T^2} \int_0^T t\,dt.$$

On integration this becomes

$$U = \tfrac{1}{2} \sum_i^n I_{0i}\Phi_{0i}. \tag{6.27}$$

The quantities Φ_{0i} can be written in terms of the self-inductances L_i of the circuits and the mutual inductances M_{ij} between the circuits:

$$\Phi_{0i} = L_i I_{0i} + \sum_{j \neq i} M_{ij}I_{0j}.$$

On substitution of this expression into Equation (6.27) we obtain

$$U = \tfrac{1}{2} \sum_i L_i I_{0i}^2 + \tfrac{1}{2} \sum_i \sum_{j \neq i} M_{ij}I_{0i}I_{0j}.$$

For just two circuits

$$U = \tfrac{1}{2}L_1 I_1^2 + M_{12}I_1 I_2 + \tfrac{1}{2}L_2 I_2^2, \tag{6.28}$$

since $M_{12} = M_{21}$. The sign of M_{12} may be either positive or negative depending on whether the current in one coil increases or decreases the flux through the other coil, i.e. on whether the fields from the two coils add up or not.

The potential energy of a coil in a field and the force on the coil

Let us now consider the torques and forces on a circuit which carries a fixed current I_1 and is placed in an external field arising from a fixed current I_2 in another coil. The same problem was discussed in section 4.6.1, where expression (4.52) was obtained for the potential energy U_P of coil 1 in the field $\mathbf{B}_2$ of coil 2.

$$U_P = -I_1 \int_{S_1} \mathbf{B}_2 \cdot d\mathbf{S}, \tag{6.29}$$

where S_1 is a surface whose rim is the contour of coil 1. In terms of the mutual inductance M_{12} between the two coils, the magnetic flux through coil 1 is given by

$$\int_{S_1} \mathbf{B}_2 \cdot d\mathbf{S} = I_2 M_{12}. \tag{6.30}$$

The sign of M_{12} is given automatically by Equation (6.30) if the direction of the vectors $d\mathbf{S}$ is given by the direction of the current I_1 and the right-hand screw rule. Combining Equations (6.29) and (6.30),

$$U_P = -I_1 I_2 M_{12}. \tag{6.31}$$

The total magnetic energy of the system, given by Equation (6.28) contains terms representing self-energies and involving self-inductances, and an interaction energy $+ M_{12} I_1 I_2$. The potential energy of one coil in the field of the other, given by Equation (6.31), is equal to $- M_{12} I_1 I_2$. The potential energy U_P is related to the mechanical work done by the system if a coil moves under the influence of the magnetic forces. If one coil undergoes a displacement $d\mathbf{r}$ under the influence of a magnetic force $\mathbf{F}$, the mechanical work done by the system dW is given by

$$dW = \mathbf{F} \cdot d\mathbf{r} = -dU_P. \tag{6.32}$$

The force $\mathbf{F}$ is given by

$$\mathbf{F} = -\operatorname{grad} U_P$$

as given in section 4.6.1. During the displacement $d\mathbf{r}$ the batteries supplying the currents I_1 and I_2 do work dW_b against the induced e.m.f.'s in order to maintain the currents constant.

$$dW_b = I_1 \, d\Phi_1 + I_2 \, d\Phi_2,$$

where $d\Phi_1$ and $d\Phi_2$ are the changes in flux through the circuits during the displacement $d\mathbf{r}$. This result follows from Equation (6.26). If the currents are constant the change in the total magnetic energy of the system is obtained from Equation (6.27),

$$dU = \tfrac{1}{2}(I_1 \, d\Phi_1 + I_2 \, d\Phi_2)$$
$$= \tfrac{1}{2} dW_b.$$

The work done by the batteries against the induced e.m.f.'s equals the sum of the mechanical work done by the magnetic force in moving the coil and the increase in total magnetic energy, therefore

$$dW_b = dW + dU.$$

Eliminating dW_b from the last two equations we obtain

$$dU = dW$$
$$= -dU_P,$$

from Equation (6.32)*. In terms of the total magnetic energy U the force on the coil $\mathbf{F}$ is now given by

$$\mathbf{F} = +\operatorname{grad} U. \tag{6.33}$$

This equation is very useful for determining forces on conductors or on magnetized materials in magnetic fields, for which it is also valid. For example the force between the pole faces of an electromagnet can be calculated by determining how the total magnetic energy varies with the separation of the pole faces, keeping the current in the magnetizing coils constant. Some examples are given in the problems at the end of the chapter. As a simple example of the application of ideas developed so far in this section we will obtain a rough estimate of the energies stored in the electric and magnetic fields existing in a hydrogen atom. This will give an idea of the relative importance of the electric and magnetic effects in determining atomic structure. We will again use a classical model of the hydrogen atom in which an electron is revolving around a proton in an orbit of radius a with speed v. Assuming that the electric field is little different from that due to two stationary charges, Equation (1.31) gives the value of about 8 eV for the energy stored in the electric field if we put $a = 10^{-10}$ m. In order to estimate the energy U stored in the magnetic field of the moving electron, let us assume that the flux through the electron's orbit is equal to $\pi a^2 B_0$, where B_0 is the field at the centre due to the current $ev/2\pi a$ flowing in the circle of radius a. This field is given by Equation (4.34) to be

$$B_0 = \frac{\mu_0 ev}{4\pi a^2}.$$

Hence

$$U \simeq \frac{\mu_0 e^2 v^2}{16\pi a},$$

from Equation (6.27). Taking the angular momentum $m_e va$ of the electron equal

---

* The situation discussed here is somewhat similar to that encountered in electrostatics when the charges on two conductors are changed whilst batteries maintain their potentials constant. The stored energy increases by an amount dU, while the batteries do work $2\,dU$, corresponding to an amount of external work dU done by the system.

to the typical value of Planck's constant h divided by 2π, we obtain

$$U \simeq \frac{\mu_0 e^2 h^2}{64\pi^3 m_e^2 a^3}.$$

This expression has a value of about 5×10^{-5} eV. Even allowing for the inadequacies in our atomic model and in the assumptions made*, one would suspect that magnetic interactions play a part only in the finer details of atomic structure. This tentative conclusion is confirmed by proper treatments of atoms.

The total magnetic energy in terms of the fields **B** and **H**

We now return to the aim of expressing the total magnetic energy in terms of the magnetic fields **B** and **H**. We apply Equation (6.27) to the simple example of a long solenoid which has N turns per unit length, each of radius r with a current I flowing. If the solenoid has a length l the energy stored in the magnetic field is

$$U = \tfrac{1}{2}LI^2$$
$$= \tfrac{1}{2}\mu_0 N^2 \pi r^2 l I^2, \tag{6.34}$$

from Equation (6.19). The fields B and H inside the solenoid (continuing to neglect end effects) are given by

$$B = \mu_0 NI,$$
$$H = NI,$$

and Equation (6.34) can be written

$$U = \tfrac{1}{2}BH\mathscr{V},$$

where $\mathscr{V}$ is the volume of the solenoid. We have neglected end effects and within this approximation the volume $\mathscr{V}$ is the only region of space in which the fields are non-zero. The above equation for the total energy U may thus be written

$$U = \tfrac{1}{2}\int BH \, d\tau,$$

where the integral is taken over all space. The fields **B** and **H** are parallel, and so the total energy may finally be expressed in the form

$$\boxed{U = \tfrac{1}{2}\int \mathbf{B} \cdot \mathbf{H} \, d\tau.} \tag{6.35}$$

This equation is generally true for linear media, and is the expression we would have anticipated, by analogy with Equation (2.26) for the energy stored in an

---

* We have ignored the fields due to the intrinsic magnetic dipole moments. The conclusion reached is still valid.

electrostatic field. It is not true in the presence of non-linear media, i.e. in those situations in which the magnetic flux is not proportional to the currents in circuits. This is because its derivation depends on Equation (6.27) which itself depends on the assumption of linearity.

The energy stored in a magnetic field may be considered to be distributed with an energy density $\frac{1}{2}\mathbf{B}\cdot\mathbf{H}$. The total energy is stored in the whole field and the integral in Equation (6.35) will normally be taken over all space.

If the relative permeability can be taken to be everywhere unity Equation (6.35) becomes

$$U = \frac{1}{2\mu_0} \int B^2 \, d\tau.$$

Superconducting magnets can produce fields of 10 T. The energy density in such fields is about 4×10^7 joules m^{-3}. A typical superconducting magnet might have an energy of 5×10^5 joules stored in its field, enough to boil 200 gm of water. The energy density in the largest electric fields in free space, about 10^7 V m^{-1}, is around 450 joules m^{-3}, much less than that in large magnetic fields.

A more general derivation of Equation (6.35) is given below. It can be omitted without loss of continuity. The work done in setting up a given charge distribution can be calculated with the aid of the inverse square law, and written down neatly, as in Equation (2.27), in terms of the charges and the electrostatic potential ϕ. Similarly the work done in setting up currents in inductive circuits can be calculated with the aid of Faraday's law, and written down in terms of the currents and the magnetic vector potential $\mathbf{A}$.

Suppose that currents I_i flow in n circuits through which are fluxes Φ_i. The flux Φ_i through the ith circuit is

$$\Phi_i \overset{\underline{\vee}}{=} \int_{S_i} \mathbf{B} \cdot d\mathbf{S},$$

where $\mathbf{B}$ is the total magnetic field. In terms of the vector potential $\mathbf{A}$

$$\mathbf{B} = \text{curl } \mathbf{A},$$

and hence

$$\Phi_i = \int_{S_i} \text{curl } \mathbf{A} \cdot d\mathbf{S}.$$

The right-hand side of this equation can be transformed to a line integral using Stokes' theorem, to give

$$\Phi_i = \int_{s_1} \mathbf{A} \cdot d\mathbf{i}.$$

The total magnetic energy U is now given by Equation (6.27),

$$U = \tfrac{1}{2} \sum_i^n \int_{s_i} \mathbf{A} \cdot I_i \, d\mathbf{l}. \tag{6.36}$$

The currents I_i flowing in separate thin circuits can be described by a current density $\mathbf{j}_f$. At a place where there is a current I_i, the current density is such that $\mathbf{j}_f \, d\tau$ equals $I_i \, d\mathbf{l}$, where $d\mathbf{l}$ is a vector in the direction of the current I_i. At places not occupied by circuits the current density is zero. The sum of line integrals in Equation (6.36) can thus be replaced by an integral over all space, with the variable $I_i \, d\mathbf{l}$ replaced by $\mathbf{j}_f \, d\tau$. We obtain

$$U = \tfrac{1}{2} \int \mathbf{A} \cdot \mathbf{j}_f \, d\tau. \tag{6.37}$$

This formula should be compared with Equation (2.27) for the energy in an electrostatic field

$$U = \tfrac{1}{2} \int \rho_f \phi \, d\tau + \tfrac{1}{2} \int_S \sigma_f \phi \, dS,$$

where ρ_f and ϕ are the electric charge density and electrostatic potential respectively, and σ_f is the surface density of free charge on conducting surfaces S. Equation (6.37) can be further transformed by noting that the free current density $\mathbf{j}_f$ and the magnetic intensity $\mathbf{H}$ are everywhere related by Equation (5.30)

$$\mathbf{j}_f = \operatorname{curl} \mathbf{H},$$

and substituting for $\mathbf{j}_f$ in Equation (6.37)

$$U = \tfrac{1}{2} \int \mathbf{A} \cdot \operatorname{curl} \mathbf{H} \, d\tau.$$

By using the vector identity (Appendix B)

$$\operatorname{div} (\mathbf{A} \wedge \mathbf{H}) \equiv \mathbf{H} \cdot \operatorname{curl} \mathbf{A} - \mathbf{A} \cdot \operatorname{curl} \mathbf{H}$$

the above equation becomes

$$U = \tfrac{1}{2} \int \mathbf{H} \cdot \operatorname{curl} \mathbf{A} \, d\tau - \tfrac{1}{2} \int \operatorname{div} (\mathbf{A} \wedge \mathbf{H}) \, d\tau.$$

The second integral on the right-hand side can be transformed into a surface integral using the divergence theorem (1.17), and

$$U = \tfrac{1}{2} \int \mathbf{H} \cdot \operatorname{curl} \mathbf{A} \, d\tau - \tfrac{1}{2} \int (\mathbf{A} \wedge \mathbf{H}) \cdot d\mathbf{S}.$$

The surface integral is to be evaluated over an infinite surface bounding all space. However, we may choose a finite surface at very great distances from the sources of the field. Over this surface the fields are negligibly small, since they fall off with distance r at least as fast as r^{-2}. The vector $(\mathbf{A} \wedge \mathbf{H})$ thus falls off at least as fast as r^{-3} and the surface integral becomes indefinitely small. We thus obtain finally

$$U = \tfrac{1}{2} \int \mathbf{H} \cdot \operatorname{curl} \mathbf{A} \, d\tau$$

$$= \tfrac{1}{2} \int \mathbf{H} \cdot \mathbf{B} \, d\tau,$$

the volume integral being taken over all space. This is the same result as Equation (6.35).

6.3.2 Hysteresis

Equation (6.35) is not valid if non-linear media like iron are present. If they are present the magnetic field $\mathbf{B}$ is not proportional to the currents which give rise to the field. Equation (6.26) however remains valid for that part of the work done (in changing the magnetic fluxes through the circuits) which goes to increase the stored magnetic energy.

Consider the simple situation in which a circuit is wound over a piece of magnetic material. Let the circuit have negligible resistance so that heating losses can be ignored. The work dW_b done by a battery to increase the flux through the circuit by an amount $d\Phi$ in time dt is given by Equation (6.26).

$$dW_b = I \frac{d\Phi}{dt} dt, \tag{6.38}$$

where I is the current in the circuit. By writing the flux Φ as $\int_S \mathbf{B} \cdot d\mathbf{S}$, where S is any surface whose perimeter is the circuit it can be shown that

$$dW_b = \int \mathbf{H} \cdot d\mathbf{B} \, d\tau, \tag{6.39}$$

where the volume integral is taken over all space. (The proof of this equation is given at the end of this section.) If the relation between the fields $\mathbf{B}$ and $\mathbf{H}$ is known, Equation (6.39) can be integrated to obtain the total work done W_b in establishing a final field B_0.

$$W_b = \int \int_0^{B_0} \mathbf{H} \cdot d\mathbf{B} \, d\tau. \tag{6.40}$$

When the field is reduced to zero again, the total work done is zero if the relation between the fields $\mathbf{B}$ and $\mathbf{H}$ is single valued, since then

$$\int \int_0^{B_0} \mathbf{H} \cdot d\mathbf{B} \, d\tau = - \int \int_{B_0}^0 \mathbf{H} \cdot d\mathbf{B} \, d\tau.$$

In such a situation, as we discussed in the previous section, the work done against induced e.m.f.'s whilst the field is being established can be considered to be stored in the field reversibly; the energy can be regained when the field is returned to zero. However, with a piece of ferromagnetic material like iron inside the circuit, the work done to establish the field cannot all be regained, and some energy is dissipated as heat in the iron. If the iron is taken around a complete hysteresis cycle (i.e. the current through the circuit, initially at a certain value, is decreased through zero to the same value in the opposite direction and then returned to its original condition), the energy dissipated as heat in unit volume of the iron can be related to the area under the hysteresis curve*. The curve labelled *abcdefa* on Figure 6.11 is the major hysteresis curve for commercial iron. If unit volume of commercial iron were taken around this cycle the current sources have to do an amount of work equal to the integral $\oint H \cdot dB$ around the cycle. This energy is about 350 joules m^{-3} as can be estimated from the figure. Iron cores are used in transformers in order to obtain maximum flux linkage

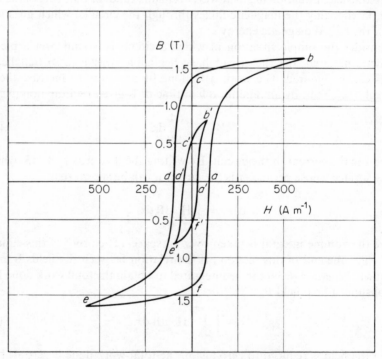

Figure 6.11. The major hysteresis curve for commercial iron. The minor hysteresis curve shown is a typical cycle around which the iron in a mains transformer is taken.

* All iron has hysteresis. In section·5.4.1 we implied that the relation between the fields **B** and **H** for soft iron was single valued. This is not a bad approximation for electromagnets, but when discussing energy losses in circuits which include iron, as now, we have to be more exact.

between primary and secondary windings. The fields in a mains transformer core are taken through a hysteresis cycle 50 times every second. Transformers are not used under conditions where the iron is near saturation. A typical situation is one in which the fields follow the minor hysteresis loop labelled $a'b'c'd'e'f'a'$ on Figure 6.11, and the area under this loop is about one sixth of the area under the major hysteresis loop. If 1000 cm^3 of commercial iron is used in a transformer which has currents flowing at 50 Hz, the power dissipation is about $(350 \times \frac{1}{6}) \times 50 \times 10^{-3}$ W, i.e. about 3 watts. Hysteresis losses are proportional to frequency, since current sources have to do an amount of work $\oint \mathbf{H} \cdot d\mathbf{B}$ per unit volume of material every cycle of the current.

Some of the hysteresis losses in ferromagnetic materials can be attributed to the friction occurring as the domains slide past each other in reorientating with the changing field. Losses of a somewhat similar kind can occur in capacitances filled with polar dielectrics used in alternating current circuits. These losses are usually much smaller than hysteresis losses, and arise when the small fraction of the molecules reorientate with the changing electric field and suffer 'friction' from their neighbours in so doing.

The following derivation of Equation (6.39) can be omitted without loss of continuity. From Equation (6.38)

$$dW_b = I \int_S \frac{\partial \mathbf{B}}{\partial t} \cdot d\mathbf{S} \, dt,$$

where S is a surface spanning the contour of the circuit. From Equation (6.17)

$$dW_b = -I \int_S \text{curl } \mathbf{E} \cdot d\mathbf{S} \, dt$$

$$= -I \oint \mathbf{E} \cdot d\mathbf{l} \, dt,$$

from Stokes' theorem (4.38). If the current I is specified by a current density $\mathbf{j}_f$, the line integral above can be replaced by a volume integral over all space, and

$$dW_b = -\int \mathbf{E} \cdot \mathbf{j}_f \, d\tau \, dt$$

$$= -\int \mathbf{E} \cdot \text{curl } \mathbf{H} \, d\tau \, dt,$$

from Equation (5.30). Using the vector identity (Appendix B)

$$\text{div} (\mathbf{E} \wedge \mathbf{H}) \equiv \mathbf{H} \cdot \text{curl } \mathbf{E} - \mathbf{E} \cdot \text{curl } \mathbf{H}$$

we obtain

$$dW_b = -\int \mathbf{H} \cdot \text{curl } \mathbf{E} \, d\tau \, dt + \int \text{div} (\mathbf{E} \wedge \mathbf{H}) \, d\tau \, dt.$$

The second integral on the right-hand side of this equation can be transformed into a surface integral using the divergence theorem (1.17). Its value is zero since the fields $\mathbf{E}$ and $\mathbf{H}$ vanish over a surface a very long way from the circuit. Hence

$$dW_b = -\int \mathbf{H} \cdot \operatorname{curl} \mathbf{E} \, d\tau \, dt$$

$$= \int \mathbf{H} \cdot \frac{\partial \mathbf{B}}{\partial t} \, d\tau \, dt$$

from Equation (6.17). We see that this is the same result as Equation (6.39),

$$dW_b = \int \mathbf{H} \cdot d\mathbf{B} \, d\tau,$$

which we used in the discussion of hysteresis.

6.3.3 The measurement of magnetic susceptibilities

Accurate ways of measuring the very small magnetic susceptibilies of paramagnetic or diamagnetic substances depend upon the measurement of the force on a sample in a non-uniform magnetic field.

Suppose a long rod of the substance, whose cross-sectional area is A, is suspended vertically between the poles of an electromagnet, as shown in Figure 6.12. When the current is passed through the coils the rod will be attracted into the field or repelled according to whether the material is paramagnetic or diamagnetic. This force in the vertical direction arises from the

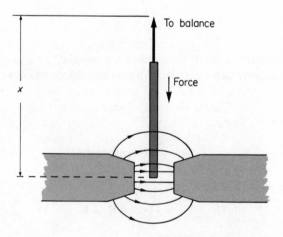

Figure 6.12. Weighing the force on a rod of paramagnetic material in a non-uniform magnetic field.

fringing field of the magnet where the field is non-uniform. However, we can find the magnitude of the force by calculating the change in potential energy of the sample when it is moved downwards through a small distance Δx. If the cross-sectional area of the rod is A, the net effect of such a displacement is to move an extra volume $A \Delta x$ into the region between the pole faces where the field $\mathbf{B}$ is uniform. Since the susceptibility χ_B of the rod is very small, it is a good approximation to assume that the field in the pole gap has the same value as in the absence of the rod, and that the rod acquires a uniform magnetization $\mathbf{M}$ equal to $\chi_B \mathbf{B}/\mu_0$. The potential energy of a dipole $\mathbf{m}$ in a field $\mathbf{B}$ is given by Equation (4.23) as $-\mathbf{m} \cdot \mathbf{B}$. Hence the work done in building up the magnetization $\mathbf{M}$ is $-\int_0^B \mathbf{m} \cdot d\mathbf{B} = -\chi_B B^2/2\mu_0$ per unit volume. The magnetic potential energy U_P of the rod is therefore changed by an amount $-\chi_B B^2 A \Delta x/2\mu_0$ when the rod moves through the distance Δx. The force $\mathbf{F}$ on the rod is given by

$$\mathbf{F} = -\operatorname{grad} U_P \tag{6.41}$$

or

$$F = \chi_B B^2 A/2\mu_0 . \tag{6.42}$$

This force is measured by weighing the rod using a sensitive balance, with and without current in the magnet coils, and the field $\mathbf{B}$ is measured in one of the ways described in the next section.

Aluminium is paramagnetic and has a susceptibility of 2.3×10^{-5}. If an aluminium rod of cross-sectional area 1 cm^2 is suspended between the poles of an electromagnet which gives a field of 0.5 T, the force on the rod, given by Equation (6.42), is about 2.3×10^{-4} N, i.e. about $\frac{1}{40}$ of the force on a mass of 1 g at the surface of the Earth.

As in the analogous electrical problem a local field correction is necessary in principle if one wishes to relate measurements of magnetic susceptibilities to atomic or molecular properties. The strength of the interaction between two typical molecular magnetic dipole moments is much smaller than that between two typical molecular electric dipole moments, and the approach adopted in section 2.2.3 to estimate the local electric field in gaseous and liquid dielectrics is valid for a wider sample of magnetic materials. The induced surface currents on the surface of a sphere cut out of a uniformly magnetized material give rise to a small reduction in the macroscopic field depending linearly on the susceptibility χ_B. However, since the susceptibility is so small the correction is negligible. For most practical purposes the magnetic fields inside volumes occupied by paramagnetic or diamagnetic materials are the same as they would have been if the materials were not there.

At the end of this section it is appropriate to point out that we have been using a different definition of magnetic susceptibility to the one usually used. The latter quantity, which we denote χ_H, is defined by relating the magnetization of a material to the magnetic intensity,

$$\mathbf{M} = \chi_H \mathbf{H}.$$

The magnetic field **B** is related to the fields **M** and **H** by

$$\mathbf{B} = \mu_0(\mathbf{H} + \mathbf{M}) = \mu\mu_0\mathbf{H},$$

where μ is the relative permeability. Hence

$$\mu = 1 + \chi_H.$$

From Equation (5.33) we may deduce the relationship between χ_H and χ_B;

$$\chi_H = \frac{\chi_B}{1 - \chi_B}.$$

For paramagnetic and diamagnetic materials the numerical difference between χ_H and χ_B is thus usually negligible. For ferromagnetic materials the difference is large; however for such materials it is not very useful to talk about magnetic susceptibility. It is more convenient to use the relative permeability μ, which itself is constant over only a very limited range of fields. The reason we choose to use χ_B instead of χ_H in this book is that it allows us to present a much more logical and satisfactory treatment of magnetic fields inside matter.

6.4   THE MEASUREMENT OF MAGNETIC FIELDS

Current is measured absolutely by measuring the force between two circuits carrying the current and relating the force and the current by the Biot–Savart law. A moving coil instrument can be calibrated in this way to give a convenient current meter. The measurement of the force or torque on a circuit carrying a known current in a magnetic field **B** provides an absolute measurement of the field. Such measurements however need sensitive equipment and are not used for everyday field determinations in the laboratory. For example a coil of area 1 cm^2 which consisted of 100 turns of thin wire could carry a current of about 1 A before overheating. If it were placed with its plane parallel to a field of 0.1 T the torque on it would be the same as that due to the weight of a mass of 10^{-7} g acting over a lever arm 1 cm long. Even if such a small torque could be accurately measured, the device would be measuring the average field over the area of the coil, and so would not be useful for fields which varied rapidly with position.

In this section we will discuss briefly some simple techniques for field measurement.

6.4.1   Search coils

Magnetic fields can be measured by determining the e.m.f.'s induced when a search coil placed in the field is removed or rotated. If a coil of N turns, each of area A, is placed in a field **B** the flux linking the coil is

$$\Phi = N \int_A \mathbf{B} \cdot d\mathbf{S}.$$

If the coil is removed from the field there is an e.m.f. V induced in the circuit containing the coil. If the total resistance of the closed circuit which includes the coil is R, the induced current I is

$$I = \frac{dq}{dt} = \frac{V}{R} = -\frac{1}{R}\frac{d\Phi}{dt}.$$

The total charge that flows around the coil circuit during the time it takes to move the coil to a place where there is no field is

$$q = -\frac{1}{R}\int_0^\infty \frac{d\Phi}{dt}\,dt$$

$$= \frac{\Phi}{R}$$

$$= \frac{N}{R}\int_A \mathbf{B}\cdot d\mathbf{S}.$$

If the coil is small, so that the field $\mathbf{B}$ is uniform over the dimensions of the coil, and if the coil is perpendicular to the field the magnetic flux is given by

$$\Phi = NBA,$$

and hence

$$q = \frac{NBA}{R}. \tag{6.43}$$

If the charge q is measured, and N and A known, the field B can be determined. One way of measuring the charge q is to use a suspended coil instrument. If the charge can be considered to have passed through the instrument before it has begun to rotate, it can be shown that the initial throw of the coil is proportional to the charge q.

A rotating search coil is often used to provide an alternating voltage signal which is used to stabilize magnetic fields produced by electromagnets. The uniform rotation of a small coil about an axis in the plane of the coil perpendicular to a magnetic field results in a sinusoidally varying induced voltage in the coil. The peak value of the voltage is constant if the field is constant. Figure 6.13 shows a coil of area A having N turns rotating at angular frequency ω about the axis CD. A uniform magnetic field $\mathbf{B}$ acts in the direction perpendicular to the axis. When the perpendicular to the plane of the coil makes an angle θ with the direction of the field the flux through the coil is

$$\Phi = NAB\cos\theta.$$

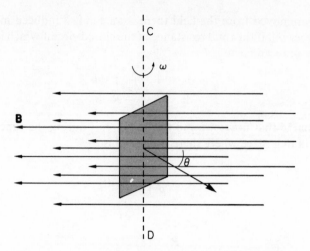

Figure 6.13. A coil rotating in a uniform magnetic field which is in a direction perpendicular to the axis of rotation CD.

The e.m.f. induced in the coil at this instant is

$$V = -\frac{d\Phi}{dt} = BNA \sin\theta \frac{d\theta}{dt}$$

$$= \omega BNA \sin \omega t,$$

choosing the zero of time such that $\theta = 0$ when $t = 0$. For a small coil, with 100 turns each of area 1 cm^2, rotating at 3600 revolutions per minute in a field of 0.05 T the peak voltage induced is about 0.2 V.

A method often used to stabilize fields from electromagnets is to compare the voltage output from a coil rotating in the field with the voltage output from a second coil. The second coil is driven off the same rotating shaft and rotates in the constant field of a permanent magnet. The difference between the voltages induced in the two coils is used to adjust the output of the current generator powering the electromagnet and so keep the field of the electromagnet constant.

6.4.2 Other methods

Any physical effect which depends upon magnetic fields can be used in principle to measure them. Only some effects have the desired linear dependence on field strength and lend themselves to simple measurement techniques. One such, the Hall effect, has already been discussed in section 4.2.2. Another quantity which depends linearly on field is the potential energy of a magnetic dipole placed in the field. The proton has an intrinsic angular momentum quantum number equal to one half, and possesses a magnetic dipole moment m equal to $+2.793$ n.m.

(One nuclear magneton, n.m., is equal to one Bohr magneton multiplied by the ratio of the mass of the electron to the mass of the proton. Its numerical value is 5.051×10^{-27} A m^2. The plus sign means that the magnetic dipole points in the same direction as the intrinsic spin.) There will thus be two energy states for a proton in a diamagnetic material in a magnetic field **B** corresponding to the two allowed projections of the magnetic moment vector in the direction of the field. The two energy states have an energy difference equal to $2mB$. The energy in a radio-frequency field of frequency v may be considered to reside in discrete units, called quanta or photons, each of which have energy hv, where h is Planck's constant. If a radio-frequency field of frequency v, given by

$$hv = 2mB,$$

is applied to the sample it will induce transitions both up and down between the two energy states. If there are more transitions up than down there will be a net absorption of energy from the oscillating field, which can be detected. The frequency at which this occurs, as already implied, is in the radio-frequency region and can be measured very accurately. Hence the field can be determined in terms of the known magnetic moment of the proton. This technique is known as nuclear magnetic resonance, or NMR. The reason why there are more transitions up than down is that random interactions between the protons and other microscopic constituents of the material maintain statistical equilibrium. This means that the number of protons occupying the lower level is very slightly greater than the number occupying the upper level, according to the Boltzmann factor. The number of transitions upwards, proportional to the number of protons in the lower state, is thus greater than the number down, proportional to the number of protons in the upper level. This of course implies some de-excitation of the upper levels by transitions that do not involve emission of photons. This is done via the same mechanism responsible for the statistical equilibrium. The difference in energy between the two levels in a field of 0.1 T is about 2×10^{-8} eV. The corresponding frequency at which absorption will occur is about 4.2 MHz. This is a frequency which can easily be measured to an accuracy of one part in 10^5. NMR is the technique most often used to measure magnetic fields accurately.

PROBLEMS 6

6.1 A rod of dielectric material is spun about its axis with angular velocity ω. A uniform magnetic field B exists in a direction along the axis of the bar. Determine a charge distribution which produces the same electric field as does the rotating rod. The electric susceptibility of the material is χ_E.

6.2 A very small piece of magnetized material, whose magnetic dipole moment is **M**, pointing in the x-direction, moves with speed v along the x-axis. A closed circular loop of thin wire, radius a, lies in the y–z plane at position $x = 0$. Determine the force opposing the motion of the particle. The speed v is much less than the speed of light, and the distance of the particle from the coil is much greater than the radius a. The resistance of the loop of wire is R, and it has negligible self-inductance.

6.3 When the armature of an electromagnetic relay is attracted, a closed iron circuit is formed, of cross-sectional area $4\,cm^2$, and length $20\,cm$. What current is required just to attract the armature if the initial gap is 1 cm, and a spring exerting a constant force of 10 newtons opposes the motion of the armature? The relay coil has 1000 turns and the permeability of the iron may be assumed to be constant at 500. To what value must the current be reduced before the armature opens again?

6.4 An iron toroid of permeability 1000 (which may be taken to be constant) has a cross-sectional area of $20\,cm^2$ and a mean radius of 50 cm. If a coil of 5000 turns is wound closely round the toroid, what would be the magnetic force between the faces of a gap 1 mm wide cut through the toroid with a current of 2 A passing through the coil?

What work must be done against the magnetic forces to increase the gap to 2 mm?

6.5 A coil of enamelled wire is wound onto a cylindrical insulating former. Describe and explain the effect on the self-inductance and the apparent resistance of the coil, when an alternating current is passed through it, of inserting into the solenoid

(i) a bar of iron

(ii) a bar of copper

(iii) a tube of copper

(iv) a tube of copper with a slit cut into it parallel to the axis of the tube.

6.6 If a steady e.m.f. is suddenly applied to a coil of self-inductance L_1, show that, by the presence nearby of a closed second coil of self-inductance L_2 which has a mutual inductance M with the first coil, the effective initial self-inductance of the first coil is diminished from L_1 to $L_1 - M^2/L_2$.

6.7 A homopolar generator consists of a large metal flywheel rotating rapidly in a uniform magnetic field which is in a direction along the axis of the flywheel. The rotation causes a voltage difference between the axis and the rim of the flywheel. Homopolar generators are capable of producing very high currents for short periods of time of the order of a few seconds duration. Such high currents are used for example in experiments with plasmas (highly ionized gases) and for providing large magnetic fields in high energy nuclear physics machines. When the large currents are required through a coil, for example, the ends of the coil are connected to the rim and axis of the flywheel and the kinetic energy of the flywheel is converted into electrical energy. If a flywheel of mass 10^4 kg and radius 3 m is rotating at 3000 revolutions per minute in a field of 0.5 T, what would be the initial current delivered to a load of $10^{-3}\,\Omega$? How long would it take for the flywheel to slow to half its speed?

6.8 The coil of a moving coil galvanometer is connected in series with a circuit of resistance R and negligible self-inductance. Describe the motion of the coil if it is given a small displacement from its equilibrium position.

6.9 A coil of area A, resistance R and self-inductance L is rotated about a vertical axis in the plane of the coil with angular velocity ω. The horizontal component of the Earth's magnetic field is B. Derive an equation for the current I in the coil and determine the mean torque required to rotate the coil.

6.10 Determine the total energy stored in the magnetic field of a long solenoid and hence show that the self-inductance is given by Equation (6.19).

6.11 Prove that the force per unit volume in the x-direction acting on a small paramagnetic particle in an inhomogeneous magnetic field is given to a good approximation by

$$\tfrac{1}{2}\chi_B\mu_0\frac{d}{dx}(H^2)$$

where χ_B is the magnetic susceptibility of the particle, and the particle is in air.

A rod of sulphur, $1.0\,cm^2$ cross-sectional area is placed in air in a non-uniform magnetic field, the direction of the field at each end of the specimen being at right angles to its axis. The strength of the field at one end a is 1.2 T and at the other end b

0.3 T. The force on the rod due to the field acts from a to b parallel to its axis and has magnitude 6.5×10^{-4} N. Find the magnetic susceptibility of sulphur.

6.12 A fluxmeter is a device sometimes used for measuring magnetic fields. It is a suspended coil instrument in which the suspension has virtually no restoring twist. The instrument is used in series with a search coil which is removed from or rotated in the field to be measured. The fluxmeter coil and the search coil form a continuous circuit of total resistance R. If the instrumental constant for the fluxmeter is k (see section 4.6.1), determine the deflection $\Delta \theta$ of the fluxmeter coil for a change of flux $\Delta \Phi$ through the search coil.

CHAPTER

Alternating currents and transients

In the last chapter it was shown how an induced e.m.f. occurs whenever the magnetic flux through a circuit is changing. By introducing the concept of inductance, we were able to use Faraday's Law to express induced e.m.f.s in terms of the rate of change of currents. We must now look at the opposite problem, and investigate the response of the currents in a circuit to known changes in the e.m.f.s driving them. This chapter is principally concerned with circuits in which both the currents and the e.m.f.s are varying sinusoidally with time. Such circuits are referred to as *alternating current* circuits, or A.C. circuits for short. There are good reasons for concentrating on the sinusoidal form of time variation. In the first place, many commonly occurring sources of e.m.f. do in fact vary sinusoidally. More importantly, in a circuit containing inductors and capacitors, sinusoidally varying e.m.f.s lead to sinusoidally varying currents, although as we shall see, the e.m.f.s and currents may not be in step with one another. The useful property of having the same functional form for e.m.f.s and current is not shared by any other type of time variation. Finally, according to Fourier's theorem, all functions may be expressed as a superposition of sine functions. It follows that we can predict the response of a circuit to any time variation of e.m.f. once we know its behaviour for alternating currents of all frequencies.

7.1 ALTERNATING CURRENT GENERATORS

Energy is dissipated when an electric current flows through a material with non-zero resistivity, and a continuous source of power is therefore needed to maintain a current in a circuit, except in the special case of superconductors.

Consider the energy dissipated when a source of e.m.f. V drives a current I through a resistor. An amount of work $V\,\mathrm{d}q$ is required to move a charge $\mathrm{d}q$ through the potential difference V, and

$$\text{power generated} = \text{rate of work} = V\frac{\mathrm{d}q}{\mathrm{d}t} = VI, \qquad (7.1)$$

since the current I is the rate at which charge passes any point in the circuit. Electrical energy is not always dissipated as heat in a resistor: it may for example be partly converted into light, or used to drive an electric motor. However, the argument used to derive Equation (7.1) remains correct no matter how the energy is expended. At any moment, the instantaneous power of a source of e.m.f. V generating a current I is always given by the product VI.

In the alternating current generator, or dynamo, electrical energy is derived from mechanical work. The simplest dynamo consists of a single square coil rotating about an axis at right angles to a uniform magnetic field $\mathbf{B}$, as shown in Figure 7.1. If the coil rotates with angular velocity ω, and is aligned at time zero so that $\mathbf{B}$ is in the plane of the coil, then the flux through the coil at time t is

$$\Phi = Ba^2 \sin \omega t.$$

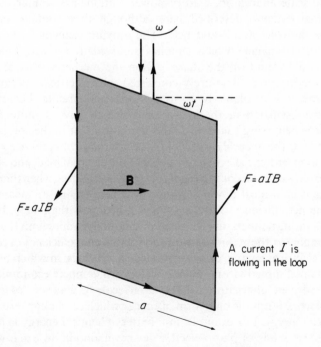

Figure 7.1. A simple dynamo made of a square loop of wire rotating in a magnetic field.

Application of Faraday's law tells us that the e.m.f. generated around the coil is

$$V = -\frac{d\Phi}{dt} = -\omega Ba^2 \cos \omega t.$$

When the coil is open-circuited, so that no current flows, no power is developed by the generator, and no work is expended in rotating the coil. If the coil is made to form part of a closed circuit through which a current I is flowing, electrical power is generated. This power is derived from mechanical work, since there is now a torque opposing the motion of the coil. We can easily check that the work done against this torque is equal to the electrical energy generated by the coil. The current in the two vertical arms of the coil in Figure 7.1 is always perpendicular to **B**, and the Lorentz force on each vertical arm has a magnitude aIB. There is no resultant force or couple on the coil due to the forces on the other arms of the coil, and the net effect of the Lorentz force is a couple of magnitude ($aIB \cdot a \cos \omega t$) opposing the motion of the coil. The power required to maintain the rotation of the coil at a steady angular velocity ω is

$$\omega \cdot aIB \cdot a \cos \omega t = VI.$$

The mechanical power consumed by the coil is thus exactly the same as the electrical power generated.

The large scale alternators used in power stations are similar in principle to the simple dynamo described here, although they contain many coils, and reverse the role of coil and field. The moving element is the *armature*, a rotating electromagnet which supplies sinusoidally varying flux through stationary coils wound on the *stator*. Again the mechanical work expended in driving the armature is entirely converted into electrical energy, but not all of this energy is available to be used in external circuits. There is always some energy dissipated in the stator coils, which have a finite resistance. If the stator is delivering a current I, and the power lost in the coil is I^2r, then r is referred to as the *internal resistance* of the alternator. In practice r is usually small compared with the effective resistance of the external load, and alternators are therefore very efficient converters of energy. Nevertheless, when the alternator is driven, as it is in most power stations, by a heat engine—a steam or a gas turbine—the net efficiency of the station is a little less than 40%, because at practical operating temperatures the laws of thermodynamics limit the efficiency of the heat engine to about this value. In spite of this poor efficiency of conversion of fuel into electrical energy, only hydro-electric schemes, in which mechanical work is obtained directly from moving water, can compete economically with heat-engine driven alternators in the large-scale generation of electricity. Mechanical work is not the only form of energy which can be converted directly into electrical energy. For example, in a battery chemical energy is the source of power; some kinds of photo-electric devices maintain an e.m.f. when they are exposed to light; in a thermocouple heat energy drives a current without the intermediary of a dynamo. For large amounts of power these alternative sources of energy are hopelessly expensive. If you work out the cost of power in a flashlight you will find it to be at least a hundred times what is charged for the mains electricity supply.

7.2 AMPLITUDE, PHASE AND PERIOD

In the last section it was shown that a coil rotating with constant angular velocity ω radians/second in a uniform magnetic field generates an e.m.f. which varies sinusoidally with time. Let us choose the zero of time at a moment when an alternator generates its maximum e.m.f., which can then be written as

$$V = V_0 \cos \omega t. \tag{7.2}$$

This voltage is drawn as a function of time in Figure 7.2. The maximum value V_0 is called the *amplitude* of the e.m.f. and the argument ωt its *phase*. The

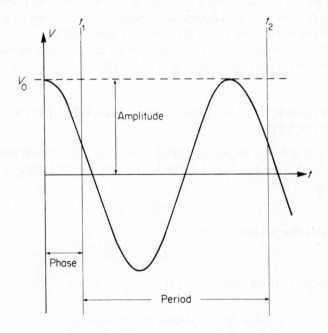

Figure 7.2. The alternating voltage waveform.

cosine pattern repeats itself every time ωt increases by 2π, and the phase is usually restricted to the range 0 to 2π. When ωt reaches an integral multiple of 2π, the phase begins again at zero. The voltages at t_1 and t_2 in Figure 7.2, for example, are said to be at the same phase. The *period* between successive repetitions of the same phase is

$$T = \frac{2\pi}{\omega} \text{ seconds.}$$

The sequence of events occurring in one complete period is called the *cycle* of the generator. The voltage pattern repeats at a frequency

$$v = \frac{1}{T} = \frac{\omega}{2\pi} \text{ cycles per second.}$$

There is a special unit of frequency called the *hertz*, after the discoverer of radio transmission:

$$1 \text{ hertz} \equiv 1 \text{ Hz} \equiv 1 \text{ cycle per second.}$$

The frequency of the main electricity supply is 50 Hz in Europe, and 60 Hz in the United States. Alternating voltages of much higher frequency occur in the tuned circuits used in telecommunications, and it is important to know how the circuits respond to the application of a sinusoidally varying e.m.f. of any frequency. We shall begin the study of A.C. circuits by considering the currents driven by an A.C. generator through single circuit elements.

7.3   RESISTANCE, CAPACITANCE AND INDUCTANCE IN A.C. CIRCUITS

Figure 7.3(*a*) shows a circuit consisting of an A.C. generator (indicated by the symbol ⊝) driving a current I through a resistor R. The generator delivers an oscillating e.m.f.

$$V = V_0 \cos \omega t,$$

and the resistor obeys Ohm's law:

$$V = IR.$$

Hence the current through the resistor is

$$I = \frac{V}{R} = \frac{V_0}{R} \cos \omega t = I_0 \cos \omega t. \tag{7.3}$$

The current is in phase with the applied voltage, and has an amplitude $I_0 = V_0/R$.

The next diagram, Figure 7.3(*b*), shows the same generator connected across a capacitor of capacitance C. Again the voltage is $V_0 \cos \omega t$, and it is now related to the charge Q carried by the capacitor through the equation

$$V = Q/C.$$

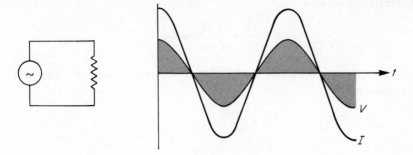

(*a*)   Resistance: current and voltage are in phase.

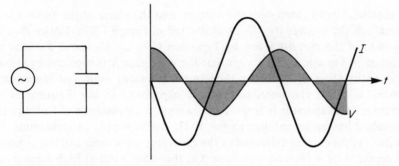

(*b*)   Capacitance: current leads voltage.

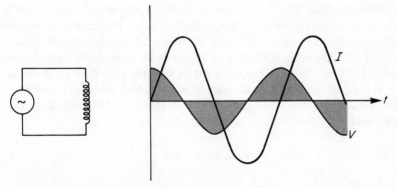

(*c*)   Inductance: current lags behind voltage.

Figure 7.3. Phase relations between current and voltage.

Differentiating we find

$$\frac{dV}{dt} = \frac{1}{C}\frac{dQ}{dt} = \frac{I}{C},$$

where I is the current flowing in the leads to the capacitor plates. Rearranging this equation,

$$I = C\frac{dV}{dt}$$

$$= -\omega C V_0 \sin \omega t$$

$$= \omega C V_0 \cos\left(\omega t + \frac{\pi}{2}\right). \tag{7.4}$$

The current is not in step with the voltage, and the phase of the current *leads* the phase of the voltage by $\pi/2$, as illustrated in Figure 7.3(b). Notice that the amplitude of the current given by Equation (7.4) has the same form as the amplitude in Equation (7.3) except that the resistance R is replaced by $(1/\omega C)$. The quantity $(1/\omega C)$, which has the same dimensions as R and is measured in ohms, is called the *impedance* of the capacitor. At low frequencies the impedance of a capacitor is large. For example, a capacitance of 1 μF (quite a large value) has an impedance to the 50 Hz mains supply of 1600 ohms: the amplitude of the current delivered by the generator is the same as if the capacitor were replaced by a 1600 ohm resistor. On the other hand at high frequencies, $1/\omega C$ becomes small, and capacitors then present a low impedance to the flow of alternating current.

Finally, in Figure 7.3(c), the circuit element across the generator is a coil with self-inductance L. This circuit is a rather unrealistic one, since coils usually have a resistance which is not negligibly small. However, it is convenient for the moment to imagine that such an idealized self-inductance exists, and then later to deal with circuits containing both inductance and resistance. According to Lenz' law the e.m.f. caused by the changing current in the inductance is $(-L\,dI/dt)$, the minus sign indicating that the e.m.f. opposes the applied voltage. In the circuit of Figure 7.3(c) this e.m.f. balances the generator voltage, and

$$V = V_0 \cos \omega t = L\frac{dI}{dt}.$$

Integrating this equation, and setting the constant of integration equal to zero, since the circuit contains no source of steady current, we obtain

$$\frac{V_0}{\omega} \sin \omega t = LI,$$

or

$$I = \frac{V_0}{\omega L} \sin \omega t$$

$$= \frac{V_0}{\omega L} \cos \left(\omega t - \frac{\pi}{2} \right). \tag{7.5}$$

As with the capacitor there is a phase difference of $\pi/2$ between current and voltage, but in the inductance the current *lags* behind the applied voltage. The amplitude of the current is $V_0/\omega L$, and here, by comparison with equation (7.3), the resistance R is replaced by the impedance ωL of the inductance. For a coil with a self-inductance of 10 mH, the impedance to a 50 Hz mains supply is only about 6 ohms, but at high frequencies the impedance of an inductance becomes large.

7.4   THE PHASOR DIAGRAM AND COMPLEX IMPEDANCE

An alternating voltage can be written as the real part of an exponential function with an imaginary argument. For a voltage with amplitude V_0,

$$V = V_0 \cos \omega t = \mathscr{R}e \left[V_0 \exp \left(j\omega t \right) \right]. \tag{7.6}$$

In this equation j stands for the imaginary number $\sqrt{-1}$; the symbol j rather than i is commonly used in the context of electromagnetism in order to avoid confusion with current. In the complex plane the point $V_0 \exp(j\omega t)$ lies at the end of a vector of length V_0 rotating about the origin with angular velocity ω, as indicated in Figure 7.4. At time zero this vector lies along the x-axis, and at time t it makes an angle ωt with the x-axis. This angle is the phase of the voltage, and from Equation (7.6), the projection of the rotating vector onto

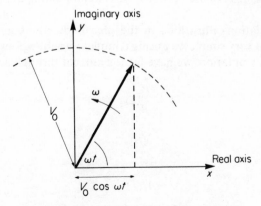

Figure 7.4. The representation of an alternating voltage in the complex plane.

the x-axis represents the magnitude of the voltage. To make the notation brief, we shall drop the prefix $\mathscr{R}e$ in Equation (7.6), and simply represent the voltage by the complex function

$$\mathbf{V} = V_0 \exp(\mathrm{j}\omega t), \qquad (7.7)$$

always remembering that only the real part of this function has physical meaning.

Alternating currents can similarly be represented by complex functions which are visualized as vectors rotating in the complex plane. The current vectors rotate at the same angular velocity as the voltage vectors, but may lead or lag by a constant phase angle. The current through an inductance, for example, always lags behind the voltage by the phase angle $\pi/2$. This is illustrated in Figure 7.5(c), in which the vectors are frozen at the moment when

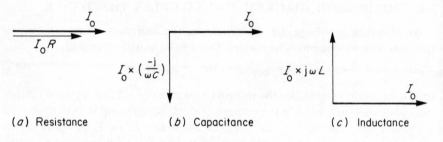

(a) Resistance   (b) Capacitance   (c) Inductance

Figure 7.5. Phasor diagrams.

the current points along the real axis. These stationary vectors whose orientation shows the relative phase between different alternating quantities are called *phasors*.

The phase relations illustrated in the phasor diagrams in Figure 7.5 can be expressed in a very simple way using complex variables. Rewriting Equation (7.3) in a complex notation, we have for the current through a resistance R

$$\mathbf{I} = \frac{V_0}{R} \exp(\mathrm{j}\omega t)$$

$$= \mathbf{V}/R,$$

or

$$\mathbf{V} = \mathbf{I}R. \qquad (7.8)$$

In this equation relating complex variables, the resistive impedance R is *real*.

Equation (7.4) for the current in the capacitative circuit becomes, when written in the complex notation,

$$\mathbf{I} = \omega C V_0 \exp\left[j\left(\omega t + \frac{\pi}{2}\right)\right]$$

$$= \omega C \mathbf{V} \exp\left(\frac{j\pi}{2}\right),$$

or

$$\mathbf{V} = \frac{\mathbf{I}}{j\omega C} = -\frac{j\mathbf{I}}{\omega C}, \tag{7.9}$$

since

$$\exp\left(\frac{j\pi}{2}\right) = \cos\frac{\pi}{2} + j\sin\frac{\pi}{2} = j.$$

Multiplication by $(-j)$ is equivalent to the rotation of a phasor through $(-\pi/2)$, and Equation (7.9) tells us both the amplitude and phase of the voltage in terms of the current. This equation has the same form as Ohm's Law if we assign to the capacitor a *complex impedance*

$$\mathbf{Z}_C = \frac{1}{j\omega C} = -\frac{j}{\omega C}.$$

Because the current phasor has been chosen to lie along the real axis, the vector representing the impedance $\mathbf{Z}_C$ in the complex plane is in the same direction as the voltage phasor along the negative imaginary axis. As they are labelled in Figure 7.5, the light arrows are measured in volts, and indicate phasors. However if you prefer you can rub out the I_0 on the label, re-scale in ohms instead of volts, and regard the light arrows as vectors representing the complex impedance.

Finally, Equation (7.5) for the current through an inductance is written as

$$\mathbf{I} = \frac{V_0}{\omega L} \exp\left[j\left(\omega t - \frac{\pi}{2}\right)\right]$$

or

$$\mathbf{V} = \mathbf{I}\omega L \exp\left(\frac{j\pi}{2}\right)$$

$$= j\omega L \mathbf{I}. \tag{7.10}$$

The inductance has a complex impedance

$$\mathbf{Z}_L = j\omega L$$

represented by a vector pointing upwards, along the positive imaginary axis.

The purely imaginary impedances we have derived apply only to idealized components. Real capacitors have a leakage resistance between their plates, and the windings of real coils have a non-zero resistance: in both instances the effect of the resistance is to reduce the phase angle between voltage and current to less than $\pi/2$. For most practical purposes capacitors closely approach the ideal, and in the analysis of an A.C. circuit it is usually adequate to represent a capacitor by an idealized component with impedance $1/j\omega C$. With inductances, however, it is often not possible to ignore the resistance of the windings.

Let us evaluate the complex impedance of a coil with a resistance R to direct current, when an alternating voltage of angular frequency ω is applied across it. At very high frequencies the behaviour of the coil is complicated because in addition to the inductive and resistive effects, it becomes necessary to take into account the capacitance between neighbouring turns on the coil. However, we shall restrict ourselves to frequencies at which the impedance of this small capacitance is extremely large compared with the inductive impedance of the coil. There is then no significant current carried by the capacitance, and it is a good approximation to assume that at any moment the current I is the same throughout the windings of the coil. Furthermore, the 'skin effect', which causes high frequency currents to concentrate near the surface of a conductor*, is not important, and the resistance of the coil is the same as for direct current.

At moderate frequencies, then, the current is uniform in the windings, and the coil has a self-inductance $L =$ (flux/current) which is independent of frequency. This inductance generates a back e.m.f. $L \, dI/dt$ opposing any change in current. The external e.m.f. V driving the current must balance the back e.m.f. and also account for the additional voltage drop IR due to the resistance. The inductive and resistive voltages add together, and V must satisfy

$$V = L\frac{dI}{dt} + IR. \tag{7.11}$$

We can imagine that the coil is replaced by a circuit consisting of an ideal inductance L in series with a separate resistance R, as shown in Figure 7.6. The voltage and current in this circuit satisfy Equation (7.11) at all frequencies. At the moderate frequencies in which we are interested, it is an *equivalent circuit* which we may use to represent the real coil.

If we choose the time-zero when the current has its maximum value, then in the complex notation

$$I = I_0 \exp{(j\omega t)}.$$

Substituting in Equation (7.11)

$$V = (j\omega L + R)I_0 \exp{(j\omega t)}$$

$$= (j\omega L + R)I.$$

---

* The skin effect is discussed in Chapter 11.

The complex impedance of the coil is $\mathbf{Z} = (j\omega L + R)$, simply the sum of the impedances of the two components of the equivalent circuit. In the phasor diagram in Figure 7.6, the voltage $\mathbf{V}$ is represented by the vector sum of the

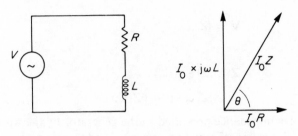

Figure 7.6. Equivalent circuit and phasor diagram for an A.C. generator driving current through a coil.

voltages across L and R. From this diagram we see that the amplitude of the voltage is

$$V_0 = I_0\sqrt{R^2 + \omega^2 L^2} = I_0|Z|,$$

and that the voltage leads the current by the phase angle

$$\theta = \tan^{-1}\left(\frac{\omega L}{R}\right). \tag{7.12}$$

For typical coils used as smoothing chokes at audio frequencies in domestic equipment, the ratio of self-inductance to resistance is of the order of 0.05 henries/ohm. At the main supply frequency, $\omega \simeq 300$ and the phase angle $\theta \simeq \tan^{-1}(15) \simeq 86°$. Even at this low frequency the voltage and current are almost 90° out of phase. Nevertheless, as we shall see in section 7.6, the resistance of a coil sometimes plays an important part in the behaviour of A.C. circuits.

We have now obtained expressions for the impedances of resistors, capacitors and inductors, which are all linear circuit elements in the sense that the amplitude of the current they carry is proportional to the amplitude of the voltage applied across them. They are called *passive* elements to distinguish them from *active* elements like dynamos or batteries which can generate e.m.f.s. When an alternating e.m.f. drives a current through a load made up of a network of passive elements, each of which is linear, then the response of the whole network must also be linear. We can define a complex impedance for such a load* in the same way as for the single circuit elements. The amplitude I_0 of the current delivered by the generator is proportional to its voltage amplitude V_0, but

---

* The problem of working out the value of the complex impedance of a network is discussed in Chapter 8.

the current and voltage are in general not in phase. If the voltage leads the current by the phase angle θ, we can write

$$\mathbf{V} = V_0 \exp{(j\omega t)} \propto I_0 \exp{(j\omega t + j\theta)}$$

or

$$\mathbf{V} = \mathbf{IZ},$$

where

$$\mathbf{Z} \equiv |Z| \exp{(j\theta)}$$
$$\equiv |Z| \cos \theta + j|Z| \sin \theta.$$

$\mathbf{Z}$ is the complex impedance presented to the generator by the whole network. The real part $|Z| \cos \theta$ is referred to as *resistive*, and the imaginary part $|Z| \sin \theta$ is called the *reactance* of the network. The reactance is inductive or capacitive, depending on whether the voltage leads the current or lags behind it. The vector diagram in Figure 7.7 shows a situation in which the reactance is capacitive.

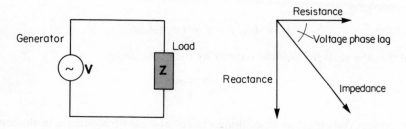

Figure 7.7. Any load made up of linear passive elements can be represented by a single impedance.

7.5 POWER IN A.C. CIRCUITS

When an alternating e.m.f. drives a current through a load made up of passive elements, it must do work in order to maintain the current. In order to calculate the power demanded from an A.C. generator, all we need to know about the load is its complex impedance. Before dealing with general loads, we shall first consider the power requirements of single idealized circuit elements. The power consumption varies during the cycle, and in the resistor in Figure 7.3(a), for example, the rate of energy dissipation at time t is

$$VI = \frac{V^2}{R} = \frac{1}{R}(V_0 \cos \omega t)^2.$$

The mean power W is proportional to the mean value of V^2, averaged over a complete cycle of period $2\pi/\omega$. We have

$$W = \frac{\omega}{2\pi} \int_0^{2\pi/\omega} \frac{V_0^2}{R} \cos^2 \omega t \, dt = \frac{V_0^2}{2R}. \tag{7.13}$$

The mean power is half the peak power dissipated at the moment when the voltage has its maximum value V_0. To avoid the occurrence of a factor $\frac{1}{2}$ in the expression for the power, it is customary to calibrate A.C. instruments in terms of root mean square (RMS) values. The root mean square voltage is defined by

$$V_{\text{RMS}}^2 = (\text{average of } V^2 \text{ over a complete cycle}) = \tfrac{1}{2}V_0^2$$

or

$$V_{\text{RMS}} = \frac{V_0}{\sqrt{2}}. \tag{7.14}$$

Similarly the RMS current I_{RMS} is defined by

$$I_{\text{RMS}} = \frac{I_0}{\sqrt{2}}. \tag{7.15}$$

The RMS voltage across a resistor R carrying a current I_{RMS} is $V_{\text{RMS}} = I_{\text{RMS}}R$. Using Equations (7.13), (7.14) and (7.15), the power dissipation can be written in the same form as for a D.C. circuit:

$$W = V_{\text{RMS}}I_{\text{RMS}} = I_{\text{RMS}}^2 R = V_{\text{RMS}}^2/R. \tag{7.16}$$

Equation (7.16) holds only for a resistor, and in an ideal capacitor or an ideal inductance no power is dissipated at all. In the capacitor circuit of Figure 7.4(b), for example, the generator delivers a voltage $V = V_0 \cos \omega t$ and a current $I = -\omega C V_0 \sin \omega t$ (Equation 7.4). The power delivered by the generator is

$$VI = -\omega C V_0^2 \cos \omega t \sin \omega t$$

$$= -\tfrac{1}{2}\omega C V_0^2 \sin 2\omega t,$$

and the total work done in a complete cycle is

$$-\tfrac{1}{2}\omega C V_0^2 \int_0^{2\pi/\omega} \sin 2\omega t \, dt = 0.$$

Energy surges to and fro between the generator and the capacitor, which absorbs its peak energy $\tfrac{1}{2}C V_0^2$ twice in every cycle. The average power output of the generator is nevertheless zero. Similarly in the ideal inductance shown in Figure 7.3(c) the stored magnetic energy builds up to $\tfrac{1}{2}L I_0^2$ twice in every cycle, and this energy is returned to the generator as the current falls to zero.

Now we can calculate the power required to drive current through any load made up of passive circuit elements. We shall discover that only the resistive part of the load absorbs power, and that the reactance simply stores energy, and later returns it. Consider the load circuit in Figure 7.7, with a complex impedance $\mathbf{Z}$. We must be rather careful in using complex numbers when working out power dissipation. Although it is permissible to add complex impedances in series, as we did for the equivalent circuit of the coil with a finite resistance, remember that only the real part of the quantities appearing in complex equations has any physical meaning. Before *multiplying* voltage and current to find the power, it is necessary to throw away the imaginary parts to avoid having a spurious contribution to the product. Again choosing the time-zero when the current is maximum,

$$\mathbf{I} = I_0 \exp{(\mathrm{j}\omega t)}$$

and

$$\mathscr{R}e\,(\mathbf{I}) = I_0 \cos \omega t$$

The voltage is

$$\mathbf{V} = \mathbf{IZ}$$
$$= I_0|Z| \exp{(\mathrm{j}\omega t + \mathrm{j}\theta)},$$

and

$$\mathscr{R}e\,(\mathbf{V}) = I_0|Z| \cos{(\omega t + \theta)}$$
$$= I_0|Z| \cos \theta \cos \omega t - I_0|Z| \sin \theta \sin \omega t.$$

The power drawn from the generator at time t is

$$\mathscr{R}e\,(\mathbf{V}) \times \mathscr{R}e\,(\mathbf{I}) = I_0^2|Z| \cos \theta \cos^2 \omega t - \tfrac{1}{2}I_0^2|Z| \sin \theta \sin 2\omega t.$$

dissipated in   exchanged between
resistance.   generator and reactance.

Averaged over a complete cycle, the power dissipated in the load is

$$W = \tfrac{1}{2}I_0^2|Z| \cos \theta = \tfrac{1}{2}V_0 I_0 \cos \theta \qquad (7.17)$$

or in terms of RMS values

$$W = I_{\mathrm{RMS}}^2|Z| \cos \theta = V_{\mathrm{RMS}} I_{\mathrm{RMS}} \cos \theta. \qquad (7.18)$$

V_{RMS} and I_{RMS} are the values registered by an A.C. voltmeter and an A.C. ammeter measuring the output of the generator. For a purely resistive load the power is given directly by their product, but for a reactive load, the power is reduced by the *power factor* $\cos \theta$. In a typical coil with $\theta = 86°$ at the main supply frequency, the power dissipation is less by a factor of ten ($\cos \theta \simeq \frac{1}{10}$) than in a non-reactive component whose impedance has the same magnitude.

7.6 RESONANCE

A capacitor presents a large impedance to low frequency alternating current, but has a small impedance at high frequency. The opposite is true for an inductive coil which resists changes in current, and generates a back e.m.f. proportional to the frequency. What happens when an inductive coil and a capacitor are connected together in series across an A.C. generator, as in the circuit diagram of Figure 7.8? The impedance is now large both at high

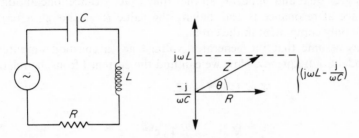

Figure 7.8. The series resonant circuit.

frequencies (because of the coil) and at low frequencies (because of the capacitor). A minimum value of the impedance occurs at some intermediate frequency called the *resonant frequency* of the circuit. We shall find that if the resistance of the coil windings is sufficiently low, then the impedance increases rapidly on either side of the minimum as the frequency moves away from the resonant frequency. A coil and a capacitor connected in series can select signals near their resonant frequency while rejecting signals of other frequencies for which the impedance is high. The circuit is said to be 'tuned' to the resonant frequency; tuned circuits are used in wireless receivers to discriminate between stations broadcasting on different wavelengths. To investigate the sharpness of tuning (i.e. the width of the frequency band over which signals are accepted), we must find the resonant frequency and work out the variation of impedance near this frequency.

The phasor diagram of the resonant circuit is shown in Figure 7.8. The complex impedance $\mathbf{Z}$ is the sum of the impedances of the components L, C and R separately. At an angular frequency ω

$$\mathbf{Z} = R + j\left(\omega L - \frac{1}{\omega C}\right). \tag{7.19}$$

The magnitude of the impedance is

$$|Z| = \left\{R^2 + \left(\omega L - \frac{1}{\omega C}\right)^2\right\}^{1/2}, \tag{7.20}$$

which has its minimum value at the resonant frequency ω_0 given by

$$\omega_0 L - \frac{1}{\omega_0 C} = 0,$$

or

$$\omega_0^2 = \frac{1}{LC}. \tag{7.21}$$

Resonance occurs at the frequency at which the inductive and capacitative phasors are equal and opposite, so that they exactly cancel one another. The impedance at resonance is real, having the value R, as though a resistor R were the only component in the circuit.

Let us assume that the generator voltage has a constant amplitude V_0, independent of frequency. Now we can find the current $\mathbf{I}$ from the relation

$$\mathbf{V} = V_0 \, e^{j\omega t} = \mathbf{IZ}.$$

Hence

$$\mathbf{I} = \frac{\mathbf{V}}{\mathbf{Z}} = \frac{V_0 \, e^{j\omega t}}{R + j(\omega L - 1/\omega C)}.$$

Multiplying top and bottom by $\{R - j(\omega L - 1/\omega C)\}$,

$$\mathbf{I} = \left\{ \frac{R}{R^2 + (\omega L - 1/\omega C)^2} - \frac{j(\omega L - 1/\omega C)}{R^2 + (\omega L - 1/\omega C)^2} \right\} V_0 \, e^{j\omega t}. \tag{7.22}$$

In the previous section we showed that there is a net demand for power from the generator only from that component of the current which is in phase with the voltage, namely the first term in Equation (7.22). If the current has amplitude I_0 and lags behind the voltage by a phase angle θ, then the component in phase with the voltage is

$$I_0 \cos \theta = \frac{V_0 R}{R^2 + (\omega L - 1/\omega C)^2}.$$

The power W, averaged over a complete cycle, is given by Equation (7.17):

$$W = \tfrac{1}{2} V_0 I_0 \cos \theta$$

$$= \frac{V_0^2}{2R} \left\{ \frac{1}{1 + (1/R^2)(\omega L - 1/\omega C)^2} \right\}. \tag{7.23}$$

Away from resonance the power falls off rapidly, as shown in Figure 7.9. Restricting attention to angular frequencies near to ω_0, we can make a Taylor expansion and find an approximate expression for W by neglecting high order terms. Writing

$$f(\omega) = \left(\omega L - \frac{1}{\omega C} \right),$$

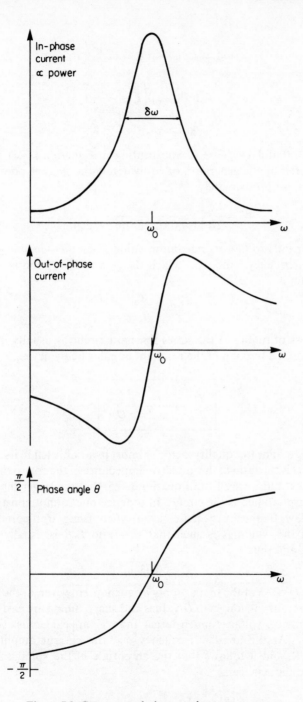

Figure 7.9. Currents and phase angles near resonance.

we have

$$f(\omega) \simeq f(\omega_0) + (\omega - \omega_0)\frac{\partial f}{\partial \omega}\bigg|_{\omega_0}$$

$$= (\omega - \omega_0)\left(L + \frac{1}{\omega_0^2 C}\right)$$

$$= 2(\omega - \omega_0)L,$$

since $f(\omega_0) = 0$ and $1/\omega_0^2 C = L$. Substituting in Equation (7.23), the average power delivered by the generator, or equivalently the average power absorbed by the circuit, is approximately

$$W = \frac{V_0^2}{2R}\left\{\frac{1}{1 + (4L^2/R^2)(\omega - \omega_0)^2}\right\}. \tag{7.24}$$

The power falls to half its maximum value when $(\omega - \omega_0) = \pm R/2L$, and the angular frequency band over which the power absorption is greater than half maximum is

$$\delta\omega = \frac{R}{L}.$$

The sharpness of tuning of the series resonant circuit is usually measured by expressing $\delta\omega$ as a fraction of the resonant angular frequency ω_0;

$$\boxed{\frac{\delta\omega}{\omega_0} = \frac{R}{\omega_0 L} = \frac{1}{Q}.} \tag{7.25}$$

The symbol Q stands for 'quality factor' (almost invariably left in its abbreviated form as 'Q'). Q is the ratio of the inductive impedance of the coil to the resistance of its windings, and we see from Equation (7.25) that the larger the value of Q, the more sharply tuned is the circuit. In practice one cannot construct high-Q circuits at low frequency because, even when using iron-cored coils, the resistance of the windings is such that the ratio L/R is usually about 0.05 henries/ohm, and thus

$$Q \simeq 0.05\omega_0.$$

For example $Q \simeq 15$ at the main supply frequency. However, at the frequencies used in radio transmission, high Q-values and sharp tuning are easily obtained.

Near resonance, voltages much larger than V_0 appear across the coil and the capacitor. At the angular frequency ω_0, the current amplitude in the circuit is V_0/R, and it follows that the amplitude of the voltage across both capacitor and inductance is

$$\frac{V_0}{R}\left(\frac{1}{\omega_0 C}\right) = \frac{V_0}{R}(\omega_0 L) = QV_0.$$

Energy is needed to build up the voltage across the capacitor, and to drive current against the inductive impedance of the coil. We shall discuss in the next section how the energy builds up when the generator is first switched on; once the alternating current is established, however, the energy is stored by the circuit and surges to and fro between capacitor and coil. To find the amount of stored energy, we drop the complex notation, writing the voltage across the capacitor as $(QV_0 \cos \omega_0 t)$, and the current, which is out of phase by $\pi/2$, as $I_0 \cos(\omega_0 t + \pi/2) = -I_0 \sin \omega_0 t$. The stored energy is

$$\tfrac{1}{2}C(QV_0 \cos \omega_0 t)^2 + \tfrac{1}{2}L(-I_0 \sin \omega_0 t)^2$$

electric energy + magnetic energy.

After replacing V_0 by $I_0 R$, a little manipulation shows that the electrostatic energy is $\tfrac{1}{2}LI_0^2 \cos^2 \omega_0 t$. The total energy stored is therefore $\tfrac{1}{2}LI_0^2(\cos^2 \omega_0 t + \sin^2 \omega_0 t) = \tfrac{1}{2}LI_0^2$. Comparing the stored energy with the energy dissipated in the resistance during one complete cycle of period $2\pi/\omega_0$, we find

$$\frac{\text{Stored energy at resonance}}{\text{Energy dissipated in one cycle}} = \frac{\tfrac{1}{2}LI_0^2}{\tfrac{1}{2}RI_0^2} \times \frac{\omega_0}{2\pi} = \frac{Q}{2\pi}. \qquad (7.26)$$

Q is equally well defined in terms of energy by this relation as by Equation (7.25).

Away from resonance, the peak energies stored by the capacitor and the coil are not equal, and the circuit must exchange the difference with the generator. This energy is supplied by the component of the current which is $\pi/2$ out of phase with the voltage, namely the second term in Equation (7.22). The magnitude of the out-of-phase current, which is plotted in Figure 7.9, changes sign at resonance, as the reactance changes from being capacitative to being inductive. Also shown in the figure is the phase angle by which the voltage leads the current. From the phasor diagram in Figure 7.8 we can see that the phase angle is

$$\theta = \tan^{-1}\left(\frac{\omega L - 1/\omega C}{R}\right).$$

Discussion of complex impedance has allowed us to work out the behaviour of the current in the series LCR circuit. The same result can also be derived by direct solution of a differential equation for the current. Writing the voltage $V_0 \cos \omega t$ of the generator as the sum of the voltages across each of the components L, C and R, we have

$$V_0 \cos \omega t = L\frac{dI}{dt} + IR + \frac{Q}{C}$$

where Q is the charge on the capacitor at time t. Differentiating with respect to time, and replacing dQ/dt by I,

$$-\omega V_0 \sin \omega t = L \frac{d^2 I}{dt^2} + R \frac{dI}{dt} + \frac{I}{C}. \tag{7.27}$$

The solution to this equation, as may be verified by substitution, is just the real part of the complex current already given in Equation (7.22):

$$I = \mathcal{R}e\,(\mathbf{I}) \propto R \cos \omega t + \left(\omega L - \frac{1}{\omega C}\right) \sin \omega t. \tag{7.28}$$

Other resonance systems always satisfy a differential equation which has the same form as Equation (7.27) near resonance. For example, the rectangular cavities discussed in Chapter 13 act as very high frequency resonators, and the current they draw is related to the applied A.C. voltage by an equation like (7.27), even though the cavity is not made up of discrete components. The power absorption of a cavity peaks at resonance, and the amplitude and phase of the current vary in the same way as for the series LCR circuit. The same differential equation describes resonance not only in electromagnetism. but also in other branches of physics*, and the solution always exhibits the same characteristics. For example, energy absorption always passes through a maximum at a resonant frequency. This explains the occurrence of sharp absorption lines in atomic spectra; how it comes about that a tenor may break a wine-glass by singing its fundamental note; and why a motor-car shudders when an unbalanced wheel rotates at its resonant frequency.

7.7 TRANSIENTS

Until now we have been considering the behaviour of electrical circuits in an equilibrium state, in which an A.C. generator maintains alternating currents of a fixed frequency everywhere in the circuit. Whenever a change is made to the e.m.f. (A.C. or D.C.) in a circuit containing capacitance or inductance, transient currents flow after the change. Let us consider, for example, the circuit shown in Figure 7.10 consisting of a coil of self-inductance L and resistance R, and a battery of voltage V_1. Initially the switch is open, and no current flows; the circuit is in an equilibrium state. When the switch is closed, current is driven through the circuit by the battery, and a new equilibrium state is approached in which the current is $I_1 = V_1/R$. But the current cannot suddenly jump to its new value, because a changing current through the coil induces an e.m.f. which opposes the change. Including the induced e.m.f., the differential equation relating current and voltage in the circuit is

$$L \frac{dI}{dt} + IR = V_1. \tag{7.29}$$

---

* A full treatment of resonance in different systems can be found in *The Physics of Vibrations and Waves* by H. J. Pain (Wiley 1968).

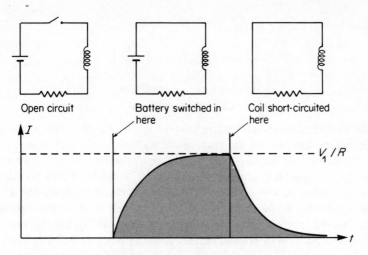

Figure 7.10. Transient currents flow in a coil whenever there is a sudden change in the voltage across its terminals.

One solution of this equation is $I_{eq} = I_1 = V_1/R$, the constant current achieved at equilibrium. This solution is the *particular integral* of Equation (7.29). However, to the particular integral we can add any solution of the equation

$$L\frac{dI}{dt} + IR = 0. \tag{7.30}$$

The solution of Equation (7.30) is the *complementary function* of Equation (7.29); it is the complementary function which describes the *transient currents* in an electrical circuit. Writing the complementary function as I_{tr}, we see by substitution in Equation (7.30) that

$$I_{tr} = k \exp\left(-\frac{Rt}{L}\right),$$

where k is any constant. The value of the constant is determined by the initial condition of the problem*, namely that the current is zero at the moment of switching. If we choose this moment to be the time-zero, then

$$I = I_1 + I_{tr} = 0 \quad \text{at } t = 0,$$

* There is an alternative method of solving transient problems which avoids the cumbersome procedure of finding the particular integral and the complementary function separately, and then combining them to satisfy the initial conditions. This method, which is called the Laplace transform method, incorporates the initial conditions from the start. A good account of it may be found in *An Introduction to the Laplace Transformation with Engineering Applications* by J. C. Jaeger and G. W. Newstead (Methuen, 1969).

leading to $k = I_1$, and

$$I = I_1 - I_1 \exp\left(-\frac{Rt}{L}\right). \qquad (7.31)$$

$$\underset{\substack{\text{equilibrium} \\ \text{current}}}{} \quad \underset{\substack{\text{transient} \\ \text{current}}}{}$$

In mathematical terms a transient current in any circuit is described by the complementary function of the corresponding differential equation. In terms of physical quantities, we can see that because reactive circuits store electro-static and/or magnetic energy, there cannot be a sudden jump between two equilibrium states of the circuit. The transient current allows the changes to occur gradually, while energy is absorbed or delivered by the circuit. For example, as the current in a coil increases towards its equilibrium value according to Equation (7.31), the battery is doing work to build up the stored energy. The work done in a circuit containing self-inductance has already been discussed in section 6.3. There it was shown that energy $\frac{1}{2}LI_1^2$ is stored by the coil as the current builds up to its equilibrium value I_1. Similarly, if the coil is suddenly short-circuited while carrying a current I_1, as shown in Figure 7.10(c), a transient current flows until the stored energy $\frac{1}{2}LI_1^2$ is dissipated. The complementary function is the same as when the battery was switched into the circuit, and the transient current dies away exponentially with time constant L/R. Now the equilibrium current is zero, and since the initial current is I_1, the current at a time t after short-circuiting is simply

$$I = I_1 \exp\left(-\frac{Rt}{L}\right).$$

As we showed in section 6.3, the rate of decay of the current in the shorted coil is such that heat losses in the resistance account for the whole of the energy $\frac{1}{2}LI_1^2$ initially stored in the coil.

Very rapid decay of the current in a coil can occur if it is suddenly open-circuited. Huge back e.m.f.s are induced, causing a spark to cross the break in the circuit; most of the energy stored in the coil is dissipated in the spark. Because of the large e.m.f.s induced, it is dangerous to break the circuit of a big coil.

The introduction of complex impedances greatly simplified the discussion of the equilibrium behaviour of A.C. circuits—i.e. the discussion of the particular integrals of their differential equations. Transients are described by the complementary functions of the same differential equations, and not surprisingly it is still helpful to use complex variables. We shall illustrate the complex variable technique by applying it first to the circuit of Figure 7.10.

To emphasize the connection with A.C. theory, let us write the transient current as

$$\mathbf{I}_{tr} = k\,e^{jpt}.$$

In this equation, $\mathbf{p}$ does not represent a real angular frequency, but is merely a parameter describing the time variation of the transient. It is allowed to have complex values, and in this example in which the transient decays exponentially, $\mathbf{p}$ is purely imaginary: substituting in Equation (7.30), we find

$$k(\mathbf{j p}L + R)\,e^{\mathbf{j p}t} = 0, \tag{7.32}$$

or

$$\mathbf{p} = -\frac{\mathbf{j}R}{L}.$$

Hence $\mathbf{I}_{tr} = k \exp(-Rt/L)$. As before the value of the proportionality constant k can only be determined by reference to the initial conditions and the equilibrium current. The quantity $\mathbf{Z} = (\mathbf{j p}L + R)$ is analogous to the impedance $(\mathbf{j}\omega L + R)$ which the coil presents to an alternating current of angular frequency ω. Because the sum of the potential differences around the circuit due to the transient is zero, we can interpret Equation (7.32) by saying that the circuit presents no impedance to the transient.

Sometimes transient currents do not die away monotonically, but show oscillations with a diminishing amplitude. Oscillatory behaviour occurs after the switch is closed in the circuit of Figure 7.11 for some values of L, C and R. For transients in this circuit the complex impedance satisfies the equation

$$\mathbf{Z} = R + \mathbf{j}\left(\mathbf{p}L - \frac{1}{\mathbf{p}C}\right) = 0,$$

or multiplying by $(-\mathbf{j p})$,

$$\mathbf{p}^2 L - \mathbf{j p}R - \frac{1}{C} = 0.$$

Solving the quadratic in $\mathbf{p}$, we find that there are two possible values of $\mathbf{p}$:

$$\mathbf{p}_1 = \frac{\mathbf{j}R}{2L} + \frac{1}{2L}\sqrt{\left\{\frac{4L}{C} - R^2\right\}}$$

and

$$\mathbf{p}_2 = \frac{\mathbf{j}R}{2L} - \frac{1}{2L}\sqrt{\left\{\frac{4L}{C} - R^2\right\}}. \tag{7.33}$$

There are *two* complementary functions $e^{\mathbf{j p}_1 t}$ and $e^{\mathbf{j p}_2 t}$, and the transient current in the circuit must be made up of a suitably weighted sum of these functions[*]. The equilibrium current in the circuit is zero, and at a time t after closing the switch the current is

$$\mathbf{I}(t) = A\,e^{\mathbf{j p}_1 t} + B\,e^{\mathbf{j p}_2 t}.$$

---

[*] In the example we have chosen here, the current turns out to be real. Sometimes the transients obtained using the complex variable technique may have imaginary components. As in A.C. theory, the imaginary components have no physical significance.

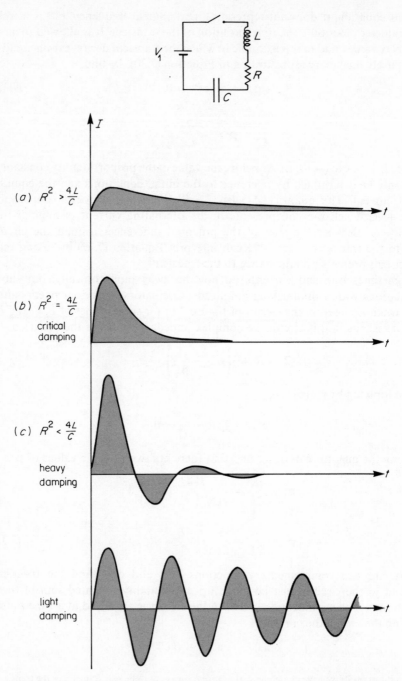

Figure 7.11. Transient currents in an *LCR* circuit as *R* changes for fixed *L* and *C*.

Two initial conditions are needed to specify the constants A and B. The first condition is that because of the presence of the inductance L, there can be no discontinuity in current, and the initial current must be zero. Secondly we note that since the initial current is zero, at the moment of closing the switch the whole of the battery voltage appears across the inductance. Hence

$$L\frac{dI}{dt} = V_1 \quad \text{at } t = 0.$$

The current which satisfies these two initial conditions is

$$I(t) = -\frac{jV_1}{\sqrt{\dfrac{4L}{C} - R^2}}\left\{e^{jp_1 t} - e^{jp_2 t}\right\}. \tag{7.34}$$

There are three different forms of this transient, depending on the value of $(4L/C - R^2)$.

(a) $R^2 > 4L/C$. Both p_1 and p_2 are then purely imaginary, and there are no oscillations. For a 'heavily damped' circuit in which $R^2 \gg 4L/C$, Equation (7.34) simplifies in first order to

$$I(t) = \frac{V_1}{R}\left\{\exp\left(-\frac{t}{CR}\right) - \exp\left(-\frac{Rt}{L}\right)\right\}.$$

The voltage drop across the inductance L quickly dies away, and the capacitance is charged up through R with a time constant CR.

(b) $R^2 = 4L/C$. Equation (7.34) then reduces to

$$I(t) = \frac{V_1}{L}t\exp\left(-\frac{Rt}{2L}\right),$$

and the circuit is described as 'critically damped'. For fixed values of L and C, the transient dies away most rapidly in a critically damped circuit.

(c) $R^2 < 4L/C$. Now the transient current is oscillatory, and

$$I(t) = \frac{2V_1}{\sqrt{\dfrac{4L}{C} - R^2}}\exp\left(-\frac{Rt}{2L}\right)\sin\left\{\left(\sqrt{\frac{1}{LC} - \frac{R^2}{4L^2}}\right)t\right\}.$$

Any circuit in which transients oscillate is said to 'ring'. Figure 7.11 illustrates the transient in a ringing circuit in which there is fairly heavy damping so that the oscillations die away rapidly, and in a lightly damped circuit for which $4L/C \gg R^2$. Under the latter condition the angular frequency of the oscillation is approximately equal to the resonant value $\omega_0 = 1/\sqrt{LC}$. The amplitude of the oscillations gradually diminishes, and successive maxima, separated by the period $2\pi/\omega_0$, are reduced by the factor $\exp(-\pi R/\omega_0 L)$. The quantity $\pi R/\omega_0 L$, which is called the *logarithmic decrement* of the circuit, is related to the sharpness of tuning, determined by $Q = \omega_0 L/R$ (see Equation (7.25)). We see that in a sharply tuned circuit, for which Q is large, the logarithmic decrement is small, and consequently transient oscillations take a long time to die away.

Transient currents for various amounts of damping are plotted in Figure 7.11. In spite of the different shapes of these plots, the final result is in each case the same. The battery has delivered a total charge CV_1 to the capacitor, i.e. the algebraic area under all the current curves is equal to CV_1, and the battery has delivered an amount of energy CV_1^2. At equilibrium, electrostatic energy $\frac{1}{2}CV_1^2$ has also been dissipated in the resistance. Further, if the battery were short-circuited, the stored energy $\frac{1}{2}CV_1^2$ would be dissipated as heat during a transient similar in form to the one occurring when the battery is switched into the circuit. Once again the transient has been shown to adjust the circuit to the requirements of stored energy in the reactive components.

PROBLEMS 7

7.1 Calculate the amplitude and phase of $\mathbf{V}_{\text{out}}$ in the circuits shown in Figure 7.12.

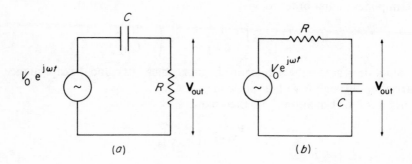

Figure 7.12.

7.2 Sketch the voltage V_{out} as a function of time after the switch is closed in the circuits shown in Figure 7.13.

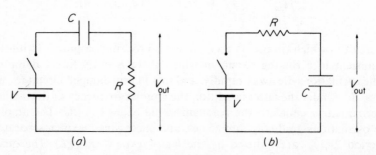

Figure 7.13.

7.3 A capacitor discharges through its leakage resistance with a time constant of 1 sec. What is the power factor of this capacitor when it is in a 50 Hz A.C. circuit? (You may represent it by an ideal capacitor in parallel with a resistor).

7.4 A 1 μF capacitor is placed in series with a 1000 Ω resistor. At what frequency is their impedance 2000 Ω? Draw the phasor diagram at this frequency.

7.5 The resistor and capacitor of the previous question are placed in series across a 240 volt 50 Hz mains supply. What is the RMS current drawn from the mains? What is the RMS voltage across the capacitor?

7.6 A coil has a self-inductance of 10 mH and a resistance of 100 Ω. What capacitance must be put in series with the coil to make a critically damped circuit? If the critically damped circuit is driven by an alternating voltage of constant amplitude, at what angular frequency is the current through the circuit a maximum? What is the Q of the circuit?

7.7 You are making a tuned circuit for a medium-wave radio receiver using an air-filled variable capacitor with the range 30–300 pF, in series with a coil wound on a toroidal ferrite core whose properties are given below.

With how many turns should you wind the ferrite core if the receiver is to cover the band from 500 kHz to 1500 kHz? If the windings are made with 25 gauge copper wire, will you be disturbed by the disc jockey on 1214 kHz when you are listening to the current affairs programme on 1151 kHz?

Data: 25 gauge wire has a diameter of 0.5 mm. The resistivity of copper is 1.7×10^6 ohm cm. The ferrite core has a relative permeability 200, a magnetic path length 3 cm and a flux area (cross-section of core) 0.1 cm^2.

7.8 For the circuit shown in Figure 7.14 the amplitude of the current drawn from the source of e.m.f. has a minimum value near the resonant angular frequency $\omega_0 = 1/\sqrt{LC}$. Find the energy stored by the coil and by the capacitor, and hence the Q of the circuit.

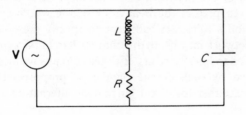

Figure 7.14.

CHAPTER 8

Linear circuits

8.1 NETWORKS

A practical circuit often contains many components linked into a network by a large number of connecting wires. The currents may not be the same in different branches of the network, and one must be able to calculate these currents in order to predict the behaviour of the whole circuit. Two rules, known as Kirchhoff's rules, are sufficient to specify these currents, however complicated the network may be. In this chapter Kirchhoff's rules are explained, and they are applied to linear circuits. By linear circuits we mean those in which the current carried by each circuit element is proportional to the voltage difference across it, either for direct current or alternating current at a fixed frequency.

8.1.1 Kirchhoff's rules

The currents in the components of a network must satisfy Kirchhoff's rules, which are:

1 *The total current arriving at any junction of the network is zero.*
2 *The sum of the voltages around any closed loop of the network is zero.*

The rules are valid both for D.C. and A.C. circuits, but we shall illustrate them first in a direct current application, later generalizing to alternating currents. The D.C. network in Figure 8.1 contains a number of sources of e.m.f. V_a, V_b... driving currents through resistors with resistances $R_1, R_2 \ldots$. The values $R_1, R_2 \ldots$ include the resistance of the leads to the junctions of the network.

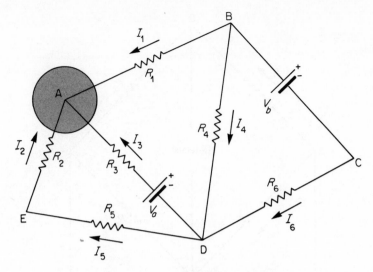

Figure 8.1. A direct current network.

The first rule is an expression of the conservation of charge. In the figure, for instance, no charge can accumulate at the junction A when the network is in a steady state, and the total current entering the shaded area is $I_1 + I_2 + I_3 = 0$. To satisfy this condition at least one of the currents must be negative. A negative value indicates that the current flows in the opposite direction to the one assumed in the diagram.

In the direct current network, only steady electric and magnetic fields are present. As we found in section 1.5.2, a steady electric field **E** can be expressed as the gradient of a scalar potential which is a function of position only. Around any closed path the potential returns to its starting value, i.e., the line integral of the field **E** is zero:

$$\oint \mathbf{E} \cdot \mathbf{dl} = 0.$$

Kirchhoff's second rule applies this equation to paths which follow the connections of the network. Along these paths potential differences occur only across the resistors and sources of e.m.f. As with the first rule, it is necessary to be careful about the signs of current and voltage. Around the loop ABDA in Figure 8.1, for example, the e.m.f. V_a balances the voltage drops across the resistors R_1, R_4 and R_3, and

$$I_1 R_1 - I_4 R_4 - I_3 R_3 + V_a = 0.$$

As an elementary example of network analysis, let us apply Kirchhoff's rule to the Wheatstone bridge shown in Figure 8.2. The value of an unknown

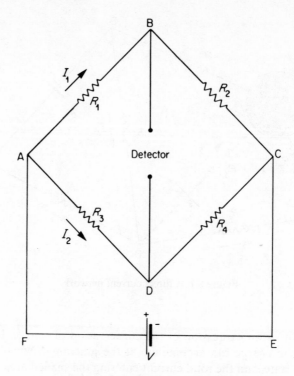

Figure 8.2. The Wheatstone Bridge.

resistor R_1 is to be determined by adjusting a variable standard resistance R_2 until there is no potential difference between points B and D. The detector used to search for the null point draws a current which is very small compared with the current flowing in the arms of the bridge even when it is off balance. Neglecting the current through the detector, we can assume that the same current I_1 flows through the resistors R_1 and R_2, and similarly a current I_2 through R_3 and R_4. To satisfy Kirchhoff's first rule at all junctions in the network, the battery must deliver a current $(I_1 + I_2)$. Now apply the second rule to the loop ABCEFA;

$$I_1R_1 + I_1R_2 - V = 0; \tag{8.1}$$

and to the loop ADCEFA;

$$I_2R_3 + I_2R_4 - V = 0. \tag{8.2}$$

The only other loop which can be made along the current-carrying branches of the network is the loop ABCDA. The sum of the voltages around this loop is

$$I_1R_1 + I_1R_2 - I_2R_3 - I_2R_4 = 0.$$

This equation can also be derived by subtracting Equation (8.2) from Equation (8.1). There are thus only two independent equations, from which solutions for the two unknowns I_1 and I_2 can be determined. For any network, Kirchhoff's rules always provide the same number of independent equations as there are unknown currents. When all the impedances in the network are linear, the set of linear simultaneous equations has a unique solution. Note, however, that if some of the circuit elements are non-linear, so that higher powers of the current appear in the equations derived from Kirchhoff's second rule, the solution may not be unique; it is quite possible for there to be more than one way of distributing currents in an electrical circuit with non-linear components.

The Wheatstone bridge contains only resistors which are linear, and the currents are uniquely determined. It follows immediately from Equations (8.1) and (8.2) that

$$I_1 = \frac{V}{R_1 + R_2},$$

and

$$I_2 = \frac{V}{R_3 + R_4}.$$

Hence the output voltage across the terminals of the detector is

$$v = I_1 R_2 - I_2 R_4$$
$$= \frac{V(R_2 R_3 - R_1 R_4)}{(R_3 + R_4)(R_1 + R_2)}, \tag{8.3}$$

and the null point occurs when

$$R_3/R_4 = R_1/R_2. \tag{8.4}$$

Suppose that a just detectable output voltage δv is caused by a change δR_2 in the variable resistor from its balance position. We have

$$\delta v = \frac{\partial v}{\partial R_2} \delta R_2,$$

and by differentiating Equation (8.3) it follows that

$$\delta R_2 = \frac{\delta v (R_1 + R_2)^2}{V R_1}.$$

Now the relative accuracy of measurement of the unknown is

$$\frac{\delta R_1}{R_1} = \frac{\delta R_2}{R_2} = \frac{\delta v}{V} \frac{(R_1 + R_2)^2}{R_1 R_2}.$$

Since δv is fixed by the sensitivity of the detector, the least error in the unknown occurs when $(R_1 + R_2)^2/R_1 R_2$ is minimum, i.e. when $R_1 = R_2$. At balance this implies $R_3 = R_4$, and the bridge is best used with equal values for the fixed resistors.

The practical limitation to the accuracy which can be attained in the measurement of resistance with the D.C. Wheatstone bridge is often not the size of the detectable output voltage, but the interference due to unwanted e.m.f.s in the circuit. The most important of these are usually thermo-electric e.m.f.s generated when the temperature of the circuit is not uniform. The thermo-electric currents are independent of the currents driven by the battery. Imagine two bridge circuits, one containing no e.m.f. other than the battery, and the other only thermo-electric e.m.f.s. Because the circuit is linear, the currents flowing in a circuit with both types of e.m.f. is the sum of the currents in the separate circuits. We can eliminate the effect of thermo-electric e.m.f.s by making a second measurement of R_2 after reversing the battery. The unwanted e.m.f.s are not reversed at the same time, and provided that they are small, the mean of the two measurements of R_1 is a good estimate of the true balance value. A still better way of eliminating the effect of thermo-electric currents by taking advantage of their independence is to replace the battery by an A.C. generator, and to use a detector sensitive only to signals of the same frequency as the generator. After generalizing Kirchhoff's rules to apply to A.C. networks, we shall show that the balance condition (8.4) still holds at low frequencies for the A.C. resistance bridge.

8.1.2   A.C. networks

Kirchhoff's two rules apply for A.C. as well as D.C. In an A.C. network carrying currents of an angular frequency ω, conservation of charge still requires that at all times the current arriving at any junction is zero. The currents in different sections of the network may have different amplitudes, and they may not be in phase with each other. Suppose, for example, that three leads meet at a junction of the network carrying currents of amplitude I_1, I_2 and I_3. At time $t = 0$ their phase angles are θ_1, θ_2, and θ_3, say, measuring the current in each lead in the direction towards the junction. Using the complex number notation for the currents, Kirchhoff's first rule requires that

$$(I_1\,e^{j\theta_1} + I_2\,e^{j\theta_2} + I_3\,e^{j\theta_3})\,e^{j\omega t} = 0.$$

$$\therefore\quad \mathbf{I}_1 + \mathbf{I}_2 + \mathbf{I}_3 = 0.$$

where $\mathbf{I}_1$, $\mathbf{I}_2$ and $\mathbf{I}_3$ are the phasors representing the currents arriving at the junction. Thus the vector sum of the phasors is zero, as sketched in Figure 8.3. Similarly the vector sum of phasors is zero for any number of leads meeting at a junction.

The changing currents in an A.C. network cause changing magnetic fields, and these lead to induced e.m.f.s. Faraday's law tells us that the e.m.f. around a closed path is $\oint \mathbf{E} \cdot d\mathbf{l} = -d\Phi/dt$, where Φ is the total magnetic flux threading the area enclosed by the path. Kirchhoff's second rule for an A.C. network applies Faraday's law to a path following the connections of the network. As for D.C., the rule states that the sum of the voltages around the loop is

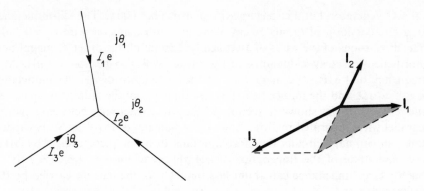

Figure 8.3. The vector sum of the phasors representing currents meeting at a junction is zero. The phasors therefore form the side of a closed polygon—like the shaded triangle in this figure.

zero: note, however, that the sum must now include the voltage drops due to induced e.m.f.s. Induced e.m.f.s are generated by the self-inductances and mutual inductances of coils, but in addition there are always 'stray' inductances associated with the leads between the components; even in an A.C. network containing no coils, changing fluxes thread the areas enclosed within the network. At moderately low frequencies the stray inductances are very small and they can usually be neglected. This is fortunate, since their magnitudes depend on the exact shape and dimensions of the circuit, and are generally almost impossible to calculate.

Not only are there always small inductances present in an A.C. network, but there is inevitably stray capacitance between different parts of the circuit, and between different parts of the same component. We have already met stray capacitance in the discussion in section 7.4 of the current through an inductive coil, which has capacitance between neighbouring windings. Neglecting this capacitance, the coil can be represented by an approximate equivalent circuit consisting of a resistance R in series with an idealized self-inductance L. Similarly, by neglecting all the stray inductances and capacitances which have little effect at low frequencies, an equivalent circuit can be built up for the whole of an A.C. network. Each component is represented by an idealized circuit element, and the idealized elements are linked together in the same way as the components of the real circuit. The leads are assumed to have no impedance, and voltage differences only occur across the idealized elements. Just as for a D.C. network, application of Kirchhoff's two rules to the equivalent A.C. network leads to a set of linear equations whose solution yields the currents in all branches of the network.

Let us estimate how good an approximation it is to neglect stray impedances in the Wheatstone bridge circuit of Figure 8.2 if the battery is replaced by an A.C. generator. Let the resistance in each arm be 1000 Ω. The self-inductance of a circular loop of radius 10 cm made of 1 mm diameter wire is $\simeq 0.7 \, \mu$H. The dimensions of the wires of a typical bridge are of this order of magnitude, and hence the stray inductance in the bridge will also be about 1 μH. At a frequency of 1 MHz the magnitude of the impedance of a 1 μH inductance is $\omega L \simeq 6 \, \Omega$, and the induced e.m.f. is less than 1 % of the voltage drop across the resistors. It was shown in section 3.5 that the capacitance between a pair of parallel wires separated by three times their diameter is 15 pF/m. Stray capacitance in ordinary circuits therefore amounts to a few picofarads. At 1 MHz the magnitude of the impedance of a 1 pF capacitance is $1/\omega C \simeq 2.10^5 \, \Omega$. Such a large impedance can shunt less than 1 % of the current carried by the resistance.

No hard and fast rules can be given for the frequency limit above which the neglect of stray capacitance and inductance ceases to be a good approximation. The example worked out above suggests that one may have to take strays into account at frequencies above 1 MHz. To some extent this can be done by adding to the equivalent circuit discrete capacitances and inductances representing the strays. It is difficult to find accurate values for the impedances of these extra circuit elements, but even a rough guide to the high frequency behaviour of a circuit may be useful. Eventually the network analysis must break down at frequencies above about 10^9 Hz. Stray inductances and capacitances cannot both be assumed to cause small corrections at these frequencies; any component is either shunted by stray capacitance or causes a smaller voltage drop than the e.m.f. generated by stray inductance. However, at low frequencies, it is usually safe to neglect stray impedances in a Wheatstone bridge with resistors in each arm. If a detector which responds only to alternating voltages is used, unwanted thermo-electric e.m.f.s are not observed, and the null-point occurs when the balance condition (8.4) is satisfied.

8.2 AUDIO-FREQUENCY BRIDGES

The Wheatstone bridge circuit is used at audio-frequencies (i.e. at frequencies up to about 20 kHz) to measure the impedance of reactive elements. The equivalent circuit for the general A.C. bridge is shown in Figure 8.4. The arms of the bridge contain circuit elements with complex impedances Z_1, Z_2, Z_3, Z_4. The currents I_1 and I_2 may not be in phase with the generator voltage or with each other. In order to obtain a balance with no alternating voltage across the detector, the voltages across Z_1 and Z_3 must not only have the same amplitude, but they must also be in phase with one another. The balance condition is found by applying Kirchhoff's second rule to the bridge. The resulting equations have the same form as they did for the D.C. bridge, with the complex impedances replacing resistances. The balance condition becomes

$$\frac{Z_1}{Z_2} = \frac{Z_3}{Z_4}. \tag{8.5}$$

This equation has real and imaginary parts, thus imposing the two conditions on the component values required to ensure that both the amplitude and the phase of the voltages across Z_1 and Z_3 are the same. An A.C. bridge must contain two variable components, and in searching for a balance these are adjusted alternately to give successive minima in the detector output until the null-point is reached.

There are many forms of the A.C. bridge, with different combinations of components in the arms. As a practical example we shall consider the Owen inductance bridge, which has the circuit diagram shown in Figure 8.5. The self-inductance L and the resistance R represent a coil whose inductance is to

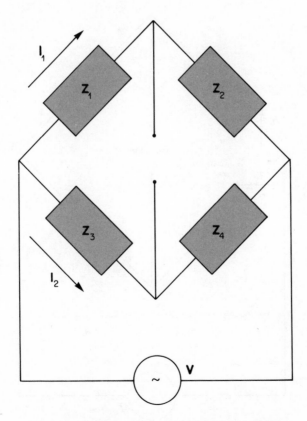

Figure 8.4. The alternating current Wheatstone Bridge.

be measured. The bridge is balanced by adjusting the variable resistors R_1 and R_3. Substituting the appropriate values in Equation (8.5), at balance we have

$$\frac{j\omega L + R + R_1}{R_2} = \frac{R_3 + 1/j\omega C_3}{1/j\omega C_4}$$

$$= \frac{C_4}{C_3} + j\omega C_4 R_3.$$

Equating real parts

$$R = \frac{R_2 C_4}{C_3} - R_1; \qquad (8.6)$$

and imaginary parts;

$$L = C_4 R_2 R_3. \qquad (8.7)$$

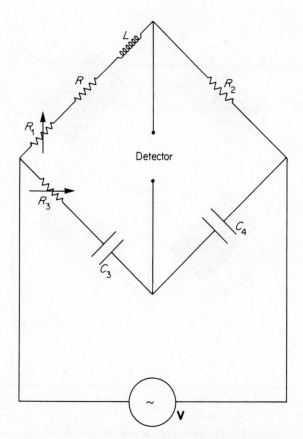

Figure 8.5. The Owen inductance bridge.

The Owen bridge has the following good features which make it convenient for routine laboratory use:

(i) The balance condition is independent of frequency. The accuracy of measurement is not affected if the time variation of the generator voltage is not exactly sinusoidal, and other angular frequencies besides ω are present in the circuit.

(ii) The resistance R_1 does not occur in Equation (8.7). The balance setting of R_3 is fixed by the inductance L of the coil, whatever its resistance R. It follows that the resistance R_3 can be calibrated to give a direct reading of inductance. In fact the inductance is proportional to R_3, with the proportionality constant known from the values of the fixed components C_4 and R_2. Similarly, since R_3 does not occur in Equation (8.6), R_1 can give a direct reading of R.

(iii) The bridge is suitable for measuring small inductances. If the inductance to be measured is much less than 100 μH, it is no longer permissible to neglect the stray inductances in components and leads. Their effect can be eliminated by repeating the measurement after removing the coil and joining together the leads which were attached to it. If the new value of the variable resistance at balance is R_3', the strays are equivalent to an inductance $C_4 R_2 R_3'$ in the same arm of the bridge as R_1. Evidently the inductance of the coil is $C_4 R_2 (R_3 - R_3')$.

8.2.1 The transformer ratio-arm bridge

At audio-frequencies stray inductances and stray capacitances cannot be simultaneously important. In the Owen bridge, for example, the voltage drop across the coil must be comparable to the voltage drops across other components if the bridge is to be sensitive. When the coil itself has such a small impedance that stray inductances are significant, then stray capacitances have much higher impedances than any of the components, and carry negligible currents. Conversely, only stray capacitances, and not stray inductances, need to be considered in an A.C. bridge measurement of a small capacitance. The most important stray capacitance is that between the leads and earth. There is invariably an appreciable capacitance to earth in a sensitive bridge, since in order to reduce interference from outside sources, the bridge is housed in a box maintained at earth potential, and leads to an unknown impedance are also shielded by an earthed braiding. Stray capacitances to earth at those junctions of the bridge joined to the generator do not matter, because they merely draw current from the generator. They do not alter the balance condition. The effects of stray capacitances to earth at the other corners of the bridge can be minimized by arranging that there is only a small voltage drop to earth at these corners. The strays then carry little current, even if they are comparable with the capacitance to be measured. One way of ensuring that the voltage drop is small is to use the transformer ratio-arm bridge, whose circuit diagram is shown in Figure 8.6. Two arms of the bridge contain the unknown impedance

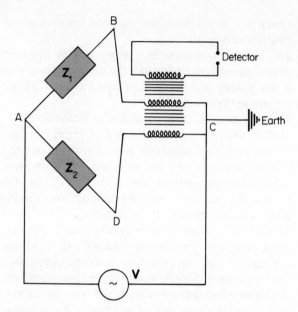

Figure 8.6. The transformer ratio-arm bridge.

Z_1 and a variable impedance Z_2 to be compared with the unknown. In the other arms are two identical coils, closely wound on the same iron core*. The detector is across yet a third coil wound on the core. The balance condition with no voltage across the detector occurs when the magnetic flux through the core is zero. We are now meeting a new situation, with e.m.f.s in the circuit generated by the mutual inductance of the coils wound on the core. These e.m.f.s cannot be dealt with in the same way as e.m.f.s generated by self-inductance, simply by adding to the equivalent circuit an idealized element with a fixed impedance. The mutual inductance generates an e.m.f. in one branch of the network due to changing current in another branch, and it must be included in the equivalent circuit as a voltage generator. Figure 8.7 shows the equivalent circuit of the transformer ratio-arm bridge, including the stray capacitances to earth; the point C on the bridge has been connected to earth. The detector coil has a self-inductance L_D and a mutual inductance M_D with each of the ratio-arm coils. Because the latter are wound in opposite directions, they cause opposing fluxes in the core, and the e.m.f. in the detector circuit is $j\omega M_D(I_1 - I_2)$. At balance the net flux is zero, and $I_1 = I_2$.

The ratio-arm coils each have self-inductance L and resistance R. Their mutual inductance is M and because the coils are closely wound, $L = M$. We have again assumed that the detector draws no current, and the detector

---

* The principle of operation of the bridge is the same if these coils have a different number of turns. For simplicity we restrict the discussion to identical coils.

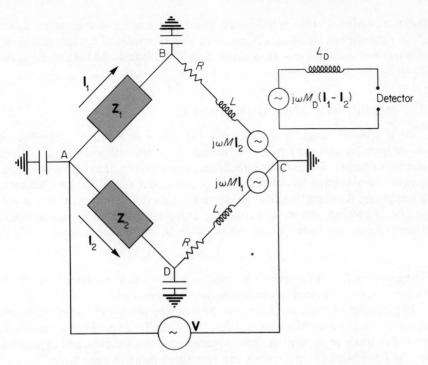

Figure 8.7. The equivalent circuit of the transformer ratio-arm bridge.

circuit therefore does not generate an induced e.m.f. in the bridge circuit. The e.m.f.s due to the self-inductance and the mutual inductance in each of the bridge coils oppose one another, and the total voltage drop across the upper coil in Figure 8.7 is

$$I_1 R + j\omega L I_1 - j\omega M I_2 = I_1 R + j\omega L(I_1 - I_2)$$

$$= I_1 R \text{ at balance.}$$

When the bridge is slightly disturbed from balance, so that I_1 and I_2 change to $(I_1 + \Delta I)$ and $(I_1 - \Delta I)$, the voltage drop is

$$I_1 R + \Delta I(R + 2j\omega L),$$

and for small changes from balance this arm of the bridge behaves like an impedance $(R + 2j\omega L)$. For maximum sensitivity, $(R + 2j\omega L)$ should have a similar magnitude to that of the impedance Z_1, and if the bridge is operated at a frequency at which $R \ll \omega L$, almost the whole of the voltage of the generator appears across Z_1. The voltage between point B on the bridge and the earthed point C is small, and stray capacitance at B carries very little current at balance. Similarly stray capacitance at D draws little current, and in working out the balance condition, it is a good approximation to neglect strays, even if Z_1 and Z_2 are themselves small capacitances. In this symmetrical bridge, application of Kirchhoff's second rule around the path ABCD yields the balance condition $Z_1 = Z_2$.

8.3   IMPEDANCE AND ADMITTANCE

At a specified angular frequency ω there is a well-defined value for the impedance between any pair of junctions in a network consisting of linear passive elements. To find this impedance, one imagines that a source of e.m.f. generating a voltage $V = V_0 \exp(j\omega t)$ is connected across the two junctions. By applying Kirchhoff's rules the currents in all the branches of the network can be calculated, and in particular the current I delivered by the generator. Since the circuit is linear, I is proportional to V, and we can write

$$V = IZ, \quad \text{or} \quad Z = V/I.$$

The quantity Z is the required impedance of the whole network between the two junctions connected to the terminals of the generator.

In general the impedance Z can only be found by solving a set of simultaneous equations derived by the application of Kirchhoff's rules. For elements connected in series or in parallel, these equations are not coupled, and expressions for the impedance of the network can be written down at once.

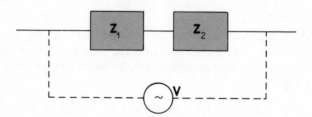

Figure 8.8. Impedances connected in series.

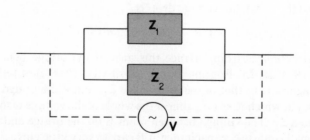

Figure 8.9. Impedances connected in parallel.

(a) *Series circuits*

The two components with impedances Z_1 and Z_2 in Figure 8.8 are connected in series, with no junction between them leading to other components. If a generator of voltage V is connected across the free ends of the components, then by Kirchhoff's first rule, the same current I flows through Z_1 and Z_2. The voltage drop across them is

$$V = IZ_1 + IZ_2 = I(Z_1 + Z_2)$$

The impedance of the pair of components is

$$Z = V/I = Z_1 + Z_2. \tag{8.8}$$

Similarly if any number of components are connected in series, the impedance between the ends of the series circuit is the sum of the impedances of each component separately.

(b) *Parallel circuits*

When the parallel circuit of Figure 8.9 is connected across a generator of voltage V, currents I_1 and I_2 flow through the components Z_1 and Z_2 respectively. By Kirchhoff's first rule the current delivered by the generator is

$$I = I_1 + I_2.$$

The whole of the generator voltage is applied across each component, and

$$V = I_1Z_1 = I_2Z_2.$$

Hence

$$I = V\left\{\frac{1}{Z_1} + \frac{1}{Z_2}\right\},$$

and

$$\frac{1}{Z} = \frac{I}{V} = \frac{1}{Z_1} + \frac{1}{Z_2}. \tag{8.9}$$

The reciprocal of impedance is called *admittance*, and is denoted by the symbol Y. For the parallel circuit,

$$Y = \frac{1}{Z} = \frac{1}{Z_1} + \frac{1}{Z_2} = Y_1 + Y_2. \tag{8.10}$$

Similarly, if any number of components are connected in parallel, the admittance of the circuit is the sum of the admittances of each component separately.

The imaginary generators, shown connected by dotted lines in Figures 8.8 and 8.9, were introduced in order to drive current through the series and parallel circuits, so that we could work out the current through each component and the voltage drop across it. But the impedances derived in Equations (8.8) and (8.9) do not depend on the voltage of the generators; these equations represent the impedances of the circuits whether or not the generators are there. Let us suppose that the impedance between two junctions of some network is $\mathbf{Z}_1$ and that instead of being connected to a source of e.m.f. the two junctions lead to other components forming part of a larger circuit. Provided that the original network has external connections only at these two junctions, it can be replaced by a single component of impedance $\mathbf{Z}$ for the purpose of working out the currents in the rest of the circuit.

When a network consists of a combination of series and parallel connections, its total impedance can easily be found by successively calculating the impedance of larger and larger sections. As an example we shall calculate the impedance at an angular frequency ω between the points A and D of the network in Figure 8.10. First we sum the impedances R_2 and $j\omega L$ in the series section BCD

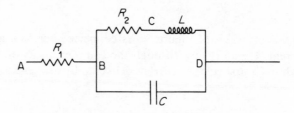

Figure 8.10. A network which is made up entirely of
series and parallel connections.

and replace them with the single impedance $(R_2 + j\omega L)$. This impedance is in parallel with the capacitor, and the total admittance $\mathbf{Y}$ between B and D is the sum:

$$\mathbf{Y} = \frac{1}{R_2 + j\omega L} + j\omega C$$

$$= \frac{1 - \omega^2 LC + j\omega C R_2}{R_2 + j\omega L}.$$

The impedance $\mathbf{Y}^{-1}$ of this section is in series with the resistor R_1, and finally the impedance of the whole network between A and D is

$$\mathbf{Z} = R_1 + \mathbf{Y}^{-1}$$

$$= R_1 + \frac{R_2 + j\omega L}{1 - \omega^2 LC + j\omega C R_2}.$$

Notice that if $\omega C R_2$ is small the denominator of the second term becomes small when $\omega^2 = 1/LC$. The network is resonant, and it has a high impedance near the resonant frequency $\omega_0^2 = 1/LC$ of the series LCR circuit considered in section 7.6. The detailed behaviour of the parallel resonant circuit is left as an exercise at the end of this chapter.

Not all networks can be broken down into series and parallel combinations. The simplest network which will not reduce in this way is the bridge circuit in Figure 8.11. If we wish to find the impedance between A and C we must

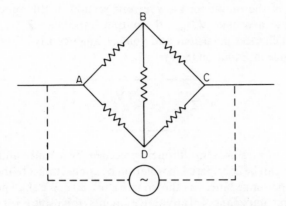

Figure 8.11. A network which cannot be broken up into sections connected in series or in parallel.

connect an imaginary generator and apply Kirchhoff's rules to the whole network, since there are no subsections which are connected to the rest of the network at only two points.

The idea of splitting up electrical circuits into sections is an important one. Most complicated circuits are in fact designed in sections which have separate functions and are more or less independent of one another. In a radio set, for example, the tuner which selects the broadcasting station, the amplifier which drives the loudspeaker and the loudspeaker itself are different sections. Indeed the loudspeaker is often mounted in a separate cabinet, joined to the amplifier only by a lead carrying the signal and a common earth. The impedance of the loudspeaker between the signal lead and earth is called its *input impedance*. Usually an input impedance can also be defined for a circuit which contains active and non-linear components. For example, if an audiofrequency amplifier is not to distort a signal, its response must be linear at the signal frequency. Throughout the amplifier, changes in current at this frequency are proportional to the signal voltage, even if they are superimposed on non-linear steady currents. The effective input impedance for the signal is

$$\mathbf{Z}_{input} = \frac{\partial \mathbf{V}_{input}}{\partial \mathbf{I}_{input}},$$ (8.11)

where V_{input} and I_{input} are the voltage and current at the input lead. For the purpose of analysing the behaviour of a second circuit connected to the input of the original one, the whole of the original circuit may be represented by a single element with impedance Z_{input}.

When we are studying a single section of a large circuit, it is not only necessary to include the input impedance of the following section, but also to account for the effect of the output of the previous section. If this output delivers a current I_{output} at a voltage V_{output}, the output impedance Z_{output} is defined in terms of small changes in current and voltage, analogously to the definition of input impedance in Equation (8.11):

$$Z_{output} = -\frac{\partial V_{output}}{\partial I_{output}}. \tag{8.12}$$

The amplifier in a radio set, although preceded by a tuner and followed by a loudspeaker, can be considered in isolation by connecting it to the appropriate output and input impedances as shown in Figure 8.12. We shall not discuss the input and output impedances of any active circuits; this topic is left to specialized books on electronics. In the remaining sections of this chapter, however, input and output impedances are calculated for some important passive circuits.

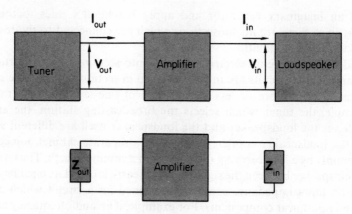

Figure 8.12. Replacing preceding and following circuits by their output and input impedances.

8.4   FILTERS AND DELAY LINES

Two sections of an A.C. network are often linked by a capacitor, which transmits alternating voltages while allowing steady voltage differences to be maintained between the sections. Let us consider the behaviour of the network in Figure 8.13 at different frequencies. Across the input terminals 1 and 2 there is a voltage V_{in} of angular frequency ω. This voltage is modified by the network and a voltage V_{out}, also of angular frequency ω, appears across the output terminals 3 and 4. The ratio $T = V_{out}/V_{in}$ is called the *voltage transfer function* of the network. The value of T depends on the input impedance of any circuit to which terminals 3 and 4 are connected: we shall assume that this input impedance is so high that a negligible current is drawn from the network, and the transfer function is practically the same as if the network were open-circuited. The network then acts as a voltage divider, and since the impedance of the capacitance C in Figure 8.13 is $1/j\omega C$, we have

$$T = \frac{V_{out}}{V_{in}} = \frac{R}{R + 1/j\omega C} = \frac{j\omega CR}{1 + j\omega CR}. \tag{8.13}$$

In passing through the network the input voltage is attenuated and also suffers a phase change. Looking at Equation (8.13), we see that at frequencies low enough so that $\omega CR \ll 1$, T is approximately $j\omega CR$. The output voltage is therefore heavily attenuated, and is in advance of the input by a phase angle $\pi/2$. At high frequencies for which $\omega CR \gg 1$, on the other hand, the transfer function is almost unity. The network thus acts as a *high-pass filter*, transmitting high-frequency signals and attenuating those of low frequency.

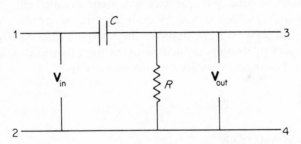

Figure 8.13. A simple high-pass filter network.

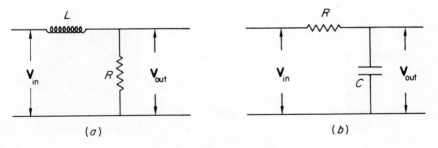

Figure 8.14. Low-pass filters.

Similarly *low-pass filters* can be made with the simple voltage divider networks shown in Figure 8.14. The open circuit voltage transfer functions for these networks are

$$\mathbf{T} = \frac{1}{1 + j\omega L/R} \qquad (8.14a)$$

for the circuit (*a*) in Figure 8.14, and

$$\mathbf{T} = \frac{1}{1 + j\omega CR} \qquad (8.14b)$$

for the circuit (*b*). Both these transfer functions have the same form of dependence on ω. At low frequencies $\mathbf{T}$ approaches unity, and the network passes an alternating voltage without modification. At high frequencies, when $\omega L \gg R$ or $\omega CR \gg 1$, as the case may be, the output is attenuated and lags behind the input by the phase angle $\pi/2$. Notice that the input impedances of the two circuits are not the same; (*a*) is inductive and (*b*) is capacitive.

Because of their simplicity, voltage dividers are very often used as filters when it is acceptable that the transfer function changes slowly with frequency. But sometimes filters must be designed to meet the more critical requirement that the transfer function should change rapidly from small values to nearly unity over a narrow band of frequencies near some specified cut-off frequency. A sharper cut-off can be achieved by adding more components to the filter network. For example, we can improve the low-pass filter in Figure 8.14(*a*) by using a capacitor to by-pass the resistor at high frequencies, as shown in Figure 8.15(*a*). The input impedance of this network is

$$\mathbf{Z} = j\omega L + \frac{1}{j\omega C + 1/R}.$$

Expanding in powers of ωCR,

$$\mathbf{Z} = R + j\omega(L - R^2 C) + \cdots$$

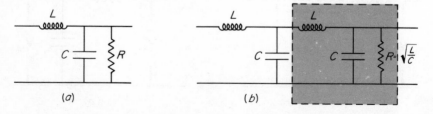

(a) (b)

Figure 8.15. A low-pass filter made with a ladder network of similar sections.

If we choose the component values so that $L = R^2C$, i.e. $R = \sqrt{LC}$, then to first order in ω the input impedance is R, the same as if the inductor and capacitor had not been included in the circuit. Now look at the circuit in Figure 8.15(b). The shaded part of the circuit within the dotted lines is just the circuit (a) and at low frequencies is equivalent to a resistance $R = \sqrt{L/C}$. Replacing the dotted section by a resistance $\sqrt{L/C}$ we are back to circuit (a), and evidently the input impedance of circuit (b) is also $\sqrt{L/C}$ at low frequency. Thus we can add more and more LC sections, whilst the low frequency input impedance remains the same. A network of repeating units like this is called a *ladder* network. When the input impedance of a ladder network at a particular frequency is independent of the number of sections, the network is said to be *matched* at this frequency, and its input impedance is called the *characteristic impedance* for this frequency. Thus at low frequencies $\sqrt{L/C}$ is the characteristic impedance of the LC ladder.

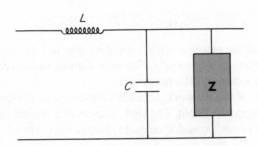

Figure 8.16. An additional section added to a low-pass filter with input impedance $\mathbf{Z}$.

To find out how efficient the ladder is as a low-pass filter, we must work out its transfer function at different frequencies. Imagine for simplicity that the ladder is matched at all frequencies, i.e. it is terminated by an impedance $\mathbf{Z}$ which is equal to the characteristic impedance of the ladder at all frequencies. The impedance $\mathbf{Z}$ is approximately equal to $\sqrt{L/C}$ at low frequencies, but it includes terms in ω^2 and higher powers of ω which become important at high frequencies. The input impedance of the network in Figure 8.16, formed by adding a single section to an existing ladder, must also be $\mathbf{Z}$:

hence
$$\mathbf{Z} = j\omega L + \frac{\mathbf{Z}}{1 + j\omega C\mathbf{Z}}. \tag{8.15}$$

The voltage transfer function of the new section is

$$\mathbf{T} = \left(\frac{1}{j\omega C + 1/\mathbf{Z}}\right)\Big/\mathbf{Z} = \frac{1}{1 + j\omega C\mathbf{Z}}.$$

Hence

$$j\omega C\mathbf{Z} = \frac{1}{\mathbf{T}} - 1.$$

From (8.15),

$$j\omega C\mathbf{Z} = -\omega^2 LC + \frac{j\omega C\mathbf{Z}}{1 + j\omega C\mathbf{Z}},$$

or

$$\frac{1}{\mathbf{T}} - 1 = -\omega^2 LC + 1 - \mathbf{T}.$$

Rearranging,

$$\mathbf{T}^2 + \mathbf{T}(\omega^2 LC - 2) + 1 = 0.$$

The solution is

$$\mathbf{T} = 1 - \tfrac{1}{2}\omega^2 LC + \sqrt{(1 - \tfrac{1}{2}\omega^2 LC)^2 - 1}. \tag{8.16}$$

(We have ignored the solution with a negative sign in front of the square root: this solution represents a reversal of the role of input and output, when a signal travels backwards through the network.)

The behaviour of the network depends on the sign of the quantity under the square root in Equation (8.16). The sign changes at a cut-off angular frequency $\omega_c = 2/\sqrt{LC}$. Let us consider angular frequencies above and below this cut off.

(i) $\omega < \omega_c$.

$$\mathbf{T} = 1 - 2\left(\frac{\omega}{\omega_c}\right)^2 - j\sqrt{1 - \left[1 - 2\left(\frac{\omega}{\omega_c}\right)^2\right]^2}.$$

or

$$\mathbf{T} = e^{-j\theta} = \cos\theta - j\sin\theta, \tag{8.17}$$

where

$$\cos\theta = 1 - 2\left(\frac{\omega}{\omega_c}\right)^2. \tag{8.18}$$

The alternating voltage is transmitted without attenuation but in passing through each section of the network its phase is changed by the angle θ.

(ii) $\omega > \omega_c$.

$$\mathbf{T} = 1 - 2\left(\frac{\omega}{\omega_c}\right)^2 + \sqrt{\left[1 - 2\left(\frac{\omega}{\omega_c}\right)^2\right]^2 - 1}. \tag{8.19}$$

The transfer function is real and it lies between (-1) and 0. The voltage signal is therefore attenuated and reversed in phase by each section.

The modulus and phase of **T** are sketched in Figure 8.17. The angular frequency range up to the cut-off at ω_c is called the *pass-band* of the network, and the range above ω_c is the *stop-band*. In practice it is not possible to construct

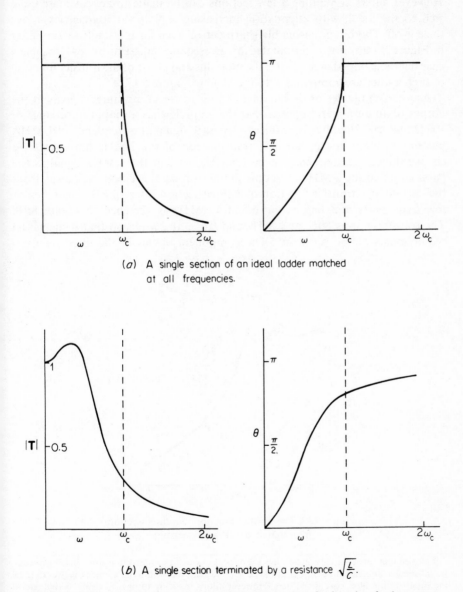

(*a*) A single section of an ideal ladder matched at all frequencies.

(*b*) A single section terminated by a resistance $\sqrt{\frac{L}{C}}$.

Figure 8.17. The magnitude and phase of the voltage transfer function for low-pass filters.

an ideal filter with no attenuation at all in the pass-band, because real coils always have some resistance, unlike the ideal inductances we have used in the equivalent circuits. Further, it is not possible to make a component with impedance equal to the characteristic impedance of the ladder network at all frequencies; such a component is required to terminate an ideal filter. However, filters containing a few sections can be made to approximate quite well to the ideal, with attenuation increasing rapidly for frequencies above the cut-off. There is a reasonably sharp cut-off even for the single-section filter in Figure 8.14(a) if it is terminated by a resistance equal to its low-frequency characteristic impedance $\sqrt{L/C}$: the modulus and phase of the transfer function of such a filter are compared with the ideal in Figure 8.17.

As an example, let us design an LC filter to reject unwanted signals at the output of an amplifier. Suppose that the amplifier has an input impedance of $100\,\Omega$ and that the filter is required to pass a signal at a frequency of 1 MHz, but not to allow more than 1% transmission of the 2 MHz harmonic. We choose the cut-off frequency to be 1.5 MHz, so that the 1 MHz signal in the pass-band and the 2 MHz harmonic in the stop-band are both well away from the cut-off. The transfer function of a three-section filter which is matched at low frequencies and has its cut-off at 1.5 MHz is sketched in Figure 8.18. The transmission is 80% at 1 MHz and 0.6% at 2 MHz, so that it meets the requirements of the problem*. The component values of the filter are fixed

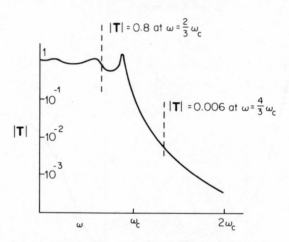

Figure 8.18. The transfer function of a three-section
filter matched at low frequencies.

* Notice that this transfer function has sharp peaks, which must lead to distortion of non-sinusoidal signals. Somewhat better performance may be achieved with more complex networks than the equal-section filter considered here. However, filters made up of passive elements (inductors, capacitors and resistors) always have undesirable peaks in their transfer functions. Such peaks may only be eliminated by the use of active filters, which incorporate amplification and feedback.

by the input impedance of the amplifier ($100\,\Omega$) and the cut-off frequency (1.5 MHz), since we have

$$\sqrt{\frac{L}{C}} = 100,$$

and

$$\omega_c = \frac{2}{\sqrt{LC}} = 2\pi f_c = 3\pi \times 10^6.$$

The solution of these two equations is $L \simeq 22\,\mu\text{H}$, $C \simeq 2200\,\text{pF}$.

Delay lines

A signal in the pass-band of an ideal matched low-pass filter is not attenuated, but it is changed in phase. The phase lag θ introduced by a single section may be found from Equation (8.18)

$$\cos\theta = 1 - 2\left(\frac{\omega}{\omega_c}\right)^2.$$

Making the expansion $\cos\theta = 1 - \frac{1}{2}\theta^2 + \frac{1}{24}\theta^4 + \cdots$ we see that provided $\frac{1}{24}\theta^4$ is small, $\theta \simeq 2\omega/\omega_c$. Each section introduces the same phase shift θ, and the alternating voltage at the input is progressively delayed in phase as it passes along the network. Associated with the phase shift is a time delay; a phase lag of 2π radians corresponds to a time delay equal to the period $2\pi/\omega$ of the alternating voltage. Because the phase shift introduced by each section of the network is proportional to angular frequency for angular frequencies well below cut-off, the time-delay

$$\frac{\theta}{\omega} = \frac{2}{\omega_c} = \sqrt{LC} \tag{8.20}$$

is constant. Any signal made up of a superposition of alternating voltages at low frequencies is delayed by the same time in passing through the filter network. Since the low frequencies are not attenuated, the network acts as a *delay line*, delaying the signal without distorting it.

In practice this kind of delay line is only satisfactory if it does not contain very many sections. The delay of each section is $2/\omega_c = 1/\pi f_c$, and the maximum delay which can be introduced by a line made up of a few sections is of the order of the period of its cut-off frequency. Looking at the transfer function for the three-section line in Figure 8.18, one sees that the transmitted amplitude begins to change rapidly with frequency at about $\omega/\omega_c = \frac{1}{2}$. Above this frequency the time delay also ceases to be constant. For this filter $1/\pi f_c = 1/(1.5\pi)\,\mu\text{s}$, and in passing through all three sections, low frequencies are delayed by $2/\pi = 0.64\,\mu\text{s}$. The delay at $\frac{1}{2}f_c = 0.75\,\text{MHz}$ is $0.67\,\mu\text{s}$, nearly the same as the low-frequency value, but the delay at 1 MHz is 0.81 μs. Non-sinusoidal signals including frequency components as high as 1 MHz will suffer considerable distortion in passing through the network.

High-pass and band-pass filters

As well as being used as low-pass filters and delay lines, ladder networks can be made up into high-pass or *band-pass* filters. (A band-pass filter is one which passes signals in a band between two cut-off frequencies, and attenuates signals outside the band.) The analysis of these filters proceeds along the same lines as for the low-pass filter, and we shall not present it here. Networks which act as high-pass and band-pass filters are shown in Figures 8.19 and 8.20.

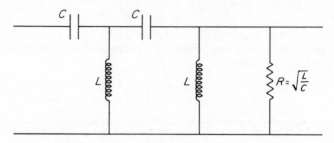

Figure 8.19. A high-pass filter with characteristic impedance $R = \sqrt{L/C}$ at high frequency.

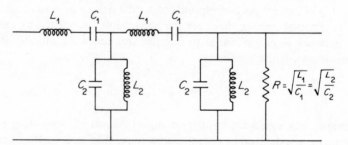

Figure 8.20. A band-pass filter. The cut-off frequencies are at

$$\omega = \sqrt{\frac{1}{L_1 C_1} + \frac{1}{L_2 C_2}} \pm \frac{1}{\sqrt{L_1 C_1}}$$

and the characteristic impedance is R near the centre of the pass-band.

8.5 TRANSFORMERS

A transformer consists of two or more coils wound on the same core. The core is made of a high permeability material such as soft iron and its function is to increase the magnetic flux linking the coils, hence increasing the mutual inductance between each pair of them. When an alternating current is passed through one of the coils, the e.m.f. induced in the other is thus made as large as possible. Depending on the number of turns in the coils, the induced e.m.f. may be many times larger or smaller than the e.m.f. driving the transformer. If the induced e.m.f. is now allowed to drive a load, electrical power can be very efficiently transferred from the generator to the load circuit. The transformer thus provides a convenient way of changing the amplitude of an alternating e.m.f.

At first sight it may seem surprising that we should think of treating a transformer as a *linear* component, since the magnetic field in a ferromagnetic material is not linearly related to the current generating the field. However, the magnitudes of the currents in the transformer coils are not very sensitive to the magnetic properties of the core material, provided only that it does have a high permeability. The behaviour of the transformer is in fact determined mainly by the ratio of the numbers of turns in its coils. We can relate the currents and voltages in the coils by approximate expressions in which the turns ratio is the only transformer parameter. These expressions—which are linear—are very simple, and since they illustrate several important properties of transformers it is useful to imagine an 'ideal' transformer for which they hold exactly. The ideal transformer is discussed in detail in the next section, some applications of transformers are described in section 8.5.2 and then in section 8.5.3 we shall explain the ways in which real transformers depart from the ideal.

8.5.1 The ideal transformer

An ideal transformer is one which has perfect magnetic coupling between its coils, and which dissipates no energy. The coils must therefore have zero resistance, and no heat losses occur within the core itself. The core of an ideal transformer is assumed to have a very high permeability indeed, yet not to exhibit saturation. This ensures that the self-inductance of each coil is enormous, and also that the core traps all the magnetic flux in the transformer. In the ideal toroidal transformer in Figure 8.21 this is represented by drawing a high density of field lines in the core and none outside. The field lines have neither beginning nor end (because there are no free magnetic poles) and each line makes a complete circuit within the core. All the lines pass through any cross-section of the core—in other words the whole of the magnetic flux is carried by the core and threads both the coils wound on it.

The left-hand coil in Figure 8.21, wound with N_1 turns, is connected to an external source of e.m.f. which generates an alternating voltage $\mathbf{V}_1$ across its terminals. It is called the *primary* coil. The other coil—the secondary coil— has N_2 turns and is connected to a load with impedance $\mathbf{Z}_2$. By applying Faraday's law to the two coils, we can now find the induced e.m.f. $\mathbf{V}_2$ appearing across the terminals of the secondary coil. Exactly the same flux, of magnitude $\mathbf{\Phi}$, say, passes through each turn of both coils. An induced e.m.f. $(-d\mathbf{\Phi}/dt)$ is induced around each turn, and since there is no resistance, the induced

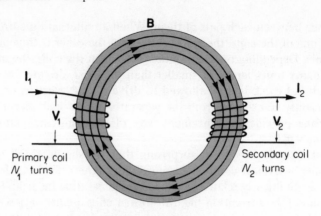

Figure 8.21. An ideal transformer.

e.m.fs balance the e.m.fs across the terminals. Thus

$$\mathbf{V}_1 - N_1 \frac{d\mathbf{\Phi}}{dt} = 0, \tag{8.21}$$

and

$$\mathbf{V}_2 - N_2 \frac{d\mathbf{\Phi}}{dt} = 0. \tag{8.22}$$

Eliminating $d\mathbf{\Phi}/dt$, it follows immediately that

$$\mathbf{V}_2 = \left(\frac{N_2}{N_1}\right)\mathbf{V}_1 = n\mathbf{V}_1, \tag{8.23}$$

writing n for the turns ratio N_2/N_1. The output voltage of the secondary is in phase with the primary voltage, and the ratio of output to input amplitudes is the turns ratio N_2/N_1.

The currents $\mathbf{I}_1$ and $\mathbf{I}_2$ flowing in the primary and secondary circuits are related by Ampère's Law to the line integral of $\mathbf{B}$ around a circuit threading the coils. Consider a circuit of length l lying entirely within the core. If the core has a uniform cross-section A, the magnetic field is nearly uniform over any cross-section, and has the average magnitude $B = \Phi/A$. In applying Ampère's Law to this circuit, we must remember that the secondary current is flowing in such a direction as to oppose the flux changes caused by the primary current. The two currents are therefore flowing in the opposite sense around the lines of $\mathbf{B}$ and

$$N_1\mathbf{I}_1 - N_2\mathbf{I}_2 = \frac{\int \mathbf{B} \cdot d\mathbf{l}}{\mu\mu_0}$$

$$= \frac{\mathbf{\Phi}l}{\mu\mu_0 A} \approx 0, \tag{8.24}$$

since μ is indefinitely large. Practically no current can flow in the primary when the secondary is open-circuited, because the self-inductance of the primary coil of an ideal transformer is so large. When the transformer is on load, the fluxes generated by the primary and secondary coils almost exactly cancel each other. This is why the transformer behaves as a linear circuit element in spite of the non-linear properties of its ferromagnetic core. Ignoring the small term on the right-hand side of Equation (8.24), we find

$$I_2 = \left(\frac{N_1}{N_2}\right)I_1 = \frac{I_1}{n}. \tag{8.25}$$

The output and input currents are in phase, and the ratio of their amplitudes is the inverse of the turns ratio N_2/N_1.

The value of the currents depends on the load impedance. In the secondary circuit,

$$V_2 = I_2 Z_2,$$

and hence from Equations (8.23) and (8.25),

$$V_1 = I_1\left(\frac{N_1^2}{N_2^2}Z_2\right) = \frac{I_1 Z_2}{n^2}.$$

To the primary circuit, the transformer looks like an impedance Z_2/n^2, i.e. it has an input impedance

$$Z_{in} = \frac{Z_2}{n^2}. \tag{8.26}$$

Similarly, if the output impedance of the circuit driving the primary coil is Z_1, the output impedance of the transformer is

$$Z_{out} = n^2 Z_1. \tag{8.27}$$

(It is left to the reader to show that Equation (8.27) is correct.)

In the analysis of the primary circuit, the transformer is simply replaced by a component with impedance Z_{in}. In the secondary it must be represented by source of e.m.f. in series with the impedance Z_{out}. Thus the equivalent circuit for an ideal transformer driven by a generator of voltage V and internal impedance Z_1 is as shown in Figure 8.22.

Note that since the ideal transformer has no losses, all the power delivered to the primary coil is transferred to the secondary circuit. The primary and secondary voltages are in phase with one another, and so are the primary and secondary currents. The power factors in the two circuits are the same, with a value $\cos \theta$ which is determined by the nature of the load. The power output

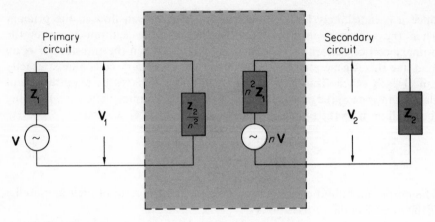

Figure 8.22. The equivalent circuit for an ideal transformer.

of the secondary is

$$V_2 I_2 \cos \theta = (n V_1)\left(\frac{I_1}{n}\right) \cos \theta$$

$$= V_1 I_1 \cos \theta$$

$$= \text{power delivered to the primary.}$$

8.5.2 Applications of transformers

The equations we have derived for an ideal transformer are good approxima-tions to the behaviour of a real one. The transformer is aptly named, since we have found that it will pass electrical power from one circuit to another, while transforming voltage, current and impedance. In this section some applications of transformers are briefly described to illustrate the usefulness of each of their transformations.

(i) *Voltage transformation.* This is the most common and obvious application of transformers. Electrical power is normally obtained from a main supply at 110 or 240 volts, and in any circuit for which the mains voltage is unsuitable, a transformer is usually to be found. For example, a mains transformer will be used to drive the rectifier in a high-voltage D.C. power supply. This trans-former requires more turns in the secondary coil than in the primary, and since it increases the voltage, is referred to as a *step-up* transformer. Similarly a *step-down* transformer is used if a 6.3 volt supply is wanted for the filament of a thermionic tube. Although the turns ratio in the 240 V–6.3 V transformer is very small, it does not follow that the secondary coil is much smaller than the primary. Power is dissipated in the windings and in the core of a real transformer: the power loss is small compared with the power transferred from primary to secondary, but the maximum permissible temperature rise of the transformer will nevertheless limit the amount of power it can transfer. In a step-down transformer the secondary has to carry a higher current than the primary, and must be wound with thicker wire to reduce the resistive heating. The current density, and hence the power density, are the same in the primary and secondary coils if they contain the same volume of conductor. Whenever transformers are to handle appreciable amounts of power, it is most economical to have equal heat losses in the primary and secondary coils, which must therefore be made about the same size.

(ii) *Current transformation.* Electrical power is transmitted over large distances by conductors carried high above the ground on their ungainly pylons. The conductors must be so high because they operate at a high voltage— 400 kV for the British grid—and the main reason for choosing a high voltage is to make the current as small as possible for a given power, in order to minimize the resistive losses in the conductors. At generating stations, huge amounts of power are converted to low current and high voltage by transformers, and it is obviously important that the transformer losses should be small. In this respect big transformers approach quite closely to the ideal, with losses of only about 0.2% of the input power.

(iii) *Impedance transformation.* There are many kinds of circuit which will only function efficiently if they are connected to a load whose impedance has a particular value. For example, the delay lines described in section 8.4 transmit all signals containing frequencies below cut-off only if they are matched by a terminating load equal to their low-frequency characteristic impedance. It may happen that the impedance of the load is such that it is impracticable to construct a delay line to match it. Thus if the load impedance were $1\,\Omega$, the inductance per section of a matching line with a cut-off at 1.5 MHz would have to be a small fraction of a microhenry, less than the stray inductances always present in the circuit. In this case a step-down transformer with a turns ratio of $1:10$ would convert the load impedance to $100\,\Omega$, and match it to the line designed in section 8.4.

Impedances also may need to be matched in circuits designed to transfer power. If an A.C. generator of R.M.S. voltage V and output impedance R_1 (purely resistive) is driving a load resistance R_2, then the R.M.S. current in the circuit is $I = V/(R_1 + R_2)$. The power delivered to the load is $I^2 R_2 = V^2 R_2/(R_1 + R_2)^2$, which has its maximum value at $R_1 = R_2$, when the same power is dissipated in the generator and the load. The output impedance of the audio-amplifier in radio or record-playing equipment is sometimes much greater than the input impedance of the loudspeaker, which is often $3\,\Omega$. The amplifier and speaker are then matched with a transformer to achieve the maximum sound output. Audio transformers are available which introduce little distortion over the whole range of audible frequencies, but in modern hi-fi equipment amplifiers and speakers are usually matched directly without an intermediate transformer. Transformers are still to be found at the other end of the recording process, where the tiny power of the microphone is generated at an inconveniently low output impedance.

Isolating transformers

Another useful property of the transformer is that the primary and secondary circuits may be completely insulated from one another. An isolating transformer can therefore transfer A.C. signals between two circuits which are at different D.C. levels. If the coils of an ideal isolating transformer have the same number of turns, it will change the D.C. level without affecting the A.C. signal at all; voltage, current and impedances are all unaltered.

★ **8.5.3 Real transformers**

None of the characteristics assumed for an ideal transformer can be achieved in practice. In a real transformer

(i) the magnetic coupling between the coils is not perfect;
(ii) there are power losses in the coil windings and in the core;
(iii) at high magnetic fields the core material exhibits saturation.

The way in which each of these factors affects the performance and design of transformers will be considered in turn.

Leakage flux

In spite of the high permeability of the soft iron used in transformer cores, the flux linkage between the coils is never quite complete. The self-inductances L_1 and L_2 of the primary and secondary coils are related to their mutual inductance M by Equation (6.25):

$$M^2 = k^2 L_1 L_2.$$

Here the coupling coefficient k represents the fraction of the flux through one of the coils which threads the other. The remaining fraction $(1 - k)$ of the flux does not link the coils and does not contribute to the e.m.f. induced in the secondary. To find what effect this leakage flux has, we shall rewrite the transformer equations in terms of the self- and mutual-inductances, still assuming that the coils have zero resistance. The current in the secondary flows so as to oppose the flux changes caused by the primary, and for the angular frequency ω, the primary and secondary voltages are

$$\mathbf{V}_1 = L_1 \frac{d\mathbf{I}_1}{dt} - M \frac{d\mathbf{I}_2}{dt} = j\omega(L_1 \mathbf{I}_1 - M\mathbf{I}_2), \qquad (8.28)$$

and

$$\mathbf{V}_2 = j\omega(M\mathbf{I}_1 - L_2 \mathbf{I}_2) = \mathbf{I}_2 \mathbf{Z}_2 \qquad (8.29)$$

when the secondary is driving a load $\mathbf{Z}_2$.

From Equation (8.29),

$$I_1 = \frac{(j\omega L_2 + Z_2)I_2}{j\omega M}.$$

Substituting for I_2 in Equation (8.28), we find the input impedance of the transformer to be

$$Z_{in} = \frac{V_1}{I_1} = j\omega L_1 - \frac{(j\omega M)^2}{j\omega L_2 + Z_2}$$

$$= j\omega L_1\left(1 - \frac{M^2}{L_1 L_2}\right) + \frac{j\omega Z_2 M^2/L_2}{j\omega L_2 + Z_2}$$

$$= j\omega L_1(1 - k^2) + \frac{j\omega k^2 Z_2 L_1}{j\omega L_2 + Z_2}.$$

In the second term we now put $L_2 = n^2 L_1$. The quantity n would be exactly equal to the turns ratio if there were no leakage flux, and in practice n hardly differs from the turns ratio. The input impedance becomes

$$Z_{in} = j\omega L_1(1 - k^2) + \frac{1}{(n^2/k^2 Z_2) + (1/j\omega k^2 L_1)}, \qquad (8.30)$$

the same as the input impedance of the equivalent circuit shown in Figure 8.23.

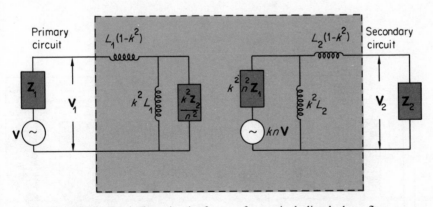

Figure 8.23. Equivalent circuit of a transformer including leakage flux.

By eliminating I_1 and I_2 from Equations (8.28) and (8.29), the secondary voltage is found to be

$$V_2 = knV_1\left\{\frac{Z_2}{Z_2 + j\omega L_2(1 - k^2)}\right\}.$$

On open circuit the output voltage is $kn\mathbf{V}_1$, smaller by the factor k than for an ideal transformer. The input voltage $\mathbf{V}_1$ itself depends on the output impedance $\mathbf{Z}_1$ of the generator driving the transformer, and if this generator develops a voltage $\mathbf{V}$, then

$$\mathbf{V}_1 = \frac{\mathbf{Z}_{in}\mathbf{V}}{\mathbf{Z}_1 + \mathbf{Z}_{in}},$$

and

$$\mathbf{V}_2 = kn\mathbf{V}\left\{\frac{\mathbf{Z}_{in}}{\mathbf{Z}_1 + \mathbf{Z}_{in}}\right\}\left\{\frac{\mathbf{Z}_2}{\mathbf{Z}_2 + j\omega L_2(1 - k^2)}\right\}. \tag{8.31}$$

The equivalent circuit which has this output voltage is shown on the right-hand side of Figure 8.23. As one would expect, the primary and secondary circuits look very similar. Their impedances are the same except that the labels 1 and 2 are interchanged, and in the secondary circuit the turns ratio n is inverted. The coupling coefficient k must be very near to unity, so that the series inductance $L_2(1 - k^2)$ in the output circuit is small compared to the load impedance $\mathbf{Z}_2$. High coupling is often achieved by winding the coils on top of one another, as in the transformer sketched in Figure 8.24, which has a core carrying flux symmetrically around both sides of the coils.

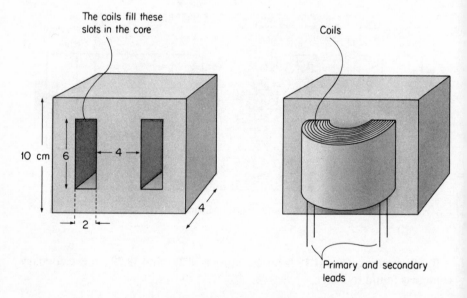

Figure 8.24. Transformer coils are wound together on the same part of the core in order to minimize leakage flux.

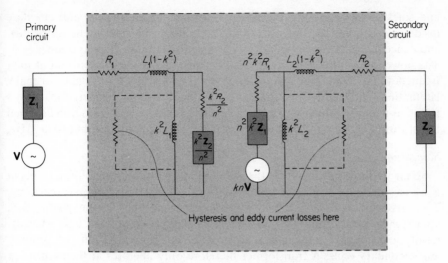

Figure 8.25. Equivalent circuit for a transformer, including losses.

Losses

(a) *Coil resistance.* The resistance of the coils is easily included in the equivalent circuit by adding series resistors R_1 and R_2 as in Figure 8.25. The primary resistance alters the output impedance of the transformer, requiring a resistor $n^2k^2R_1$ to be included in the secondary: similarly a resistor k^2R_2/n^2 appears in the primary circuit. These resistors affect the low-frequency perform-ance of the transformer. Even on open-circuit, the output voltage of the trans-former falls unless the angular frequency $\omega \gg R_1/L_1$ and $\omega \gg R_2/L_2$. The resistance per unit inductance of the coils of an audio-frequency transformer is typically about 20 ohms/henry, leading to a frequency $R_1/2\pi L_1$ of about 3 Hz. At 3 Hz the performance is very poor, but a good audio transformer will operate at 30 Hz with a power loss of no more than 3 dB.

(b) *Hysteresis losses.* Once in every cycle of the alternating current, the iron of the transformer core is taken around its hysteresis curve. An amount of energy proportional to the area of the hysteresis curve is dissipated during each cycle. The loss is not simply proportional to the voltage, since the shape of the hysteresis curve changes as the maximum flux is altered. However, at a fixed frequency the hysteresis loss can be roughly represented in the equivalent circuit by a resistor in parallel with the transformer inductance.

(c) *Eddy currents.* There are induced e.m.f.s in the transformer core, which cause currents—called eddy currents—to flow within the core, leading to a further dissipation of energy. The eddy current losses can be reduced by dividing the core into thin laminations which are insulated from one another, and stacked so as to stop the flow of current. Small currents still circulate inside each lamination; for a fixed flux amplitude the rate of change of flux, and hence the eddy current density, is proportional to the angular frequency ω. The power losses thus vary as ω^2, and become increasingly important at high frequency. For very high frequency operation it is not practicable to make laminations sufficiently thin, and cores are then made of powdered iron or of ferrite, a material which has a much higher resistivity than iron. Small-signal transformers with ferrite cores can be used at frequencies up to about 1000 MHz.

Saturation

So far we have assumed that the self- and mutual-inductances of the coils in a transformer are constant, independent of the currents they are carrying. This assumption breaks down completely when the ferromagnetic core approaches saturation. The incremental permeability then falls sharply, resulting in smaller inductances and poorer coupling between the primary and secondary coils. A transformer therefore only operates at high efficiency when carrying currents below those which saturate the core. Suppose that B_m is the maximum field which is to be allowed in a particular transformer core. If the core has a cross-sectional area A_c, then the maximum flux is $\Phi_m = A_c B_m$. The flux is related to the voltage $\mathbf{V}_1$ on the primary through Equation (8.21):

$$\mathbf{V}_1 - N_1 \frac{d\mathbf{\Phi}}{dt} = 0.$$

For an angular frequency ω, the maximum permissible R.M.S. value of the primary voltage is therefore

$$V_m = \frac{1}{\sqrt{2}} \omega N_1 \Phi_m = \frac{1}{\sqrt{2}} \omega N_1 A_c B_m.$$

For a given core size, the value of V_m can be increased by increasing the number of turns on the primary, at the expense of increasing the coil resistance and hence its power dissipation. The power dissipation is limited by the temperature rise which can be tolerated, and in practice this leads to the stipulation that the current density in the windings should not exceed some value j_m. If the total area of conductor in a cross-section of the primary coil is A_p, then the maximum current it can carry is

$$I_m = \frac{j_m A_p}{N_1}.$$

Hence

$$V_m I_m = \frac{1}{\sqrt{2}} \omega j_m B_m A_c A_p. \tag{8.32}$$

The same argument applies to the secondary coil, and since the VI products are almost equal in primary and secondary, the secondary coil should have about the same area of conductor as the primary.

Equation (8.32) relates the size of a transformer to its power rating $V_m I_m$. We shall illustrate this by working out the rating of a 50 Hz transformer wound on the core shown in Figure 8.24. The core is made of iron with the hysteresis curve drawn in Figure 6.11, and it is to be operated within the minor hysteresis loop in the figure, so that $B_m \sim 1\,T$. We shall choose j_m to be 3 amps/mm^2, a typical value for transformers about this size. Allowing a third of the area of the window in the transformer core to be taken up by insulation, $A_p = 4\,cm^2$, and substituting in Equation (8.32),

$$V_m I_m = \sqrt{2}\pi \times 50 \times 3 \times 10^6 \times 1 \times 16 \times 10^{-4} \times 4 \times 10^{-4}$$

$$\simeq 250 \text{ volt-amperes.}$$

(The rating of a transformer is always quoted in volt-amperes instead of watts.)

The resistivity of copper is $1.8 \times 10^{-8}\,\Omega\,m$, and the power density in the windings at full current is $\rho j_m^2 \simeq 0.16$ watts/c.c. Assuming rectangular winding of the coils, the total volume of copper is 128 c.c., leading to a maximum power dissipation of 20 watts. This is much bigger than the hysteresis losses, which from the area of the hysteresis loop in Figure 6.11 are 3×10^{-3} watts/cm^3 at 50 Hz, or just over one watt for this transformer. The efficiency of the transformer when operated at its maximum rating is thus a little over 90%, a typical value for this power range. Higher efficiencies can be achieved where necessary by increasing the size of the transformer, and in generating stations handling thousands of megawatts of power, the transformers have efficiencies of about 99.8%.

PROBLEMS 8

8.1 Twelve resistors, each of resistance one ohm, are joined to form the edges of a cube. What is the resistance between opposite corners of the cube? (First decide which of the resistors are equivalent when a current is flowing between the two corners, and then apply Kirchhoff's rules.)

8.2 The network in Figure 8.26 is an equivalent circuit to the valve circuit called the anode follower. Express the output voltage V_{out} appearing across the load resistance R_L

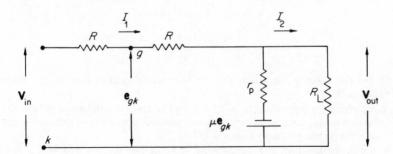

Figure 8.26. Equivalent circuit for the anode follower. The e.m.f. μe_{gk} generated by the battery is related to the voltage e_{gk} appearing between the points g and k.

as a function of the load current I_2, and hence find the output impedance of the anode follower.

8.3 The circuit in Figure 8.27 represents a coil and a capacitor placed in parallel across an alternating voltage **V**. Show that the circuit exhibits resonance and that if $L \gg CR^2$ the amplitude of the current drawn from the source of e.m.f. goes through a minimum value near the angular frequency $\omega_0 = 1/\sqrt{LC}$. What is Q for this circuit?

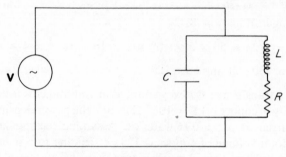

Figure 8.27.

8.4 The bridge in Figure 8.28 is a phase-shifting network. Show that if the generator voltage **V** has a constant amplitude, the voltage across XY has constant amplitude but varies in phase as R alters. Sketch the phase variation as R changes from 0 to ∞.

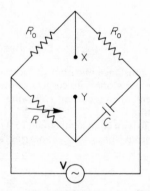

Figure 8.28.

8.5 Derive the balance conditions for the Maxwell inductance-capacitance bridge in Figure 8.29.

8.6 At resonance the impedance of an LCR circuit is real, and a balance can be found in a resonance bridge containing only resistors in three of its arms. What are the balance conditions for the bridge in Figure 8.30?

8.7 In the network shown in Figure 8.31, R is a standard resistor and M_{12} a standard mutual inductance, whose value is known accurately from its dimensions. Show that the null condition for the detector leads to an expression giving resistances in terms

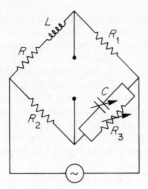

Figure 8.29.

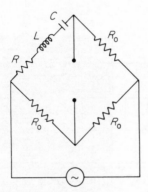

Figure 8.30.

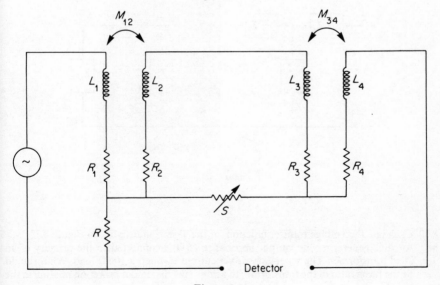

Figure 8.31.

of M_{12}, M_{34} and the angular frequency ω of the generator. (Relative values of mutual inductance and relative values of resistance can be measured precisely, and this network is used in the absolute calibration of standard resistors—see Appendix A for a discussion of electrical standards.)

8.8 An A.C. generator of voltage $\mathbf{V}$ and internal impedance $(X + jY)$ drives current through a load of impedance $\mathbf{Z}$. Show that maximum power transfer to the load occurs when $\mathbf{Z} = X - jY$. (Hint: the power is dissipated by the real part of the load impedance. For a fixed value of the real part $\mathbf{Z}$, choose the imaginary part to maximize the current amplitude, then vary the real part for maximum power transfer.)

8.9 The ladder networks described in section 8.4 may be split up into symmetrical 'π-sections' or 'T-sections' as indicated in Figure 8.32. Show that the matching impedances of the π- and T-sections are given by

$$\mathbf{Z}_\pi^2 = \mathbf{Z}_1\mathbf{Z}_2/(1 + \mathbf{Z}_1/4\mathbf{Z}_2),$$

$$\mathbf{Z}_T^2 = \mathbf{Z}_1\mathbf{Z}_2(1 + \mathbf{Z}_1/4\mathbf{Z}_2).$$

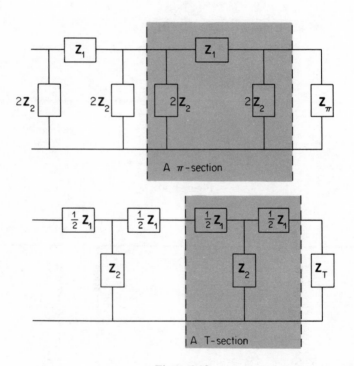

Figure 8.32.

8.10 Calculate the voltage transfer function for the T-section network in Figure 8.33.

8.11 An A.C. generator with output impedance $16\,\Omega$ is connected to the primary of an ideal transformer. The secondary drives current through a $100\,\Omega$ load. What should be the turns ratio of the transformer to achieve the maximum power dissipation in the load?

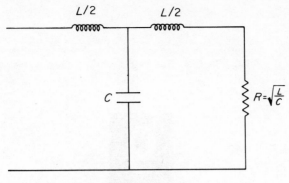

Figure 8.33.

8.12 The primary of a transformer presents an inductance of 1 H when the secondary is open-circuited. If a generator is connected to the primary and a 1000 pF capacitor across the secondary, the current drawn from the generator has a resonant maximum at a frequency $(10^4/2\pi)$ Hz. The secondary has ten times as many turns as the primary. What is the coupling coefficient of the transformer coils?

8.13 Use Faraday's law to show that the eddy current density in one of the soft iron sheets making up a laminated transformer core is proportional to the distance from the central plane of the sheet. Hence, assuming that the magnetic field $B = B_m \cos \omega t$, derive the expression $\sigma B_m^2 \omega^2 a^2/24$ for the power density in sheets of thickness a and conductivity σ. If the 50 Hz mains transformer in Figure 8.24 has laminations 1 mm thick with conductivity $2.5 \times 10^6 \, (\Omega\,\text{m})^{-1}$, what is the eddy current power loss when $B_m = 1$ T?

9

Transmission lines

In discussing A.C. circuits, we have assumed previously that current and voltage changes are transmitted instantaneously along the leads connecting different components in a network. At a given moment, a single value has been assigned to the current throughout the length of any branch of the network. This cannot be exactly true, since we know that no signal can travel faster than the speed of light. What has not been taken into account is the stray impedance associated with the leads (i.e. the self-inductance of the leads and their mutual inductance and capacitance with other parts of the circuit). When the stray impedances are included, they introduce phase changes and time delays, just as coils and capacitors introduce phase changes and time delays between the sections of the low-pass filter described in Section 8.4.

The value of stray impedances depends on the shape of each particular circuit, but their effect is always such that voltage and current changes move about the circuit at a speed not more than the speed of light. The time delays caused by strays are negligible at low frequencies, because they are then much shorter than the period of the alternating voltages and currents. At what frequency can the strays no longer be ignored? This question has already been raised in the discussion of networks in section 8.1.2. It was pointed out there that in ordinary laboratory circuits, network analysis breaks down at a frequency of about 1000 MHz. At such a high frequency, the magnitude of the impedances of stray capacitances and stray inductances are similar. All components are therefore either shunted by current carried by stray capacitance, or cause voltage drops no bigger than the voltage drops due to the stray inductance of leads. Although this argument may not seem to have anything to do with the speed of light, it is roughly equivalent to saying that at 1000 MHz the period of alternating voltages is comparable to the time taken for a light signal to cross the circuit. In a time of 10^{-9} sec, equal to one complete period at 1000 MHz, light travels 30 cm, about the size of the circuits we had in mind.

It must be emphasized that there is no well-defined frequency limit below which lead impedances may be neglected. If a signal is to be carried over a distance of a hundred metres, for example, then propagation delays are important at only 1 MHz. High frequency signals can be carried over such distances along a *transmission line*. A transmission line consists of a pair of parallel conductors which have the same cross-section anywhere along the length of the line. Between the conductors there is a capacitance which, apart from end effects, is proportional to the length of the line. A transmission line with a particular cross-section therefore has a definite capacitance per unit length between its conductors. Similarly, if the conductors are joined together at both ends of the line, a closed circuit is formed which has a self-inductance proportional to the length of the line. The line thus has a definite self-inductance per unit length. Because of the time taken to accumulate charge on the capacitance and to build up current across the inductance, delays are introduced, and signals propagate along the line at a constant speed: it turns out that this speed equals the speed of light in the medium in which the line is embedded.

As they travel along the line, signals are attenuated because of energy losses in the conductor (which has a finite resistance) and in any dielectric materials which may be present. In practice the attenuation is usually small and it is often a very good approximation to assume that a transmission line is *lossless*, so that it causes no attenuation at all. The properties of a lossless line, which are entirely determined by its capacitance per unit length and self-inductance per unit length, are first discussed in a general way in section 9.1. Then in section 9.2 some commonly used types of line are described. When a line is used to join two different circuits together, signals may sometimes be transmitted unaltered from one circuit to the other, but in general they will be partially reflected at the ends of the line. The amount of reflection depends on the values of the impedances terminating the line: this is explained in section 9.3, and the input impedance of a mismatched line is derived in section 9.4. Finally, the attenuation caused by resistive and dielectric losses is evaluated in section 9.5.

9.1 PROPAGATION OF SIGNALS IN A LOSSLESS TRANSMISSION LINE

A lossless line is made up of conductors which have zero resistance, and which are separated by a perfect insulator. No energy is dissipated in the line, whatever the frequency of the voltage across the conductors. Suppose that a lossless line has a capacitance per unit length C and a self-inductance per unit length L. Let us terminate a section of the line by joining an impedance $\mathbf{Z}$ across the conductors at one end. Viewed from the other end of the line, the circuit formed by the line plus its terminating impedance $\mathbf{Z}$ has an input impedance $\mathbf{Z}_{in}$, say. If an extra length δx is added to the line, the network has an additional self-inductance $L\,\delta x$, and there is an additional capacitance $C\,\delta x$ shunting the conductors, as shown in the equivalent circuit in Figure 9.1. This figure is reminiscent of the ladder networks discussed in the previous chapter. A transmission line is a limiting case of a low-pass filter network in which the inductance and capacitance of each section has been made infinitesimally small; the cut-off frequency is therefore indefinitely large, that is to say, the line will transmit signals of any frequency without attenuation. At low frequencies the low-pass filter has a characteristic impedance $Z_0 = \sqrt{L/C}$ (see section 8.4), and each section introduces a time delay $\sqrt{LC}$. By analogy we should expect the line to have a characteristic impedance $Z_0 = \sqrt{L/C}$ at all frequencies, so that in Figure 9.1, if the line is terminated by an impedance Z_0, then for all frequencies $\mathbf{Z}_{\text{in}} = Z_0$, no matter what is the length of the line. Also the time delay introduced by an extra length δx of the line should be $\delta x \sqrt{LC}$, at any frequency, i.e. signals are propagated along the line at a speed $1/\sqrt{LC}$. These values are correct, as we shall now verify by setting up the differential equations for the voltage and current on the line.

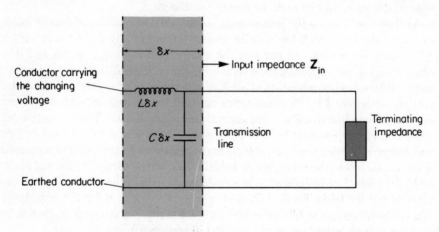

Figure 9.1. The equivalent circuit for a length δx of transmission line.

Consider the length δx of the line shown in Figure 9.2. At a time t the voltage and current at the position x on the line are $V(x, t)$ and $I(x, t)$ respectively. Over the distance δx the voltage changes to $V(x + \delta x, t)$ because there is a back e.m.f. due to the self-inductance $L\,\delta x$ of the intervening section of the line, and we have

$$V(x + \delta x, t) - V(x, t) = \frac{\partial V}{\partial x}\,\delta x = -L\,\delta x\frac{\partial I}{\partial t},$$

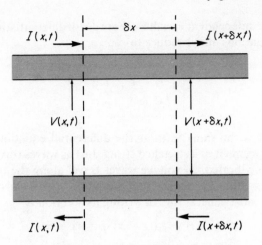

Figure 9.2. The voltage and current on a transmission line are functions of position and of time.

or

$$\frac{\partial V}{\partial x} = -L\frac{\partial I}{\partial t}. \tag{9.1}$$

The current flowing in the two conductors also varies with x, because of the capacitance between them. (Notice that the symmetry of the problem implies that in a lossless line equal and opposite currents are flowing in the conductors at any x.) The capacitance $C\,\delta x$ in the length δx carries a current $C\,\delta x(\partial V/\partial t)$, and hence

$$I(x + \delta x, t) - I(x, t) = \frac{\partial I}{\partial x}\delta x = -C\,\delta x\frac{\partial V}{\partial t},$$

or

$$\frac{\partial I}{\partial x} = -C\frac{\partial V}{\partial t}. \tag{9.2}$$

Differentiating Equation (9.1) with respect to x and Equation (9.2) with respect to t,

$$\frac{\partial^2 V}{\partial x^2} = -L\frac{\partial^2 I}{\partial x\,\partial t},$$

and

$$\frac{\partial^2 I}{\partial t\,\partial x} = -C\frac{\partial^2 V}{\partial t^2}.$$

Since x and t are independent variables, the order of differentiation is immaterial, and the two equations above reduce to

$$\frac{1}{LC}\frac{\partial^2 V}{\partial x^2} = \frac{\partial^2 V}{\partial t^2}. \tag{9.3}$$

This equation has the same form as the differential equation describing the transverse displacement of a stretched string. Just as waves travel along a string when one end is vibrated, so voltage waves travel along the transmission line when a changing voltage is applied to it. Equation 9.3 is a *wave equation* in one dimension. The most general way of writing its solution is

$$V(x, t) = f_1(vt - x) + f_2(vt + x),$$

where f_1 and f_2 are any functions whatever, and $v = 1/\sqrt{LC}$. This can be verified by substituting for V in Equation (9.3). The function f_1 represents a wave travelling along the line from left to right at a speed v in the positive

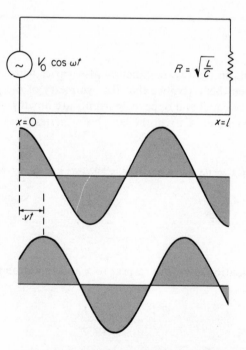

Figure 9.3. A wave travelling along a matched length of transmission line satisfies Ohm's law for the resistor at the end of the line at all times.

x-direction, and f_2 represents a wave travelling from right to left*. For example, if $V(x, t) = V_0 \cos{(vt - x)}$, then at $t = 0$ the voltage on the line has the sinusoidal variation shown in Figure 9.3. Later on, at time t, the voltage pattern has the same shape, but it has been shifted to the right through the distance vt.

Instead of eliminating the current from Equations (9.1) and (9.2), the voltage V can be eliminated, leading to a wave equation for the current:

$$\frac{1}{LC} \frac{\partial^2 I}{\partial x^2} = \frac{\partial^2 I}{\partial t^2}. \tag{9.4}$$

The current must also be a superposition of waves travelling with speed $1/\sqrt{LC}$ in either direction along the line. For the moment we will restrict attention to waves of a single angular frequency ω, travelling from left to right. Consider the line of length l shown in Figure 9.3. At the end where $x = 0$ a generator maintains an alternating voltage of angular frequency ω and amplitude V_0. The other end of the line at $x = l$ is terminated by a resistance $\sqrt{L/C}$. Waves travel away from the generator, from left to right, and the voltage and current on the line are

$$V = V_0 \cos{(\omega t - kx)},$$

$$I = V_0 \sqrt{C/L} \cos{(\omega t - kx)}. \tag{9.5}$$

Here the *wave-number* $k = \omega \sqrt{LC}$, since the solution must represent waves travelling away from the generator at a speed $1/\sqrt{LC}$. This solution satisfies both the line Equations (9.1) and (9.2), and also satisfies Ohm's Law $V = I\sqrt{L/C}$ at the end of the line $x = l$ terminated by the resistor. The input impedance of the line is

$$\frac{V(0, t)}{I(0, t)} = \sqrt{\frac{L}{C}},$$

whatever the length of the line, i.e. the characteristic impedance of the line is $\sqrt{L/C}$. Notice that although the line has inductance and capacitance, its characteristic impedance is purely resistive, so that current and voltage are in phase with each other when the line is matched. Clearly, if the line were terminated at its input end by its characteristic impedance, the impedance looking at the line from its right hand end is also $\sqrt{L/C}$. In other words, the output impedance of a transmission line matched at the input is $\sqrt{L/C}$.

The above argument holds for any angular frequency. We have shown that any alternating voltage is propagated along the line at a speed $1/\sqrt{LC}$, and that the characteristic impedance of the line is real and equal to $\sqrt{L/C}$ at all frequencies. Since the amplitudes of the voltage and current in Equation (9.5) are

---

* See Chapter 1 of the Manchester Physics Series volume on *Optics* by Smith and Thomson for a more detailed treatment of the wave equation. We shall be discussing waves in electromagnetism more fully in Chapter 11 of this book.

independent of x, the lossless line transmits all signals without attenuation and without distortion.

9.2 PRACTICAL TYPES OF TRANSMISSION LINE

In this section we calculate the distributed parameters—the capacitance per unit length and the inductance per unit length—of some commonly used transmission lines. From these parameters are derived the characteristic impedances of the lines and the speeds at which signals are propagated along them.

9.2.1 The parallel wire transmission line

Figure 9.4 is a cross-section of a transmission line made of a pair of similar cylindrical wires, each of radius a and with their centres separated by a distance $2d$. The wires are embedded in a medium of relative permittivity ε and relative permeability μ. The capacitance per unit length of a pair of wires has already been calculated in section 3.5. If $2d \gg a$, it is approximately

$$C = \frac{\pi \varepsilon \varepsilon_0}{\ln (2d/a)}. \tag{9.6}$$

To find the inductance per unit length, we make a closed circuit by joining the ends of a section of length l. The self inductance of this whole circuit is the magnetic flux through the circuit when it is carrying a current of 1 amp. The field outside a long straight wire carrying I amps, at a distance r from its centre, is $\mu \mu_0 I/(2\pi r)$ (Equation (4.32)). If the direction of the current is as shown in Figure 9.5 both wires contribute a field $\mathbf{B}$ pointing into the figure, and the total magnetic flux through the circuit is

$$2l \int_a^{2d-a} \frac{\mu \mu_0 I \, dr}{2\pi r} \simeq \frac{\mu \mu_0 l I}{\pi} \ln \left(\frac{2d}{a} \right), \qquad \text{if } 2d \gg a,$$

and the self-inductance per unit length is

$$L = \frac{\mu \mu_0}{\pi} \ln \left(\frac{2d}{a} \right). \tag{9.7}$$

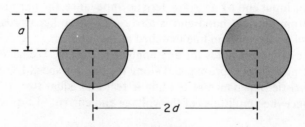

Figure 9.4. Cross-section of a parallel wire transmission line.

(There is actually a contribution to the self-inductance from the flux passing through the conductors. This is negligible when $2d \gg a$.) From Equations (9.6) and (9.7) we find that the characteristic impedance of the line is

$$Z_0 \text{ (parallel wire)} = \sqrt{\frac{L}{C}} = \left(\frac{\mu\mu_0}{\pi^2\varepsilon\varepsilon_0}\right)^{1/2} \ln\left(\frac{2d}{a}\right), \qquad (9.8)$$

and the speed of propagation of signals

$$\frac{1}{\sqrt{LC}} = \frac{1}{(\varepsilon\varepsilon_0\mu\mu_0)^{1/2}} = \frac{c}{(\varepsilon\mu)^{1/2}},$$

a constant, independent of the dimensions of the line. As we shall find in chapter 11 when we come to discuss radiation, *the constant c is the speed of light in vacuo*, and $c/(\varepsilon\mu)^{1/2}$ the speed of light in the medium in which the line is embedded.

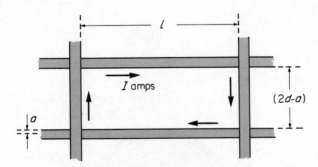

Figure 9.5. To find the self-inductance per unit length of a transmission line, a closed circuit is made by connecting the conductors at two points along the line.

It turns out that the velocity of propagation is equal to the speed of light for any lossless line, whatever the size and shape of its conductors.

Parallel wire lines are most often used *in vacuo* or in air, for which we may take $\varepsilon = \mu = 1$. For a line with $2d/a = 6$, Equation (9.8) then gives for the characteristic impedance

$$Z_0 = 120 \ln\left(\frac{2d}{a}\right) \simeq 200\ \Omega.$$

9.2.2   The coaxial cable

The coaxial cable consists of two coaxial cylindrical conductors with cross-section as in Figure 9.6. The conductors are separated by a material of relative permittivity ε and relative permeability μ. The capacitance per unit length has already been found to be (see section 2.3.2)

$$C = \frac{2\pi\varepsilon\varepsilon_0}{\ln(b/a)}. \tag{9.9}$$

Suppose that the conductors each carry a current of I amps. Applying Ampère's Law to the circuit of radius r in Figure 9.6.

$$\oint \mathbf{B} \cdot d\mathbf{l} = \mu\mu_0 I.$$

Because the cable has cylindrical symmetry, the magnitude B of the field depends only on r, and

$$\oint \mathbf{B} \cdot d\mathbf{l} = 2\pi r B,$$

leading to

$$B = \frac{\mu\mu_0 I}{2\pi r}.$$

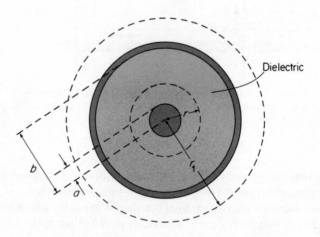

Figure 9.6. Cross-section of a coaxial cable.

The flux of the field $\mathbf{B}$ through a closed circuit formed by joining the conductors at the end of a section of cable of length l is

$$l \int_a^b \frac{\mu\mu_0 I \, dr}{2\pi r} = \frac{\mu\mu_0 I l}{2\pi} \ln\left(\frac{b}{a}\right),$$

and the self inductance per unit length is

$$L = \frac{\mu\mu_0}{2\pi} \ln\left(\frac{b}{a}\right). \tag{9.10}$$

From equations (9.9) and (9.10) we check that the speed of propagation is again $c/(\varepsilon\mu)^{1/2}$, and obtain for the characteristic impedance of the cable

$$Z_0 \text{ (coaxial cable)} = \sqrt{\frac{L}{C}} = \left(\frac{\mu\mu_0}{4\pi^2\varepsilon\varepsilon_0}\right)^{1/2} \ln\left(\frac{b}{a}\right). \tag{9.11}$$

Usually the insulator between the conductors of a coaxial cable is a solid with relative permeability very nearly unity, and we may write

$$Z_0 = \frac{60}{\sqrt{\varepsilon}} \ln\left(\frac{b}{a}\right).$$

For example, if the insulator is polyethylene with relative permittivity 2.3 then $Z_0 = 50\,\Omega$ for $b/a = 3.5$. Fifty ohms is often chosen as the characteristic impedance of coaxial cables for laboratory use. Many items of commercially available electronic equipment are standardized to have input and output impedances of fifty ohms. This allows signals to be transmitted along $50\,\Omega$ cable from one piece of equipment to another without reflections, as is explained in section 9.3.

The outer conductor of a coaxial cable is normally maintained at earth potential. There are then no electric fields outside the cable. Nor are there any magnetic fields outside it. Applying Ampère's Law to the circuit of radius r_1 in Figure 9.6, $\oint \mathbf{B} \cdot \mathbf{dl} = 0$, since equal and opposite currents flow in the two conductors of the cable: hence by symmetry $\mathbf{B} = 0$ outside. The confinement of the fields within the outer conductor gives the cable a two-fold advantage over the parallel wires. In the first place, the absence of external magnetic fields means that there can be no unwanted inductive coupling between the cable and other circuits. Secondly, there is no radiation from a coaxial cable. In chapter 11 we shall learn that energy must be radiated from a pair of parallel wires carrying high frequency currents. This leads to attenuation of signals even when there are no resistive losses.

9.2.3   Parallel strip lines

Figure 9.7 shows the cross-section of a transmission line made of two parallel strips of conductor attached to the opposite faces of a slab of dielectric material. Neglecting edge effects*, as we did in deriving the capacitance of a pair of parallel plates in section 1.5.5, the capacitance per unit length is

$$C = \frac{\varepsilon\varepsilon_0 a}{d}.$$

---

* In practical strip lines edge effects are *not* negligible, but the simplified expressions given here show roughly the way the line impedances vary with dimensions.

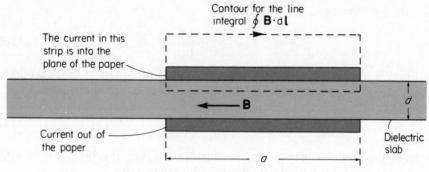

Figure 9.7. Cross-section of a strip line.

For a strip carrying current I, the magnetic field in the gap between the strips is found by applying Ampère's Law to the dotted contour in Figure 9.7. Except near the edge of the strip, there is no field outside the transmission line, since a contour right round the whole line encloses no current. Along the dotted contour, the line integral

$$\oint \mathbf{B} \cdot d\mathbf{l} = Ba = \mu\mu_0 I,$$

and the magnetic field $\mathbf{B}$ is uniform within the transmission line, having a magnitude

$$B = \frac{\mu\mu_0 I}{a}.$$

When a closed circuit is formed by connecting the ends of a length l of the line, a flux $\mu\mu_0 I dl/a$ threads it, and the self-inductance per unit length is therefore

$$L = \frac{\mu\mu_0 d}{a}. \tag{9.12}$$

Once again the speed of propagation $1/\sqrt{LC}$ is $c/(\varepsilon\mu)^{1/2}$.

The characteristic impedance of the line is

$$Z_0 \text{ (strip line)} = \sqrt{\frac{L}{C}} = \left(\frac{\mu\mu_0}{\varepsilon\varepsilon_0}\right)^{1/2} \frac{d}{a}. \tag{9.13}$$

If the strips are separated by a material with relative permittivity ε and relative permeability unity, this expression reduces to

$$Z_0 = \frac{120\pi}{\sqrt{\varepsilon}} \left(\frac{d}{a}\right).$$

A property of the strip line which is often useful is that the characteristic impedance varies directly as the ratio d/a of spacing to width, in contrast to the slow logarithmic dependence on dimensions for both the parallel wire and co-axial lines. Low impedances can be achieved by choosing $d \ll a$. It is particularly easy to connect together strip lines of different characteristic impedances, simply by fixing different widths of conductor to the same slab of dielectric material. Although strip lines are not totally enclosed like the coaxial cable, their external fields are much less than those of parallel wires, and strip lines perform satisfactorily at frequencies up to several hundred megahertz.

9.3 REFLECTIONS

If a transmission line is terminated by a resistance equal to its characteristic impedance Z_0 then an alternating voltage $V_0 \cos \omega t$ applied to the left hand end of the line generates a wave described by Equations (9.5). Rewriting these equations in the complex notation:

$$\mathbf{V} = V_0 \, e^{j(\omega t - kx)}$$

$$\mathbf{I} = \frac{V_0}{Z_0} e^{j(\omega t - kx)}.$$

The wave travels forwards from left to right, and is completely absorbed by the terminating resistor. There is no wave travelling in the opposite direction, from right to left. However, if the line is terminated by an impedance $\mathbf{Z}$ which is different from the characteristic impedance Z_0, the voltage and current no longer have the form given above, since these expressions do not meet the condition $\mathbf{V} = \mathbf{IZ}$ at the end of the line. In order to satisfy this condition, the solution of the differential equations of the line (Equations (9.1) and (9.2)) must include a term representing a wave travelling back towards the generator. The wave arriving at the terminating impedance is partly reflected, and the line is said to be mismatched. Let us choose the origin of the x-coordinate to be at the end of the line which is terminated by the impedance $\mathbf{Z}$. We write the forward-going wave as $A \exp[j(\omega t - kx)]$. This represents a wave travelling from negative values of x towards $x = 0$, where it makes a contribution $A \exp(j\omega t)$ to the voltage across the terminating impedance. Suppose that the reflected voltage is $\mathbf{K}A \exp(j\omega t)$ at $x = 0$, whence a wave $\mathbf{K}A \exp[j(\omega t + kx)]$ sets out leftwards towards the generator. The quantity $\mathbf{K}$ is then the voltage reflection coefficient at the terminating impedance $\mathbf{Z}$. The total voltage on the line is

$$\mathbf{V}(x, t) = A \exp[j(\omega t - kx)] + \mathbf{K}A \exp[j(\omega t + kx)]. \tag{9.14}$$

The expression for the current is similar except that the current reflection coefficient is $(-\mathbf{K})$. The current associated with the voltage (9.14) as a solution of the coupled equations (9.1) and (9.2) is

$$\mathbf{I}(x, t) = \frac{1}{Z_0}\{A \exp[j(\omega t - kx)] - \mathbf{K}A \exp[j(\omega t + kx)]\}. \tag{9.15}$$

At the end of the line, at $x = 0$, we have

$$\mathbf{V}(x = 0) = A\,e^{j\omega t}(1 + \mathbf{K}),$$

$$\mathbf{I}(x = 0) = \frac{A}{Z_0}\,e^{j\omega t}(1 - \mathbf{K}).$$

At $x = 0$ the ratio of voltage to current must equal the terminating impedance $\mathbf{Z}$ of the line,

$$\frac{\mathbf{V}(x = 0)}{\mathbf{I}(x = 0)} = \mathbf{Z} = Z_0 \frac{1 + \mathbf{K}}{1 - \mathbf{K}}. \tag{9.16}$$

Hence the reflection coefficient

$$\mathbf{K} = \frac{\mathbf{Z} - Z_0}{\mathbf{Z} + Z_0}. \tag{9.17}$$

The only way to avoid reflections is to terminate the line with an impedance Z_0. All the power carried by the forward-going wave on the line is then absorbed at the termination. At the other extreme, total reflection occurs when the line is open circuited ($\mathbf{Z} = \infty$, $\mathbf{K} = 1$) or short-circuited ($\mathbf{Z} = 0$, $\mathbf{K} = -1$). There is no current at the end of an open-circuited line, and no power can be transmitted; on the other hand, no power crosses the end of the shorted line either, since at this point there is no voltage across the conductors. When no power is transmitted, there are *standing waves* on the line, analogous to the standing sound waves in closed or open-ended pipes. At a node of the standing wave, the voltage across the conductors of the line is zero at all times, and a shorted line thus ends at a node. Nodes are equally spaced along the line with a spacing π/k equal to half the wavelength of the travelling waves. Similarly there are antinodes with a spacing π/k where the voltage has a maximum amplitude, and an open-circuited line ends at an antinode.

No power is transmitted whenever a line is terminated in a purely reactive impedance. The reflected wave has the same amplitude as the forward wave, and there are again standing waves on the line. However, at the end of the line the voltage across the terminating impedance is not in phase with the current. This means that there must be a phase difference between the forward and reflected waves at the end of the line. If this phase difference is δ, then the reflection coefficient is $\mathbf{K} = \exp(j\delta)$. From Equation (9.14) we see that the nodes of the standing waves occur at position x_n where the voltage amplitude

$$A\,e^{-jkx_n} + \mathbf{K}A\,e^{jkx_n} = 0.$$

Hence

$$\mathbf{K} = e^{j\delta} = -e^{-2jkx_n},$$

and

$$x_n = -\left\{\frac{\delta + (2n + 1)\pi}{2k}\right\}. \tag{9.18}$$

If a line is terminated by an impedance which causes partial reflection, standing waves are superimposed on the part of the travelling wave which is absorbed. The amplitude of the alternating voltage across the conductors varies along the line in the way shown in Figure 9.8, with maxima and minima each having a

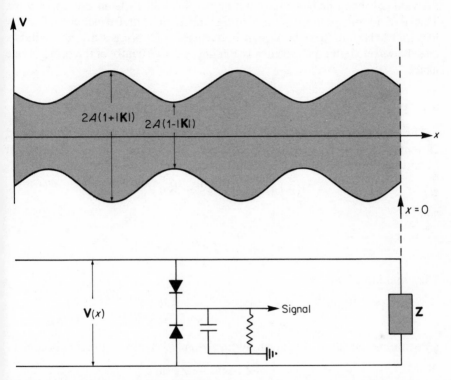

Figure 9.8. Measurement of the voltage standing-wave ratio on a mismatched transmission-line.

spacing of half a wavelength. Inspecting Equation (9.14) we see that the position x_n of the minima is still related to the phase of the reflected wave by Equation (9.18), while the maximum amplitude of the voltage is $A(1 + |\mathbf{K}|)$, and the minimum is $A(1 - |\mathbf{K}|)$. Their ratio is called the *voltage standing wave ratio*:

$$\text{Voltage standing wave ratio} = \frac{1 + |\mathbf{K}|}{1 - |\mathbf{K}|}. \tag{9.19}$$

The voltages on a parallel wire transmission line can be detected by putting a rectifying circuit across the line as shown in Figure 9.8. The current drawn by this circuit must be very small so that the line voltage is not disturbed. The magnitude of the rectified current depends on the shape of the characteristic curve of the diodes, but provided the current is small enough, it can be shown to be proportional to the square of the voltage across the line. Both the positions of the voltage minima and the voltage standing wave ratio can therefore be found from measurements of the rectified current. The magnitude and phase of the reflection coefficient **K** are then known from Equations (9.18) and (9.19), and the value of the impedance terminating the line follows from Equation (9.16). This is a useful method of measuring impedances at frequencies of a few hundred MHz: at these very high frequencies A.C. bridges are not suitable, and the wavelength on the transmission line is conveniently of the order of one metre.

★ 9.4 THE INPUT IMPEDANCE OF A MISMATCHED LINE

The input impedance of a mismatched line is found from the voltage and current at its input end. For a line of length l, the input is at $x = (-l)$, and from Equations (9.14) and (9.15) the voltage and current at this point are

$$\mathbf{V}(-l, t) = A\, e^{j\omega t}(e^{jkl} + \mathbf{K}\, e^{-jkl}),$$

and

$$\mathbf{I}(-l, t) = \frac{A}{Z_0}\, e^{j\omega t}(e^{jkl} - \mathbf{K}\, e^{-jkl}).$$

The input impedance is

$$\mathbf{Z}_{in} = \frac{\mathbf{V}(-l, t)}{\mathbf{I}(-l, t)} = Z_0 \left\{ \frac{e^{jkl} + \mathbf{K}\, e^{-jkl}}{e^{jkl} - \mathbf{K}\, e^{-jkl}} \right\}.$$

Substituting the value of **K** given by Equation (9.17), this expression becomes

$$\mathbf{Z}_{in} = Z_0 \left\{ \frac{\mathbf{Z} \cos kl + jZ_0 \sin kl}{Z_0 \cos kl + j\mathbf{Z} \sin kl} \right\}. \tag{9.20}$$

Three cases are of special interest:

(i) $\mathbf{Z} = 0$. For the shorted line equation (9.20) reduces to

$$\mathbf{Z}_{in} = jZ_0 \tan kl. \tag{9.21}$$

The input impedance is purely reactive, because, as we have already pointed out above, no power can be dissipated at the shorted end of the line.

(ii) $\mathbf{Z} = \infty$. Similarly the input impedance of an open-circuited line is purely reactive, and

$$\mathbf{Z}_{in} = -jZ_0 \cot kl$$
$$= jZ_0 \tan (kl - \pi/2). \tag{9.22}$$

The open-circuited line is thus equivalent to a shorted line differing in length by a quarter of a wavelength.

(iii) $kl = \pi/2$. Equation (9.20) now reduces to

$$\mathbf{Z}_{in} = \frac{Z_0^2}{\mathbf{Z}}. \tag{9.23}$$

The input impedance of the quarter-wave line is real when it is terminated by any resistor. It may appear to be paradoxical that the input impedance becomes very small when the line is terminated by a large resistance. For a fixed input voltage, according to Equation (9.23) the power delivered by the generator becomes indefinitely larger and larger as the terminating resistance R increases. However, we have not taken into account the output impedance of the generator. The voltage at the input of the line falls when the current demand becomes high and the voltage at the input end of an open-circuited quarter-wave lossless line must in fact always be zero, so that no power at all is delivered from the generator to the line.

Impedance matching

The power output of a generator depends on its output impedance, and it is easily shown that a generator delivers maximum power to a load which is equal to the complex conjugate of its own output impedance $\mathbf{Z}_g$. If $\mathbf{Z}_g$ and the load R are both purely resistive and the value of $\sqrt{RZ_g}$ is in the range of the characteristic impedances of practical transmission lines, then a quarter-wave line with $Z_0 = \sqrt{RZ_g}$ has an input impedance Z_g when joined to the load, and may be used to maximize the power transfer from the generator.

The quarter-wave line is here acting as an impedance transformer. It can be used in a similar way to match a load to another line of different characteristic impedance. Suppose, for example, that a long transmission line of characteristic impedance Z_0' is being used to transmit power to a load of resistance $R \neq Z_0'$. Reflections can be eliminated by inserting before the load a quarter-wave-length section of a line with characteristic impedance Z_0 satisfying $Z_0^2 = RZ_0'$. The input impedance of the quarter-wave section is $Z_0^2/R = Z_0'$, i.e. the long line is matched and no reflected wave travels along it. Notice that the matching is only achieved at the particular frequency for which the additional section of line is exactly a quarter-wavelength long.

Transmission lines as high-frequency circuit elements

Generally there are reflections whenever transmission lines of different characteristic impedances are connected together, except in special cases like the example above where the two lines are matched at a single frequency. But it is possible to make a low-pass filter transmitting a wide range of frequencies by connecting together alternate sections of two different transmission lines. The result is a ladder rather like the ladder network made of lumped capacitors and inductors discussed in section 8.4, Figure 9.9 is a plan view of such a ladder constructed of strip lines mounted on a single slab of dielectric, with alternate sections of narrow and broad strips, all of equal length l. The narrow strips have a

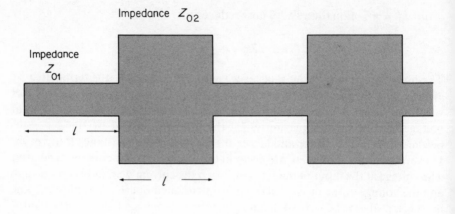

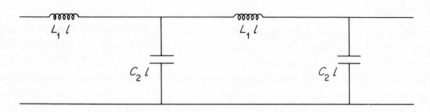

Figure 9.9. A strip line filter can be made with conductors consisting of thin sheets of the shape shown by the shaded area: the conductors are separated by a dielectric slab of constant thickness. Signals travelling from left to right meet alternate narrow and broad sections with characteristic impedances Z_{01} and Z_{02}. The approximate equivalent circuit represents the narrow sections as idealized inductances and the broad sections as lumped capacitors.

small capacitance per unit length, and hence a high characteristic impedance Z_{01}, while the wide strips have a lower characteristic impedance Z_{02}. Applying Equation (9.20) we can find the characteristic impedance of the whole ladder by requiring that the input impedance remains unchanged if another pair of narrow and broad strips is added to an existing matched ladder. It turns out that the input impedance of a ladder starting with a narrow section is

$$\mathbf{Z}_1 = \tfrac{1}{2}\mathrm{j}(Z_{01} - Z_{02})\tan kl + \sqrt{Z_{01}Z_{02} - \tfrac{1}{4}(Z_{01} - Z_{02})^2 \tan^2 kl}.$$

This impedance becomes purely imaginary if

$$Z_{01}Z_{02} - \tfrac{1}{4}(Z_{01} - Z_{02})^2 \tan^2 kl < 0.$$

When this inequality is satisfied there is total reflection, and an input signal is attenuated as it passes along the ladder. On the other hand, when the quantity under the square root sign is positive, the input impedance is partly resistive, implying that as for the LC ladder network, the input signal is transmitted without attenuation. There is a series of pass-bands because of the cyclic behaviour of $\tan kl$, and there are cut-off angular frequencies ω_c satisfying the condition

$$\tan^2(k_c l) = \tan^2\left(\frac{\omega_c l \sqrt{\varepsilon}}{c}\right)$$

$$= \frac{4Z_{01}Z_{02}}{(Z_{01} - Z_{02})^2}.$$

where ε is the relative permittivity of the insulator, and $c/\sqrt{\varepsilon}$ the propagation velocity on the line.

Let us concentrate on the lowest pass-band, for which $kl < \pi/2$, and assume that the wide strips are much wider than the narrow strips. Because the characteristic impedance is proportional to the strip width, we can put $(Z_{01} - Z_{02}) \simeq Z_{01}$, and the cut-off condition becomes

$$\tan^2\left(\frac{\omega_c l \sqrt{\varepsilon}}{c}\right) \simeq \frac{4Z_{02}}{Z_{01}},$$

or, making the approximation $\tan\theta \simeq \theta$,

$$\omega_c^2 \simeq \frac{4Z_{02}c^2}{Z_{01}l^2\varepsilon}. \tag{9.24}$$

For example, if $Z_{01} = 4Z_{02}$, the length of each section of the filter is 5 cm and the relative permittivity $\varepsilon = 2.3$, the cut-off frequency $\omega_c/2\pi$ is roughly 400 MHz. Although this frequency is so high, the transmission line filter really works in a very similar way to the LC filter discussed in section 8.4. A section of the wide strip has a much higher shunt capacitance than a section of the narrow strip, and conversely, referring back to Equation (9.12), one sees that the narrow strip has the larger series self-inductance. The sections of transmission line can be pictured as the components of a network, the wide strips representing capacitors and the narrow ones inductances, as shown in Figure 9.9. In the spirit of this picture, let us rewrite the cut-off condition (9.24) in terms of the capacitance per unit length C_2 of the wide strip and the self-inductances per unit length L_1 of the narrow strips. Signals travel along both strips at the same speed $c/\sqrt{\varepsilon}$, given in terms of the line parameters by the equations

$$\frac{c}{\sqrt{\varepsilon}} = \frac{1}{\sqrt{L_1 C_1}} = \frac{1}{\sqrt{L_2 C_2}}.$$

Using these equations to eliminate C_1 and L_2, we may put

$$Z_{01} = \sqrt{\frac{L_1}{C_1}} = \frac{L_1 c}{\sqrt{\varepsilon}},$$

and

$$Z_{02} = \sqrt{\frac{L_2}{C_2}} = \frac{\sqrt{\varepsilon}}{C_2 c}$$

The cut-off condition equation (9.24) becomes

$$\omega_c \simeq \frac{2}{\sqrt{(L_1 l)(C_2 l)}},$$

exactly the condition obtained in section 8.4 for the lumped parameter filter, since $(L_1 l)$ and $(C_2 l)$ are respectively the total inductance and total capacitance of the neighbouring sections of transmission line. This is a good example of how at high frequencies the role of the components used at low frequencies can be taken over by lengths of transmission line.

★ 9.5 LOSSY LINES

For most purposes the assumption that a transmission line is lossless is a very good approximation. At a frequency of 1 MHz an alternating voltage applied to the input of a typical laboratory coaxial cable is attenuated by considerably less than 0.1 % per metre of cable. Attenuation can safely be ignored when signals at about this frequency are transmitted from one piece of apparatus to another along a few metres of cable. However, attenuation may become important in the following circumstances:

  (i) when signals are to be transmitted for great distances along the cable.
 (ii) at very high frequencies, since it turns out that the losses increase with frequency.
(iii) when the line is handling large amounts of power. The power lost in transmission is dissipated as heat, and cooling the line may be a problem.

Attenuation arises because of the non-zero resistance of the conductor and the finite resistance of the insulator from which the transmission line is constructed. As we did when setting up the differential equations for the lossless line, let us consider the series and parallel impedances of a section of line of length δx, now including the effects of resistance. As before we write the self-inductance per unit length as L and the capacitance per unit length as C. If the resistance per unit length of the conductor is R, a resistance $R\,\delta x$ must be added to the self-inductance $L\,\delta x$, and the total series impedance to a signal of angular frequency ω is $(R + j\omega L)\,\delta x$. Similarly if the conductance per unit length of the insulator is G (conductance is the inverse of resistance), a resistance $1/(G\,\delta x)$ is added in parallel to the capacitance $C\,\delta x$. The admittance between the two conductors of the section δx is $(G + j\omega C)\,\delta x$. The equivalent circuit of the section δx is shown in Figure 9.10.

The analysis of the propagation of signals along the line is exactly the same as for the lossless line except that the impedance $j\omega L$ is replaced by $(R + j\omega L)$, and the admittance $j\omega C$ is replaced by $(G + j\omega C)$. Now the characteristic

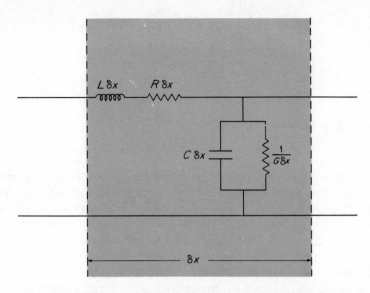

Figure 9.10. The equivalent circuit for a short length of lossy transmission line.

impedance of a lossless line is

$$Z_0 = \sqrt{\frac{L}{C}} = \sqrt{\frac{j\omega L}{j\omega C}};$$

in a lossy line the characteristic impedance becomes complex and we write

$$\mathbf{Z}_0 = \sqrt{\frac{R + j\omega L}{G + j\omega C}}. \tag{9.25}$$

The characteristic impedance approaches its lossless value $\sqrt{L/C}$ at *high* frequencies, although it is at high frequencies that the losses cause most attenuation. Usually reflections are not important even on a lossy line if it is terminated by an impedance $\sqrt{L/C}$: typically the imaginary part of the characteristic impedance is less than 10% of the real part at 1 MHz, and is decreasing with increasing frequency. For frequencies below 1 MHz the wavelength of the signals on the transmission line are so long that unless the line is very long the phase changes introduced by the line are negligible. Ordinary A.C. theory in which the line is regarded simply as a lead carrying currents is then adequate, and it does not matter if the line is mismatched.

For a matched lossless line, the voltage of a wave of angular frequency ω travelling from left to right is

$$\mathbf{V}(x, t) = V_0 \exp[j(\omega t - kx)],$$

where

$$jk = j\omega\sqrt{LC} = \sqrt{(j\omega L)(j\omega C)}.$$

In the lossy line this quantity is replaced by

$$\sqrt{(R + j\omega L)(G + j\omega C)} = \alpha + j\kappa, \quad \text{say}, \tag{9.26}$$

and the voltage becomes

$$\mathbf{V}(x, t) = V_0 \exp\left[j(\omega t - \kappa x)\right] \exp\left(-\alpha x\right). \tag{9.27}$$

This represents a damped wave with wave-number κ and attenuation constant α.

In the low frequency limit as $\omega \to 0$, $\alpha \to \sqrt{RG}$ and the voltage amplitude becomes $V_0 \exp\left(-\sqrt{RG}x\right)$. For a coaxial cable the attenuation length at zero frequency $1/\sqrt{RG}$ is very long. The resistance between the conductors was calculated in section 4.1.2 for a one metre length of typical cable, and found to be $3 \times 10^{11}\,\Omega$. The series resistance to direct current is mostly associated with the inner conductor, and for a 1 mm diameter copper wire is about 0.1 ohms per metre, leading to an enormous attenuation length of more than 10^6 m.

As the frequency increases the attenuation becomes more rapid. When $R \ll j\omega L$ and $G \ll j\omega C$, Equation (9.26) can be expanded to first order to give

$$\alpha + j\kappa \simeq j\omega\sqrt{LC} + \omega\sqrt{LC}\left\{\frac{R}{2\omega L} + \frac{G}{2\omega C}\right\}$$

$$\simeq j\omega\sqrt{LC} + \frac{R}{2Z_0} + \tfrac{1}{2}GZ_0. \tag{9.28}$$

since $\sqrt{L/C}$ is the high frequency value of Z_0. Equation (9.28) shows that to first order the wave number $\kappa = \omega\sqrt{LC}$, the same value as for the lossless line. Hence all frequencies propagate at the same speed $1/\sqrt{LC}$. Using the values given above for R and G in a coaxial cable, Equation (9.28) gives an attenuation length of 1 km for a characteristic impedance of $50\,\Omega$. This is too optimistic, because R and G both increase with frequency. There are losses in the dielectric at high frequency because of a hysteresis effect in materials with a permanent dipole moment. In the frequency range of interest for transmission lines, a roughly constant amount of work is done in taking the material once round the hysteresis loop, and the dielectric losses are therefore proportional to frequency. We shall not discuss dielectric losses further, because the attenuation of a transmission line is usually dominated by conductor losses.

The effective resistance per unit length of the conductors increases because of the skin effect which causes high frequency currents to flow in a shallow layer on the surface of the conductor. The skin effect is discussed in Chapter 11, where it is shown that at an angular frequency ω the current in a material of conductivity σ is effectively carried in a surface layer of thickness $\delta = \sqrt{2/\mu_0\sigma\omega}$. The high frequency resistance per unit length of a coaxial cable, for example, is roughly the same as the D.C. resistance of unit length of two cylinders of thickness δ and radii a and b equal to the inner and outer radii of the cable, i.e.

$$R = \frac{1}{2\pi}\sqrt{\frac{\mu_0\omega}{2\sigma}}\left(\frac{1}{a} + \frac{1}{b}\right)$$

$$\simeq 1.7 \times 10^{-8}\sqrt{\omega}\left(\frac{1}{a} + \frac{1}{b}\right) \text{ ohms per metre for copper.}$$

(9.29)

Most of the resistance is in the inner conductor, and if its diameter is 1 mm, $R \simeq 3.4 \times 10^{-5}\sqrt{\omega}\,\Omega\,\text{m}^{-1}$. For a cable with characteristic impedance 50 Ω, the attenuation length $2Z_0/R \simeq 3 \times 10^6/\sqrt{\omega}$, or about 40 metres at a frequency of 1000 MHz.

PROBLEMS 9

9.1 A 50 MHz signal propagates along a transmission line which has a characteristic impedance of 200 Ω. The voltage across the line is zero at a point 75 cm from the end. What is the impedance terminating the line?

9.2 The central conductors of three 50 Ω coaxial cables are joined together as shown in Figure 9.11. What must be the value of R if signals passing along any of the cables are to suffer no reflections? (The outer conductors are all earthed.)

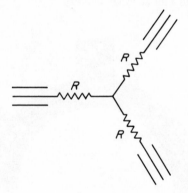

Figure 9.11.

9.3 Equal capacitors join the conductors at each end of a piece of 50 Ω coaxial cable which is a quarter wavelength long at 10 MHz. If a source of e.m.f. of this frequency is applied across one of the capacitors, for what value of capacitance will the current drawn from the source be a minimum?

9.4 A 200 Ω transmission line is terminated by a 600 Ω resistor. What is the input admittance of the line at one-sixth of a wave-length from the end? A short-circuited stub of line is added in parallel at this point, as shown in Figure 9.12. What should be its length in order to minimize the reflected signal travelling back towards the source?

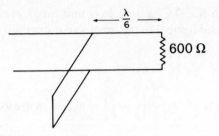

$$\frac{\lambda}{6}$$

600 Ω

Figure 9.12.

9.5 A 50 Ω cable is joined to a piece of 100 Ω cable, and the 100 Ω cable is terminated by an impedance of 50 Ω. What is the reflection coefficient for an incident signal travelling along the 50 Ω cable at the frequency for which the length of the 100 Ω cable is a quarter of a wavelength?

9.6 Figure 9.13 shows a schematic plan view of the conductors in a strip line arrangement known as a 'hybrid ring'. The ring joins together the four strip lines represented by the arms 1, 2, 3 and 4. Each arm has the same characteristic impedance Z_0, and the characteristic impedance of the strip line forming the actual ring is $\sqrt{2}Z_0$. The hybrid ring has the property that there is no reflection for signals arriving along any of the arms if each arm is matched at the end away from the ring. The ring is very nearly symmetrical, since the extra half wavelength in one section simply introduces a phase reversal. By considering the relative phases of signals travelling around the ring, deduce the outputs along the arms 2, 3 and 4 for an input on arm 1.

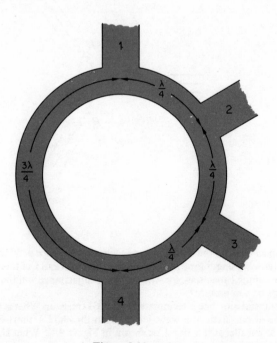

Figure 9.13.

10

Maxwell's equations

Four basic laws of electromagnetism have been discussed in the earlier chapters of this volume. These were Gauss' law; a law expressing the absence of magnetic poles; Ampère's law relating the magnetic field to the steady currents producing it; and Faraday's law of induction. We shall show that it is necessary to modify Ampère's law when the currents change with time and that, when this is done, the four laws make up a set of equations which any possible electromagnetic field must satisfy everywhere. These equations are called *Maxwell's equations*. Much of electromagnetism hereafter in this book consists of finding solutions to Maxwell's equations which fit the boundary conditions and distributions of charge and current appropriate to different problems. Often one's physical insight enables approximate solutions to be found which illustrate the most important features of a particular situation. This is the concern of Chapters 11, 12 and 13. In this chapter we will modify Ampère's law, summarize the earlier parts of the book whilst discussing Maxwell's equations, and show how the equations are able to predict and describe electromagnetic radiation.

10.1 THE EQUATION OF CONTINUITY

There is a large amount of experimental evidence suggesting that electric charge is a conserved quantity, i.e. that the total electric charge of an isolated system is constant. The best evidence comes from nuclear reactions. For example if calcium nuclei, each with charge $20e$, are bombarded with high energy α-particles (with charge $2e$), many possible nuclear reactions can take place. However the sum of the charges on any set of reaction products is always equal to $22e$. There are no known violations of the principle of conservation of charge.

The principle is on a par with the principles of conservation of momentum and energy. On a macroscopic level we have already invoked charge conservation in deducing Kirchhoff's first law in Chapter 8. We will now use the principle to derive an equation relating the macroscopic charge and current densities at each point. This equation is called the *equation of continuity*.

Consider an elementary volume $d\tau$, with sides dx, dy and dz, centred on the point (x, y, z). If the charge density ρ varies with time, the charge within the volume element at time t is $\rho(x, y, z, t)d\tau$, and at time $t + dt$ it is $(\rho + \partial\rho/\partial t \, dt)d\tau$, i.e. the change in the amount of charge within the volume element during the time dt is $\partial\rho/\partial t \, dt \, d\tau$. The charge flowing into the volume element during the time interval dt over the two faces of area $dy \, dz$ parallel to the y–z plane is

$$\{j_x(x - \tfrac{1}{2}dx, y, z, t) - j_x(x + \tfrac{1}{2}dx, y, z, t)\} \, dy \, dz \, dt$$

$$= -\frac{\partial j_x}{\partial x} dx \, dy \, dz \, dt$$

$$= -\frac{\partial j_x}{\partial x} d\tau \, dt,$$

where j_x is the x-component of the current density $\mathbf{j}$. The total charge flowing into the volume element over all six sides during the time dt is

$$-\left(\frac{\partial j_x}{\partial x} + \frac{\partial j_y}{\partial y} + \frac{\partial j_z}{\partial z}\right) d\tau \, dt$$

$$= -\operatorname{div} \mathbf{j} \, d\tau \, dt.$$

Since charge is conserved this charge must equal the change in the total charge within the volume element. Hence

$$\frac{\partial \rho}{\partial t} dt \, d\tau = -\operatorname{div} \mathbf{j} \, d\tau \, dt$$

and

$$\boxed{\frac{\partial \rho}{\partial t} + \operatorname{div} \mathbf{j} = 0.} \tag{10.1}$$

The derivation of this equation is independent of the position of the volume element, and so the relation (10.1) holds at all points and at all times. Equation (10.1) is the equation of continuity, expressing the conservation of charge in differential form. We shall now use it to show that the differential form of Ampère's law (as obtained for steady currents) is incompatible with the principle of conservation of charge when the currents are changing with time.

10.2   DISPLACEMENT CURRENT

Let us first consider the magnetic field due to currents flowing in conductors in free space and deal later with the complications introduced by the presence of matter. The differential form of Ampère's law, Equation (4.40), relates the field **B** at each point to the current density **j** at that point, and

$$\frac{1}{\mu_0} \operatorname{curl} \mathbf{B} = \mathbf{j}. \tag{10.2}$$

For currents changing with time the divergence of the right hand side of this equation may be non-zero at some points. For example, if a capacitor is being discharged the charge on the plates is decreasing, and over the surface of the plates $\partial\rho/\partial t$ is non-zero. Hence from Equation (10.1) div **j** is non-zero. As another example, if an electron beam is scanning across the front face of a television tube, the volume density of charge within the tube is changing with time. If the divergence of the right hand side of Equation (10.2) can be non-zero, the equation is clearly inadequate for changing currents since the divergence of the left hand side is identically zero everywhere. The function obtained by taking the curl of another vector function can easily be shown to have zero divergence at all points. We thus see that Ampère's law, as we have used it so far, is not true for currents varying with time. The law has to be modified by adding to the right hand side of Equation (10.2) a quantity which will make the divergence of the new right hand side everywhere zero. This requirement is satisfied, according to the equation of continuity, if the divergence of the quantity we add is equal to $\partial\rho/\partial t$.

Let us examine the differential form of Gauss' law obtained in section 1.4.3,

$$\operatorname{div} \mathbf{E} = \rho/\varepsilon_0. \tag{10.3}$$

If the charge density ρ changes with time the changing magnetic fields produced by the varying currents will induce an electric field. However the divergence of the induced electric field is everywhere zero, and so Equation (10.3) remains valid for fields that change with time. Taking the derivative with respect to time of this equation,

$$\frac{\partial}{\partial t} \operatorname{div} (\varepsilon_0 \mathbf{E}) = \frac{\partial\rho}{\partial t}.$$

The order of differentiation on the left hand side can be changed, since the operator div does not depend on time, giving

$$\operatorname{div} \varepsilon_0 \frac{\partial \mathbf{E}}{\partial t} = \frac{\partial\rho}{\partial t}.$$

Hence

$$\operatorname{div} \varepsilon_0 \frac{\partial \mathbf{E}}{\partial t} = -\operatorname{div} \mathbf{j},$$

from the equation of continuity. Hence adding the quantity $\varepsilon_0\,\partial\mathbf{E}/\partial t$ to the right hand side of equation (10.2) gives the required modification to Ampère's law in free space, which becomes

$$\frac{1}{\mu_0}\operatorname{curl}\mathbf{B} = \mathbf{j} + \varepsilon_0\frac{\partial\mathbf{E}}{\partial t}.$$

We will now derive the corresponding equation for situations where material is present. The total current density $\mathbf{j}$ at each point may now include a contribution $\mathbf{j}_M$ from the magnetization currents, in addition to the contribution $\mathbf{j}_f$ from free currents, and Equation (10.2) becomes

$$\frac{1}{\mu_0}\operatorname{curl}\mathbf{B} = \mathbf{j} = \mathbf{j}_f + \mathbf{j}_M. \tag{10.4}$$

The magnetization current density is related to the magnetization $\mathbf{M}$ by equation (5.19), and we have

$$\mathbf{j}_M = \operatorname{curl}\mathbf{M}.$$

Inserting this into Equation (10.4) and rearranging we obtain

$$\operatorname{curl}\left(\frac{\mathbf{B}}{\mu_0} - \mathbf{M}\right) = \mathbf{j}_f,$$

or

$$\operatorname{curl}\mathbf{H} = \mathbf{j}_f, \tag{10.5}$$

where $\mathbf{H}$ is the magnetic intensity defined by Equation (5.27). This equation is again inadequate for changing currents, just as was Equation (10.2), and we have to add to the right hand side a quantity whose divergence is everywhere equal to $\partial\rho_f/\partial t$. In the presence of polarized atoms or molecules the total charge density in Equation (10.3) includes a contribution ρ_p from the polarization charges, as well as the contribution ρ_f from the free charges. Equation (10.3) can now be written in terms of the electric displacement $\mathbf{D}$, as in section 2.3.2, to give

$$\operatorname{div}\mathbf{D} = \rho_f.$$

Differentiation with respect to time gives

$$\operatorname{div}\frac{\partial\mathbf{D}}{\partial t} = \frac{\partial\rho_f}{\partial t} = -\operatorname{div}\mathbf{j}_f.$$

Adding the quantity $\partial\mathbf{D}/\partial t$ to the right hand side of Equation (10.5) gives the required modification, and the final generalization of Ampère's law is

$$\boxed{\operatorname{curl}\mathbf{H} = \mathbf{j}_f + \frac{\partial\mathbf{D}}{\partial t}.} \tag{10.6}$$

The term $\partial \mathbf{D}/\partial t$ is called the *displacement current density*. It can easily be visualized as a current flowing back and forth between the plates of a capacitor filled with a polarizable medium as the charge distribution within the atoms varies with the field across the plates. The displacement current is still present however even when there is no material between the plates of the capacitor. In this case its value is $\varepsilon_0 \, \partial \mathbf{E}/\partial t$ as in the free space example we dealt with first.

Displacement current was introduced by Maxwell in 1862. With the benefit of hindsight it seems surprising perhaps that people should have thought Ampère's original law would apply to non-steady currents. However the original law correctly described the results of all the experiments made with the slowly changing currents which were met in the laboratory at the time of Maxwell. 'The form of the rules we write down is more general than the experience from which they are culled, and it is quite normal to apply them to situations we have not directly experienced. The act of genius is to realize precisely where it is that we are going too far*.' Where nowadays are we making the same sort of error as that to which Maxwell drew attention in electromagnetism?

The inclusion of displacement current in the electromagnetic field equations restores a degree of symmetry to electricity and magnetism. A changing magnetic field produces an electric field according to Faraday's law. Now we see that a changing electric field produces a magnetic field. This fact has far-reaching consequences.

Displacement currents are not usually important compared to the currents arising from the motion of free charges when investigating the behaviour of the continuous circuits met in the laboratory and discussed in Chapters 7 and 8. In order to obtain an idea of the relative size of the two types of current in conductors, let us calculate them for a wire in which there is an alternating electric field $\mathbf{E}_0 \cos \omega t$, where ω is the angular frequency. The current density $\mathbf{j}$ is given by $\mathbf{j} = \sigma \mathbf{E}_0 \cos \omega t$, where σ is the electrical conductivity, which is to a high degree of accuracy equal to the steady current conductivity up to the highest frequencies used. If we assume that there is no contribution to the macroscopic electric field from the atoms of the metal the relative permittivity of the metal is unity and the electric displacement $\mathbf{D}$ inside the wire is given by

$$\mathbf{D} = \varepsilon_0 \mathbf{E}_0 \cos \omega t,$$

and

$$\frac{\partial \mathbf{D}}{\partial t} = -\omega \varepsilon_0 \mathbf{E}_0 \sin \omega t.$$

Therefore

$$\left| \mathbf{j} \middle/ \frac{\partial \mathbf{D}}{\partial t} \right| = \frac{\sigma}{\omega \varepsilon_0}.$$

* The quotation is taken from *An Introduction to the Meaning and Structure of Physics*, by Leon N. Cooper. Harper and Row, New York, 1968.

For copper we obtain

$$\left| \mathbf{j} \middle/ \frac{\partial \mathbf{D}}{\partial t} \right| \sim \frac{10^{19}}{\omega},$$

and the ratio of the conduction current to the displacement current in a wire is very large at all frequencies used in electrical circuits.

Since the displacement current density is non-zero outside the conductor the above calculation is not directly relevant to the question of the relative importance of conduction and displacement currents for the magnetic fields produced by changing currents. In order to investigate this point we will discuss the particular example of a long solenoid carrying an alternating current $I = I_0 \sin \omega t$, where ω is the angular frequency at which the current, of amplitude I_0, oscillates. The solenoid has N turns per unit length and radius a. Let us first neglect the displacement current. Inside the solenoid the magnetic field $\mathbf{B}$ can be determined as in section 4.4.2, and is to a good approximation parallel to the axis and everywhere constant. Its magnitude is given by

$$B = \mu_0 NI = \mu_0 NI_0 \sin \omega t, \tag{10.7}$$

from Equation (4.31). The magnetic field changes with time and hence there is an induced electric field inside the solenoid. Inside, by symmetry, all points on a circle with centre on the axis and plane perpendicular to the axis are indistinguishable. The induced electric field must therefore have constant magnitude at all points on the circle and be in a direction tangential to the circle, as shown in Figure 10.1. Its direction is given by Lenz's law. Its magnitude E can be determined by applying Faraday's law to a circle of radius r.

$$2\pi r E = -\frac{\partial}{\partial t}(\pi r^2 \cdot \mu_0 NI_0 \sin \omega t).$$

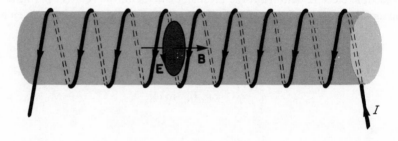

Figure 10.1. The electric and magnetic fields inside a solenoid carrying an alternating current. If the field $\mathbf{B}$ is decreasing in the direction shown, the electric field is anticlockwise looking into the solenoid from the right. The amplitude of the electric field is proportional to the distance from the centre (Equation (10.8)).

Hence

$$E = -\tfrac{1}{2}\omega\mu_0 NI_0 \cos \omega t \cdot r. \tag{10.8}$$

The electric displacement D is given by

$$D = -\tfrac{1}{2}\omega\varepsilon_0\mu_0 NI_0 \cos \omega t \cdot r,$$

where we have put the relative permittivity of the air in the solenoid equal to unity. Differentiating this equation with respect to time we obtain

$$\frac{\partial D}{\partial t} = \tfrac{1}{2}\omega^2\varepsilon_0\mu_0 NI_0 \sin \omega t \cdot r. \tag{10.9}$$

We are now in a position to determine whether the initial neglect of the displacement current was justified. The magnetic field inside the solenoid was worked out by applying Ampère's integral relationship to a rectangular loop like the one shown in Figure 10.2. The current through the surface S of the loop i.e. the surface integral $\int_S \mathbf{j} \cdot d\mathbf{S}$, is equal to NIl. The displacement current, which was neglected, is equal to $\int_S (\partial \mathbf{D}/\partial t) \cdot d\mathbf{S}$, and an estimate of the error involved in its neglect is given by the ratio

$$\int_S \frac{\partial \mathbf{D}}{\partial t} \cdot d\mathbf{S}/NIl.$$

From equation (10.9),

$$\int_S \frac{\partial \mathbf{D}}{\partial t} \cdot d\mathbf{S} = \tfrac{1}{2}\omega^2\varepsilon_0\mu_0 NIl \int_0^a r\, dr,$$

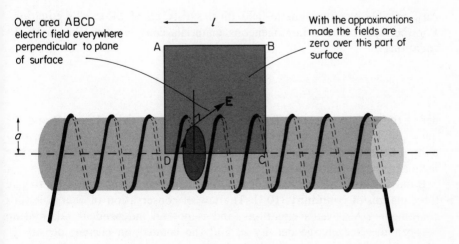

Over area ABCD electric field everywhere perpendicular to plane of surface

A $\overset{\longleftarrow\ \ l\ \ \longrightarrow}{}$ B

With the approximations made the fields are zero over this part of surface

a

E

D   C

Figure 10.2. The current due to the motion of free charges and the displacement current through a surface partly inside a solenoid.

since over the surface of the loop the vector $\partial\mathbf{D}/\partial t$ is everywhere parallel to the vector d$\mathbf{S}$. Hence the ratio of the displacement current through the loop to the conduction current is equal to $(\omega^2 a^2 \varepsilon_0 \mu_0/4)$. If the current alternates at a frequency of 1 MHz and the radius is 10 cm, the numerical value of this ratio is $\sim 10^{-6}$. This shows that the contribution from the displacement current to the magnetic field inside the solenoid is very small. It does not mean that there are no important effects due to displacement currents at frequencies of the order of 1 MHz, neither does it mean that the electric field inside the solenoid is negligible. If the amplitude of the current is 1 A and the coil has 50 turns per metre, the amplitude of the electric field near to the coil, given by Equation (10.8), is about 20 V m^{-1} at 1 MHz.

10.3 MAXWELL'S EQUATIONS

We can now summarize the equations which the field vectors $\mathbf{E}$, $\mathbf{D}$, $\mathbf{B}$ and $\mathbf{H}$ everywhere satisfy. These equations are called *Maxwell's equations*. They are:

$$\text{div } \mathbf{D} = \rho_f \tag{10.10}$$

$$\text{div } \mathbf{B} = 0 \tag{10.11}$$

$$\text{curl } \mathbf{E} = -\frac{\partial \mathbf{B}}{\partial t} \tag{10.12}$$

$$\text{curl } \mathbf{H} = \mathbf{j}_f + \frac{\partial \mathbf{D}}{\partial t}. \tag{10.13}$$

Any possible electromagnetic field must satisfy *all* of Maxwell's equations. For the isotropic and homogeneous materials considered in this book the equations

$$\mathbf{D} = \varepsilon\varepsilon_0\mathbf{E},$$

$$\mathbf{H} = \mathbf{B}/\mu\mu_0,$$

define the auxiliary vectors $\mathbf{D}$ and $\mathbf{H}$ in terms of the fields $\mathbf{E}$ and $\mathbf{B}$ and the quantities ε, ε_0, μ, and μ_0. The relative permittivity ε and the relative permeability μ depend on frequency, as in principle do all parameters which describe the macroscopic behaviour of media in electric and magnetic fields.

It may be noted that Equations (10.10) and (10.13) can be combined to give the equation of continuity (10.1). The law of conservation of charge is thus contained in Maxwell's equations, and is not an independent relationship between the free charge density ρ_f and the conduction current density $\mathbf{j}_f$. Maxwell's equations, together with the equation which gives the force on a charge q moving with velocity $\mathbf{v}$ in fields $\mathbf{E}$ and $\mathbf{B}$, viz.

$$\mathbf{F} = q\mathbf{E} + q\mathbf{v} \wedge \mathbf{B},$$

form a complete statement of the macroscopic theory of electromagnetism. Each equation is a generalization of a fairly simple experimental result and below we comment on the physical background to each one in turn.

$$\text{(i)} \quad \operatorname{div} \mathbf{D} = \rho_f.$$

The inverse square law of force between stationary charges in free space leads to Gauss' theorem. In differential form this theorem states

$$\operatorname{div} \mathbf{E} = \rho/\varepsilon_0,$$

where ρ is the volume density of charge and ε_0 is the permittivity of free space. If dielectric materials are present they acquire a polarization $\mathbf{P}$ and at each point are equivalent, for the purpose of calculating the electric field to which they give rise, to a volume density of charge $(-\operatorname{div} \mathbf{P})$. This volume density of charge must be added to the free charge density ρ_f, and we have

$$\operatorname{div} \mathbf{E} = (\rho_f - \operatorname{div} \mathbf{P})/\varepsilon_0. \tag{10.14}$$

Rather than work with the polarization it is more convenient to use a vector $\mathbf{D}$ defined by

$$\mathbf{D} = \varepsilon_0 \mathbf{E} + \mathbf{P}.$$

The electric displacement $\mathbf{D}$ is then related to the density ρ_f by

$$\operatorname{div} \mathbf{D} = \rho_f.$$

This equation, first discussed in electrostatics, is true for the total electric displacement from all sources, including induced fields and electric fields from moving charges. The divergence of the induced electric field is everywhere zero, and hence the field $\mathbf{E}$ in Equation (10.14) can be written as the sum of electrostatic and induced electric fields. Gauss' theorem is true for moving charges as well as for stationary charges, and Equation (10.14) also applies to the electric fields caused by moving charged bodies.

$$\text{(ii)} \quad \operatorname{div} \mathbf{B} = 0.$$

This equation is a consequence of the observation that all magnetic fields $\mathbf{B}$ have field lines that are continuous. There are no free magnetic 'poles' or 'charges'.

$$\text{(iii)} \quad \operatorname{curl} \mathbf{E} = -\frac{\partial \mathbf{B}}{\partial t}.$$

Changing magnetic fields produce an electric field. This equation is the differential form of Faraday's law for the induced electric field. The integral form of the law expresses the experimental observation that the e.m.f. induced in a circuit is equal to the rate of change with time of the total magnetic flux through the circuit, no matter how the flux changes. The fields $\mathbf{E}$ and $\mathbf{B}$ are the total electric fields from all sources at each point.

$$\text{(iv)}\quad \text{curl } \mathbf{H} = \mathbf{j}_f + \frac{\partial \mathbf{D}}{\partial t}.$$

Ampère's law for steady currents relates the line integral of the field $\mathbf{B}$ around a closed loop to the total current flowing through the loop. The law summarizes the results of experiments which show that two small coils carrying steady currents exert forces and torques on each other which show the same variation with separation, orientation, etc., as the forces and torques between two small electric dipoles. Ampère's law in differential form relates the field $\mathbf{B}$ to the current density $\mathbf{j}$ at each point.

$$\text{curl } \mathbf{B} = \mu_0 \mathbf{j}$$

for currents flowing in free space. μ_0 is the permeability of free space. If matter is present it acquires a magnetization $\mathbf{M}$. At each point the magnetized material, for the purpose of calculating the magnetic field to which it gives rise, is equivalent to a current density curl $\mathbf{M}$. This magnetization current density must be added to the free current density $\mathbf{j}_f$, and we have

$$\text{curl } \mathbf{B} = \mu_0(\mathbf{j}_f + \text{curl } \mathbf{M}). \tag{10.15}$$

It is more useful to introduce the magnetic intensity $\mathbf{H}$ than to work with the magnetization. The magnetic intensity is defined by the equation

$$\mathbf{H} = \mathbf{B}/\mu_0 - \mathbf{M},$$

and, rearranging Equation (10.15), is related to the free current density by the equation

$$\text{curl } \mathbf{H} = \mathbf{j}_f. \tag{10.16}$$

If the currents change with time the displacement current density has to be included, as discussed in section 10.2, and Equation (10.16) becomes

$$\text{curl } \mathbf{H} = \mathbf{j}_f + \frac{\partial \mathbf{D}}{\partial t}.$$

Much of the remainder of this book deals with mathematical solutions to Maxwell's equations in different situations, i.e. with different boundary conditions and source distributions. There is thus a lot of mathematics in the study of certain parts of electricity and magnetism. However, many important features in most situations can be understood by using physical insight to suggest approximations to simplify the mathematics.

The development of the macroscopic theory of electromagnetism—Maxwell's equations—beginning around the year 1785 with the work of Coulomb on the forces between charged bodies, and ending around 1867 with Maxwell, is an elegant example of the scientific method. The many practical applications of electromagnetic theory are a constant reminder of the benefits which may result from scientific curiosity.

10.4   ELECTROMAGNETIC RADIATION

If the electric and magnetic fields do not vary with time Maxwell's equations reduce to the equations of electrostatics and magnetostatics discussed in Chapters 1 to 5. These equations are:

$$\text{div } \mathbf{D} = \rho_f$$

$$\text{curl } \mathbf{E} = 0,$$

and

$$\text{div } \mathbf{B} = 0$$

$$\text{curl } \mathbf{H} = \mathbf{j}_f.$$

If the fields vary slowly with time, so that the displacement current is negligible, we have what is often called a quasi-static situation, and the fields obey similar equations to those above, except that the second is replaced by Faraday's law,

$$\text{curl } \mathbf{E} = -\frac{\partial \mathbf{B}}{\partial t}.$$

For fields which vary rapidly with time, typically if they have frequency components of 100 kHz or higher, the displacement current has to be included. This gives rise to new solutions to Maxwell's equations corresponding to *electromagnetic radiation.*

A simple example in which the fields change rapidly is provided by a spark breakdown between two electrically charged spheres. The large, but short-lived, current in the spark produces a changing magnetic field which induces an electric field near to the spheres. Prior to the addition of the displacement current term in Equation (10.13) one would have believed that the electric and magnetic fields would quickly die away to zero, as they do to all intents and purposes when a battery supplying a current through an inductive coil is suddenly removed. With the addition of displacement current however, the rapidly changing electric field can now in its turn induce a magnetic field. Solutions to Maxwell's equations are now possible in which the changing electric and magnetic fields combine in such a way as to produce new fields further out from the spheres as the old ones die away. This process is continually repeated, and the result is that a combination of related electric and magnetic fields travels outwards from the spheres with a speed which is determined by the electromagnetic field equations (10.10)–(10.13). The fields have energy, and some of the energy dissipated in the spark moves outwards from the spheres with the radiated fields. The spark is said to have radiated energy, which is carried away in the electromagnetic radiation field. In the next chapter we derive wavelike solutions to Maxwell's equations in free space that describe electromagnetic disturbances which are propagated at a fixed speed. In this section, however, we simply note that the quantity $(\varepsilon_0\mu_0)^{-1/2}$ has the dimensions of velocity, and that this is the speed of travel of the electromagnetic radiation field in free space. The numerical value of $(\varepsilon_0\mu_0)^{-1/2}$ is very nearly equal to $3 \times 10^8 \text{ m s}^{-1}$, which is the same as the observed speed of propagation of visible light in free space.

Using spark discharges between two spheres, Hertz in 1887 first detected electromagnetic radiation by inducing an electric field in a circuit some distance away from the spheres. He showed that the speed of the radiated field was roughly $3 \times 10^8 \text{ m s}^{-1}$, and also that the radiated field had many other properties similar to visible light. For example it could be reflected, refracted and polarized. The conclusion that visible light is an electromagnetic wave in which electric and magnetic fields vary rapidly in time was an early spectacular result of Maxwell's proper formulation of the laws of electromagnetism.

★ 10.5 THE MICROSCOPIC FIELD EQUATIONS

Throughout this book we have attempted to keep sight of the fact that the macroscopic electric and magnetic fields inside media are averages of microscopic fields. The microscopic fields are produced by charges on electrons and protons moving in free space and by the intrinsic magnetic dipole moments of electrons, protons and neutrons. In this section we will give expressions for the fields and parameters used in the macroscopic theory in terms of microscopic quantities. For convenience we will use the symbols $\mathbf{E}_a$ and $\mathbf{B}_a$ to denote the atomic fields which describe forces on microscopic charges. The force on an electron of charge $-e$, moving with velocity $\mathbf{v}$ is given by

$$\mathbf{F} = -e\mathbf{E}_a - e\mathbf{v} \wedge \mathbf{B}_a.$$

It is possible to formulate the laws of electricity and magnetism in terms of the fields $\mathbf{E}_a$ and $\mathbf{B}_a$ alone, but it is easier on a macroscopic scale to use the macroscopic fields $\mathbf{E}$, $\mathbf{D}$, $\mathbf{B}$ and $\mathbf{H}$.

The atomic fields obey relationships similar to Maxwell's equations:

$$\text{div } \mathbf{E}_a = \rho_a/\varepsilon_0 \tag{10.15}$$

$$\text{div } \mathbf{B}_a = 0 \tag{10.16}$$

$$\text{curl } \mathbf{E}_a = -\frac{\partial \mathbf{B}_a}{\partial t} \tag{10.17}$$

$$\text{curl } \mathbf{B}_a = \mu_0(\mathbf{j}_a + \mathbf{j}_{M_a}) + \varepsilon_0\mu_0 \frac{\partial \mathbf{E}_a}{\partial t}. \tag{10.18}$$

In these equations ρ_a is the microscopic volume density of charge and $\mathbf{j}_a$ is the microscopic current density. If $\mathbf{v}_a$ is the average velocity of the charges constituting the charge density ρ_a, the current density is given by $\mathbf{j}_a = \rho_a \mathbf{v}_a$. $\mathbf{j}_{M_a}$ is the current density equivalent to the microscopic magnetization $\mathbf{M}_a$, which is due to the distribution of intrinsic magnetic dipole moments. Equations (10.15) to (10.18) are in a sense more fundamental than Maxwell's equations for the macroscopic fields, since the atomic field equations hold at every point, whereas Maxwell's equations involve averages over very many atoms. The macroscopic electric and magnetic fields are defined by averaging the microscopic fields over volumes very small on a macroscopic scale but which contain very many atoms. We write the macroscopic fields as

$$\mathbf{E} = \bar{\mathbf{E}}_a$$

$$\mathbf{B} = \bar{\mathbf{B}}_a.$$

Maxwell's equations for the fields $\mathbf{E}$ and $\mathbf{B}$ can be written in terms of the polarization $\mathbf{P}$ and the magnetization $\mathbf{M}$ by substituting for the fields $\mathbf{D}$ and $\mathbf{H}$ in Equations (10.10) to (10.13), using Equations (2.20) and (5.27):

$$\text{div } \mathbf{E} = (\rho_f - \text{div } \mathbf{P})/\varepsilon_0$$

$$\text{div } \mathbf{B} = 0$$

$$\text{curl } \mathbf{E} = -\frac{\partial \mathbf{B}}{\partial t}$$

$$\text{curl } \mathbf{B} = \mu_0 \left(\mathbf{j}_f + \text{curl } \mathbf{M} + \frac{\partial \mathbf{P}}{\partial t} \right) + \varepsilon_0 \mu_0 \frac{\partial \mathbf{E}}{\partial t}.$$

It can be proved that averaging the microscopic field equations over volumes which are very small on a macroscopic scale gives a set of four equations for the fields $\mathbf{E}$ and $\mathbf{B}$ which can be cast in a form similar to the above equations. The quantities $\mathbf{P}$, $\mathbf{M}$, ρ_f, and $\mathbf{j}_f$ can then be identified with the following averages of microscopic quantities:

$$\mathbf{P} \equiv N\langle \mathbf{p} \rangle$$

$$\mathbf{M} \equiv N\langle \mathbf{m} \rangle$$

$$\rho_f \equiv -N_c e + N q_m$$

$$\mathbf{j}_f \equiv -N_c e \bar{\mathbf{v}}_c.$$

In these equations N is the number of atoms or molecules of the substance per unit volume, and N_c the number of free electrons per unit volume with average velocity $\bar{\mathbf{v}}_c$, i.e. $\bar{\mathbf{v}}_c$ is their drift velocity as used in section 4.1. q_m is the average charge on each atom or molecule; $\langle \mathbf{p} \rangle$ is the average electric dipole moment of the molecules in the direction of the field $\mathbf{E}$, and $\langle \mathbf{m} \rangle$ the average magnetic dipole moment of the molecules* in the direction of the field **B**. The above definitions of **P**, **M**, ρ_f and $\mathbf{j}_f$ are the same as the ones introduced in the earlier chapters when these quantities were first discussed. They enable us to define the auxiliary vectors **D** and **H** in terms of averages of microscopic quantities in the following way:

$$\mathbf{D} = \varepsilon_0 \overline{\mathbf{E}}_a + N\langle \mathbf{p} \rangle$$

$$\mathbf{H} = \frac{\overline{\mathbf{B}}_a}{\mu_0} - N\langle \mathbf{m} \rangle.$$

This concludes our correlation between macroscopic and microscopic quantities. The microscopic magnetization current density $\mathbf{j}_{Ma}$ is not always included in equation (10.18). The intrinsic dipole moments may be considered to be due to the motion of microscopic charges within the particle itself, e.g. due to the electron spinning about an internal axis. In this case the intrinsic moments are considered to be included in the term $\mathbf{j}_a$†. We follow the procedure outlined above because it leads naturally to an expression for the magnetic dipole moment of an atom or molecule (see Problem 10.3) which is taken over directly and used in quantum mechanics.

PROBLEMS 10

10.1 Positive charge leaves one plate of a parallel plate capacitor which is discharging through a resistor. At a certain time the rate of discharge is I amps. Calculate the displacement current flowing at this instant through a surface S which encloses one plate of the capacitor (see Figure 10.3). Show that the magnitude of the displacement current is equal to the conduction current I.

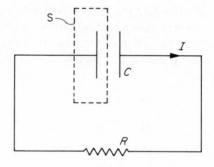

Figure 10.3.

* The average magnetic dipole moment of the molecules is here taken to include a contribution from the free electrons if there are any.

† See for example *Electric and Magnetic Susceptibilities*, by J. H. Van Vleck, Oxford University Press, 1932.

10.2 Suppose that a volume distribution of free charge density exists within a metallic conductor and is specified by the function $\rho(r, t)$. This function rapidly becomes zero as the charge redistributes itself over the surface of the conductor. If the direct current electrical conductivity of the metal is σ and it has N free electrons per unit volume, determine how $\rho(r, t)$ varies with time. How long does it take for the charge density at a certain point to drop to one half of its initial value in the case of copper? (The changes with time are so rapid that it is no longer a good enough approximation to use a simple linear relation between current density j and field E).

10.3 Obtain expressions for the electric and magnetic dipole moments of an atom or molecule in terms of the microscopic quantities ρ_a, v_a and M_a defined in section 10.5. If the molecule is considered to consist of point particles obtain expressions for the moments in terms of the charges, velocities and intrinsic magnetic dipole moments of the point particles.

11

Electromagnetic waves

We have already outlined in section 10.4 how a spark discharge can produce changing electric and magnetic fields that propagate outwards from the changing charge and current distributions in the spark. Any charge distribution which changes with time, or any varying current, can give rise to a radiated electromagnetic field. However it is only when the changes are rapid that the radiated fields are large enough to carry away an appreciable fraction of the power dissipated in the system. In practice radiated fields associated with varying currents are usually unimportant unless the variations contain frequencies higher than 100 kHz. This is about the lowest frequency used for commercial transmission of radio waves.

In this chapter we will consider some general properties of electromagnetic waves propagating in free space or in homogeneous isotropic media. We will also consider what happens to simple waves at a plane boundary between two media. The electromagnetic field will be treated classically. This means that we will assume the energy in the waves to be distributed continuously throughout the space occupied by them, and the energy to depend in some way on the macroscopic fields E, D, B and H. The same approach was adopted in the discussion of the energies in static electric and magnetic fields. The media in which the waves are travelling will also be treated classically, as if they were continuous, and will be characterized by macroscopic quantities such as relative permittivity ε, relative permeability μ, and electrical conductivity σ. These quantities vary with frequency, although the electrical conductivity of almost all substances is effectively constant for most frequencies that we will consider. The interpretation of the parameters ε, μ and σ throws much light on the microscopic description of matter, but these problems will rarely be mentioned in this chapter. The classical theory provides a good description of most electromagnetic phenomena for frequencies below about 10^{11} Hz, and for some phenomena at higher frequencies.

11.1   ELECTROMAGNETIC WAVES IN FREE SPACE

In this section we will show formally that there exist solutions to Maxwell's equations in free space corresponding to waves which always travel with the same speed c.

Any electromagnetic field must satisfy *all* of Maxwell's equations. Let us assume that there is a changing charge and current distribution within a certain region of space, and let us consider the fields produced by this source of radiation. In the free space outside the region, the charge and current densities are everywhere zero, and the following relations hold between the vectors $\mathbf{E}$ and $\mathbf{D}$, and between $\mathbf{B}$ and $\mathbf{H}$:

$$\mathbf{D} = \varepsilon_0 \mathbf{E}$$

$$\mathbf{H} = \mathbf{B}/\mu_0.$$

Maxwell's Equations (10.10) to (10.13) reduce to

$$\text{div } \mathbf{E} = 0 \tag{11.1}$$

$$\text{div } \mathbf{B} = 0 \tag{11.2}$$

$$\text{curl } \mathbf{B} = \varepsilon_0 \mu_0 \frac{\partial \mathbf{E}}{\partial t} \tag{11.3}$$

$$\text{curl } \mathbf{E} = -\frac{\partial \mathbf{B}}{\partial t}. \tag{11.4}$$

We will manipulate Equations (11.3) and (11.4) in order to derive equations obeyed by the fields $\mathbf{E}$ and $\mathbf{B}$ separately. Taking the curl of both sides of Equation (11.3) we find

$$\text{curl curl } \mathbf{B} = \varepsilon_0 \mu_0 \text{ curl} \left(\frac{\partial \mathbf{E}}{\partial t} \right).$$

Since the operator curl does not include the time variable, curl $(\partial \mathbf{E}/\partial t)$ is the same function as $\partial(\text{curl } \mathbf{E})/\partial t$, and

$$\text{curl curl } \mathbf{B} = \varepsilon_0 \mu_0 \frac{\partial}{\partial t}(\text{curl } \mathbf{E}).$$

Hence from Equation (11.4)

$$\text{curl curl } \mathbf{B} = -\varepsilon_0 \mu_0 \frac{\partial^2 \mathbf{B}}{\partial t^2}.$$

The function curl curl **B** can be replaced using the identity (B.20) given in Appendix B,

$$\text{curl curl } \mathbf{B} = \text{grad div } \mathbf{B} - \nabla^2 \mathbf{B}.$$

The divergence of the field **B** is everywhere zero and so we find finally

$$\nabla^2 \mathbf{B} = \varepsilon_0 \mu_0 \frac{\partial^2 \mathbf{B}}{\partial t^2}. \tag{11.5}$$

The magnetic field **B** obeys this equation everywhere in free space where the charge and current densities are zero.

By a procedure similar to that used above it can be shown that the electric field **E** must satisfy the equation

$$\nabla^2 \mathbf{E} = \varepsilon_0 \mu_0 \frac{\partial^2 \mathbf{E}}{\partial t^2} \tag{11.6}$$

at all points in free space where there is no charge and current. Equations (11.5) and (11.6) are called the *wave equations* for the fields **B** and **E**.

In order to investigate what sort of field the wave equation (11.6) describes let us use Cartesian coordinates and write the vector **E** in terms of its components as $\mathbf{E} = \mathbf{e}_x E_x + \mathbf{e}_y E_y + \mathbf{e}_z E_z$. Equation (11.6) then separates into three equations obeyed by the components separately. The equation obeyed by the component E_x is

$$\nabla^2 E_x = \varepsilon_0 \mu_0 \frac{\partial^2 E_x}{\partial t^2}.$$

Consider the simple situation when the field component E_x does not vary in the x- or y-directions. The above equation then reduces to

$$\frac{\partial^2 E_x}{\partial z^2} = \varepsilon_0 \mu_0 \frac{\partial^2 E_x}{\partial t^2}.$$

One solution to this equation has the form

$$E_x = E_0 f(z - vt), \tag{11.7}$$

where E_0 is a constant, f is any function, and v is given by

$$v = \frac{1}{\sqrt{\varepsilon_0 \mu_0}},$$

as can readily be verified by substitution. The expression (11.7) describes how the field component E_x varies with distance z and time. Suppose that at time t_1

the field E_x is as shown in Figure 11.1 i.e. the function f corresponds to the profile or waveform shown. At a later time t_2 it can be seen from expression (11.7) that the profile has moved along the z-axis, unchanged in shape, by an amount $v(t_2 - t_1)$. Hence Equation (11.7) represents a waveform moving

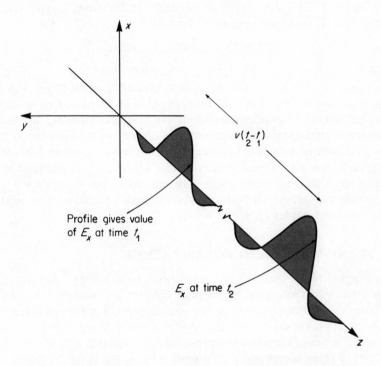

Figure 11.1. A waveform moving along the z-axis, unchanged in shape, with speed v. The component E_x of the electric field in the wave is plotted against coordinate z. E_x is independent of the coordinates x and y.

unchanged in shape with speed v along the positive Z-axis. A similar waveform moving in the direction of the negative z-axis is given by

$$E_x = E_0 f(z + vt).$$

The three dimensional Equation (11.6) also represents waves travelling with speed equal to $(\varepsilon_0 \mu_0)^{-1/2}$. *All electromagnetic waves in free space move with the same speed.* This speed is given the special symbol c, and

$$c = \frac{1}{\sqrt{\varepsilon_0 \mu_0}}. \tag{11.8}$$

The speed c is a fundamental constant which has been measured in a variety of ways with radiation of widely different frequencies. The frequency range extends from the lowest radio frequencies (~ 100 kHz) to x-ray and γ-ray frequencies in excess of 10^{18} Hz, and all the measurements are consistent with a constant value for c. The most precise measurements are made with visible light ($\sim 6 \times 10^{14}$ Hz) and very high frequency radio waves ($\sim 10^{10}$ Hz), and the value of c currently adopted as the best is

$$c = (2.997925 \pm 0.000001) \times 10^8 \text{ m s}^{-1}.$$

We will hereafter assume the value of c to be 2.998×10^8 m s^{-1}.

The constant c has a very special place in present physical theories. A basic postulate of the theory of special relativity is that all observers in uniform relative motion in free space measure the speed of light to have the same value c. This suggests that Maxwell's equations are already in a form obeyed by all observers in uniform motion in free space. The people in the different laboratories moving uniformly with respect to each other, if observing the same phenomena, would observe different electric and magnetic fields, but Equations (11.1) to (11.6) would still be obeyed by the fields. Maxwell's equations and relativity are discussed in Chapter 14.

11.2   PLANE WAVES AND POLARIZATION

The simplest type of wave that is a solution to Equation (11.5) and (11.6) is a plane wave. Plane waves will be our main concern in this chapter. Far enough away from any source of radiation, the electromagnetic wave is a plane wave, to a good approximation, and the study of plane waves illustrates many important features of the electromagnetic field. We will first review some simple properties of plane waves and then discuss what is meant by the polarization of an electromagnetic wave.

Suppose that an electromagnetic field is produced by a current in a coil, and that the current oscillates continuously with constant amplitude and constant* angular frequency ω. In the steady state the field strengths everywhere will oscillate at the same angular frequency. The electromagnetic waves produced are said to be *monochromatic*. (The name is taken from the visible part of the spectrum, but it is usually applied to waves that oscillate sinusoidally at any fixed frequency.) The power radiated by the coil, averaged over a complete cycle of the current, is constant in the steady state. The power is carried away by the radiated waves, and the same average power crosses any sphere whose centre is at the position of the coil. If we consider two such spheres, one closer in than the other, the energy crossing unit area per second (in a direction along a particular line drawn away from the coil) over the inner sphere is greater than the energy crossing unit area per second in the same direction over the

---

* This is not strictly attainable in practice. However the idea helps us to introduce some of the basic concepts of waves.

outer sphere. The average energies depend on the amplitudes of the electric and magnetic fields at the appropriate points on the two spheres, and hence the field strengths decrease with increasing distance from the source.

The fields at two points are said to be *in phase* if their oscillations occur in unison. A *wave front* is a surface over which the fields are in phase. Wave fronts in which the fields have a constant phase spread out from the coil and become weaker, as the fields decrease, like the ripples spreading out from a disturbance in the middle of a pond. Far away from the coil, in a region whose dimensions are small compared with the distance from the coil, the wave fronts become very nearly plane surfaces. These planes are perpendicular to the direction of propagation, which is the line joining the region of interest to the far-off coil. The diminution in the amplitudes of the fields within the region of interest is negligible, and the wave is said to be a *plane wave*. A plane wave has its wave fronts parallel to each other, and the electric field in a plane monochromatic electromagnetic wave propagating along the positive direction of the z-axis, with the electric field lying in the direction parallel to the x-axis, can be represented by the real part of the expression

$$\mathbf{E} = \mathbf{e}_x E_0 \exp \left[j(\omega t - kz) \right]. \tag{11.9}$$

The field has the same amplitude E_0 everywhere, and at all points with the same z-coordinate, i.e. all points in any plane parallel to the x–y plane, the field has the same phase. It is oscillating with angular frequency ω. It can readily be shown that the expression (11.9) is a solution to the wave equation (11.6), and hence the speed of propagation of the wave, c, is given by

$$c = \frac{\omega}{k}. \tag{11.10}$$

k is called the *wavenumber*, and is measured in m^{-1}. Figure 11.2 shows how the field varies with time at a fixed position $z = 0$, and how it varies with position at time $t = 0$. The variation with distance is sinusoidal, and the distance between equivalent points on successive cycles is equal to the *wavelength* λ of the wave.

$$\lambda = \frac{2\pi}{k}. \tag{11.11}$$

The third diagram on Figure 11.2 shows how the field varies with z at time δt. The waveform has the same shape, but the maxima have moved forward a distance δz, such that

$$\frac{\delta z}{\delta t} = \frac{\omega}{k}$$

Combining Equations (11.10) and (11.11) we obtain a relation between the wavelength λ and the frequency $v \, (= \omega/2\pi)$ of electromagnetic waves *in free space*,

$$\boxed{\lambda v = c.} \tag{11.12}$$

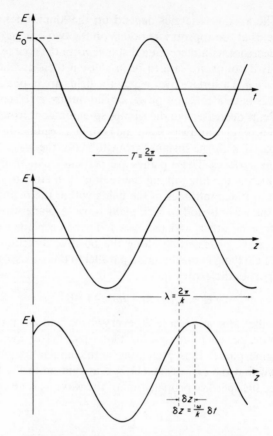

Figure 11.2. The top figure shows how the electric
field in the plane wave (11.9) varies with time at
all points with coordinate z equal to zero. The
field is independent of coordinates x and y.
The centre figure shows how the field varies with
z at time $t = 0$. The bottom figure shows the
variation with z at a later time δt. The wave
crests have moved forward a distance δz equal to
$\omega\, \delta t/k$.

We will now discuss what is meant by the polarization of a plane wave.
Expression (11.9) represents a *linearly polarized* plane wave. The electric field
vector always lies parallel to the x-axis, pointing either in the positive or
negative direction depending on the instant at which we observe the oscillating
field. A linearly polarized monochromatic plane wave in which the electric field
is always parallel to some other line in the x–y plane can be expressed as

$$\mathbf{E} = (\mathbf{e}_x E_{0x} + \mathbf{e}_y E_{0y}) \exp\left[j(\omega t - kz)\right]. \tag{11.13}$$

This corresponds to the sum of two oscillations in phase and polarized at 90° to each other. The resultant is a vector which oscillates along a line making an angle θ with the x-axis, where

$$\tan \theta = \frac{E_{0y}}{E_{0x}},$$

as shown on Figure 11.3.

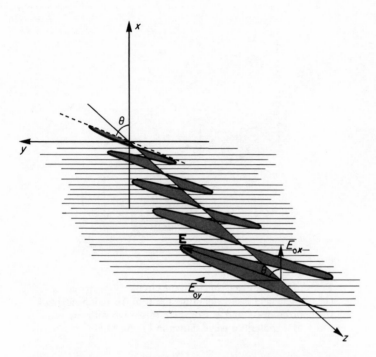

Figure 11.3. The electric field in a linearly polarized plane wave in which the electric field vector oscillates along a line making an angle θ with the x-axis.

The plane wave

$$\mathbf{E} = \mathbf{e}_x E_{0x} \exp \left[j(\omega t - kz + \phi) \right], \tag{11.14}$$

where ϕ is an arbitrary constant phase angle, is also a solution to the wave equation (11.6). The sum of waves given by expressions (11.14), (11.13) and (11.9) is also a plane wave solution to the wave equation. By adding monochromatic plane waves linearly polarized in different directions and out of phase we can construct expressions which represent plane waves in more complicated states of polarization. For example the field

$$\mathbf{E} = \mathbf{e}_x E_0 \exp\left[j(\omega t - kz)\right] + \mathbf{e}_y E_0 \exp\left[j(\omega t - kz + \pi/2)\right] \quad (11.15)$$

represents a *circularly polarized* plane wave. It is a wave progressing along the z-axis in which the vector $\mathbf{E}$ has a constant length E_0 but is continuously changing its direction. At any fixed point the vector $\mathbf{E}$ rotates* in a circle with angular frequency ω. The field as a function of position at the fixed time $t = 0$ is illustrated in Figure 11.4.

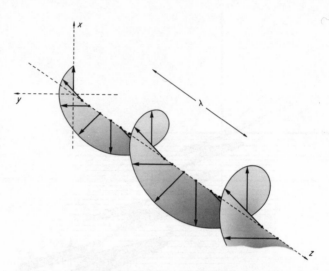

Figure 11.4. The electric field in the circularly polarized wave (11.15). The field is shown at time $t = 0$, and is independent of the coordinates x and y. Compare this wave with the plane polarized wave shown in Figure 11.5.

A *randomly polarized* plane wave (often called an unpolarized plane wave) is one in which the direction of oscillation of the field is changing randomly with time. A plane wave of visible light contains contributions from very many individual atomic or molecular de-excitations, and is randomly polarized. A plane wave at a frequency of 1 MHz which arises from the controlled sinusoidal variation of current along a wire has a fixed polarization. Only when the changing charge and current distributions producing the waves are controlled in a regular manner will such sources emit radiation of a fixed state of polarization.

---

* Looking in the direction of propagation of the wave, i.e. looking in the $+z$-direction the electric field vector, at a fixed point, rotates anticlockwise as time progresses. The wave (11.15) is said to be a left handed circularly polarized wave. Note that the $\mathbf{E}$ vector rotates clockwise as z increases if the wave is observed at a fixed time, looking in the direction of propagation.

11.2.1 Plane waves in free space

In this section we will first show that the electric and magnetic field vectors in a plane wave lie in directions perpendicular to the direction of propagation of the wave. We will afterwards derive a relation between the electric and magnetic fields in the wave.

Consider a monochromatic plane wave solution to the wave equation (11.6) in which the electric field is given by

$$\mathbf{E} = \mathbf{E}_0 \exp[j(\omega t - kz)], \tag{11.16}$$

where $\omega/k = c$, and $\mathbf{E}_0$ is an arbitrary constant vector. This expression represents a linearly polarized plane wave with the electric field direction arbitrarily chosen. We will now show that the vector $\mathbf{E}_0$ must lie in a plane parallel to the x–y plane. The choice of the positive z-axis to be the direction of propagation of the wave involves no loss of generality, since we are free to choose the coordinate axes in whatever directions are most convenient.

The field given by Equation (11.16) must satisfy *all* of Maxwell's equations. Hence from Equation (11.1)

$$\frac{\partial E_x}{\partial x} + \frac{\partial E_y}{\partial y} + \frac{\partial E_z}{\partial z} = 0 \tag{11.17}$$

at all points, and at all times. For the plane wave (11.16) the field is independent of the coordinates x and y, and hence

$$\frac{\partial E_x}{\partial x} = \frac{\partial E_y}{\partial y} = 0.$$

Equation (11.17) now becomes

$$\frac{\partial E_z}{\partial z} = 0,$$

and this tells us that the component E_z is independent of the coordinate z. The only way this statement can be reconciled with the expression (11.16) for the field $\mathbf{E}$, is for the component E_z to be everywhere zero. Hence the vector $\mathbf{E}_0$ must be perpendicular to the direction of propagation.

This conclusion is true for any plane wave, not only a linearly polarized plane wave. A wave in a more complex state of polarization can be expressed as the sum of linearly polarized waves with different amplitudes and phases, and the proof given above holds for each term in the sum, and hence for the whole wave. The electric field vector in a randomly polarized plane wave must also lie in directions perpendicular to the direction of propagation of the wave. The average time interval between the random changes in direction of the polarization of the wave is much greater than the period. Thus one is justified in using the expression (11.16) in the time intervals between changes, and the proof given above remains valid.

Let us now turn to the problem of determining the magnetic field **B** in a plane wave in terms of the electric field **E**. Let the electric field be given by expression (11.16), in which the vector $\mathbf{E}_0$ lies in the x–y plane. For simplicity we will choose the direction of the field **E** to be along the x-axis, i.e.

$$\mathbf{E} = \mathbf{e}_x E_0 \exp\left[j(\omega t - kz)\right].$$

On substituting this into Equation (11.4) we obtain

$$-\frac{\partial \mathbf{B}}{\partial t} = \operatorname{curl}\left\{\mathbf{e}_x E_0 \exp\left[j(\omega t - kz)\right]\right\}$$

$$= -\mathbf{e}_y jk E_0 \exp\left[j(\omega t - kz)\right].$$

Integrating this equation with respect to time we find

$$\mathbf{B} = \mathbf{e}_y \frac{k}{\omega} E_0 \exp\left[j(\omega t - kz)\right].$$

There is no constant of integration in the expression for the field **B** in the absence of a steady field. From Equation (11.10) the ratio of the wavenumber k of the wave to its angular frequency ω is the reciprocal of the speed of the wave

$$\frac{k}{\omega} = \frac{1}{c}.$$

Hence, finally,

$$\mathbf{B} = \mathbf{e}_y \frac{E_0}{c} \exp\left[j(\omega t - kz)\right].$$

If a similar analysis is pursued with the wave progressing in the negative z-direction, the field **E** is given by

$$\mathbf{E} = \mathbf{e}_x E_0 \exp\left[j(\omega t + kz)\right],$$

and the field **B** can be shown to be

$$\mathbf{B} = -\mathbf{e}_y \frac{E_0}{c} \exp\left[j(\omega t + kz)\right].$$

The general expression for the linearly polarized plane wave solution to Equation (11.6) is of the form

$$\mathbf{E} = \mathbf{E}_0 \exp\left[j(\omega t - \mathbf{k} \cdot \mathbf{r})\right], \tag{11.18}$$

where the vector **k** is in the direction of propagation of the wave and the vector $\mathbf{E}_0$ is perpendicular to **k**. The relationship between the fields **B** and **E** in a plane wave can be summarized in the equation

$$\mathbf{B} = \frac{1}{c}(\hat{\mathbf{k}} \wedge \mathbf{E}), \tag{11.19}$$

where $\hat{\mathbf{k}}$ is a unit vector in the direction of propagation of the wave. Figure 11.5 illustrates the relative directions of the electric and magnetic fields in a linearly polarized plane wave.

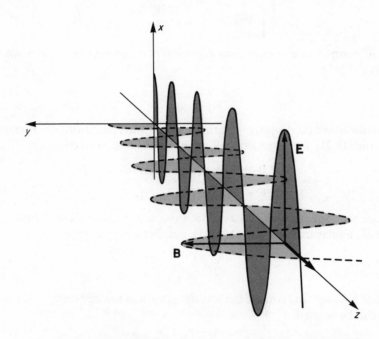

Figure 11.5. The relative directions of the electric and magnetic fields in a plane wave. The wave is linearly polarized and the electric field vector oscillates along the x-axis. The wave propagates in the z-direction. The magnetic field in the wave is along the y-axis and in phase with the electric field.

The amplitude of the magnetic field in a plane wave in free space is usually quite small. The ratio of the electric field to the magnetic field is equal to the speed of light in free space. If the amplitude of the electric field in a plane wave is $0.1 \, \text{V m}^{-1}$, as might be encountered in a radio wave some kilometres from a transmitter, the amplitude of the magnetic field is $0.1 \times (3 \times 10^8)^{-1}$ or $3.3 \times 10^{-10} \, \text{T}$.

11.2.2   Plane waves in isotropic insulating media

Let us now consider waves travelling in an isotropic, non-conducting medium of relative permittivity ε and relative permeability μ. The wave equations for the fields $\mathbf{E}$ and $\mathbf{B}$ can be derived from Maxwell's equations in a manner similar to that given in section 11.1. They are

$$\nabla^2 \mathbf{B} = \varepsilon\varepsilon_0\mu\mu_0 \frac{\partial^2 \mathbf{B}}{\partial t^2} \tag{11.20}$$

$$\nabla^2 \mathbf{E} = \varepsilon\varepsilon_0\mu\mu_0 \frac{\partial^2 \mathbf{E}}{\partial t^2}. \tag{11.21}$$

The solutions to these equations represent waves travelling with a speed v given by

$$v = \frac{1}{\sqrt{\varepsilon\varepsilon_0\mu\mu_0}}. \tag{11.22}$$

The monochromatic, linearly polarized, plane wave solutions to Equations (11.20) and (11.21), for a wave travelling in the positive z-direction, are

$$\mathbf{B} = \mathbf{B}_0 \exp[\mathrm{j}(\omega t - kz)]$$

$$\mathbf{E} = \mathbf{E}_0 \exp[\mathrm{j}(\omega t - kz)].$$

These are the same expressions as for a wave in free space. Now however, for a wave of angular frequency ω, the wave number k is given by

$$k = \frac{\omega}{v} \tag{11.23}$$

instead of $k = \omega/c$. In terms of the wavelength λ and the frequency v, Equation (11.23) can be written

$$\lambda v = v. \tag{11.24}$$

The wavelength of a wave of fixed frequency is thus different when the wave is travelling in a medium to the wavelength when the wave is travelling in free space. If the product of the relative permittivity and the relative permeability of the medium is greater than unity, as is usually the case, the speed v of the wave in the medium is less than the speed c in free space, from Equation (11.22). The wavelength λ of the wave is thus decreased, as can be seen from Equation (11.24).

By arguments similar to those used in section 11.2.1 above, it can be shown that the electric and magnetic fields in plane waves travelling in isotropic media must also lie in directions perpendicular to the direction of propagation of the wave. If the direction of propagation is given by the unit vector $\hat{\mathbf{k}}$ it may also be shown in the same way that the relationship between the electric and magnetic fields in the wave in the medium is

$$\mathbf{B} = \frac{1}{v}(\hat{\mathbf{k}} \wedge \mathbf{E}). \tag{11.25}$$

The wavelength of a wave of fixed frequency changes as we go from one medium to another, as we have seen. Equation (11.22) shows that the speed of the wave in a medium depends on the frequency, since the quantities ε and μ vary with frequency. The ratio of the speed of an electromagnetic wave in free space to its speed in a medium is equal to the *refractive index n* of the medium for radiation of the frequency of the wave.

$$n = \frac{c}{v} = \sqrt{\varepsilon\mu}. \qquad (11.26)$$

Thus the refractive index varies with frequency. The variation with frequency of the refractive index of the glass of a prism gives rise to the *dispersion* observed as the different colours of the visible spectrum are separated by being refracted through different angles.

In order to show that the definition (11.26) of refractive index is the same as the definition in terms of angles of incidence and refraction of a wave at a plane boundary, consider a plane wave incident on the flat interface shown in Figure 11.6. Let the wave be incident in free space onto a medium with relative

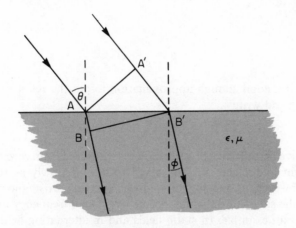

Figure 11.6. A plane wave refracted at the plane boundary between two media. The directions of propagation of the wave in air and in the medium are given by the directions of the rays A'B' and AB respectively.

permittivity ε and relative permeability μ. The wave front of the incident wave is AA' and the wave front of the reflected wave is BB'. Along the surfaces AA' and BB' the phase of the wave is constant, hence the change in phase over the distance AB equals the change over A'B'. The frequency of the wave does not change as it goes into the medium. Any changes in the properties of the wave are due to the oscillations of the electrons in the atoms or molecules of the medium as they interact with the incident field. The electrons in turn radiate a field which modifies the original wave, but they radiate at the same frequency and so the final product is a wave of the original frequency. The wavelength and speed of the wave change but the frequency does not. Hence the phases at the points B and B′ are the same if the time taken for the radiation to go from A to B and from A′ to B′ are the same. For this to be so we require that

$$\frac{A'B'}{c} = \frac{AB}{v}$$

or

$$\frac{\sin\theta}{\sin\phi} = \frac{c}{v}$$

But

$$\frac{\sin\theta}{\sin\phi} = n$$

and hence

$$n = \frac{c}{v} = \sqrt{\varepsilon\mu}.$$

Usually it is a good enough approximation to put the relative permeability equal to unity and write

$$n = \sqrt{\varepsilon}. \tag{11.27}$$

If the relative permittivity of a substance is measured from zero frequency up to visible light frequencies it is observed to fall off slowly as the frequency increases. However, at the higher frequencies, there exist anomalies superimposed on the slow fall off which occurs over small frequency ranges. Let us take water as an example. In static fields and in alternating fields at low frequencies the relative permittivity of water is about 80 at room temperature. As the frequency of the alternating electric field increases above 100 MHz, the relative permittivity begins to decrease. It falls throughout the far infrared region of the spectrum, and is down to about 8 at 10 GHz. At low frequencies the chief contribution to the polarization arises from the preferred orientation along the field of the permanent dipole moment of the water molecules. As the frequency increases, the molecules, which are restricted by frictional torques from other nearby molecules, are unable to turn fast enough to follow the rapid variations of the field direction. The contribution to the relative permittivity from the permanent dipole moments thus falls off to zero. As the frequency increases further, into the near infrared and visible parts of the spectrum, the *induced* dipole moment of each molecule in turn fails to keep up with the field alternations, and the relative permittivity falls even further. The anomalies in the curve showing how the relative permittivity varies with frequency occur at frequencies corresponding to the internal modes of excitation of the water molecule. At a frequency of 5.1×10^{15} Hz (yellow light from a sodium vapour lamp) the refractive index of water at room temperature is down to 1.33, corresponding to a relative permittivity of 1.77.

11.3   ENERGY IN ELECTROMAGNETIC WAVES

In this section we will obtain an expression giving the energy flowing across unit area per second at any point in an electromagnetic wave. This quantity is a vector which depends on the electric and magnetic fields, and is called the *Poynting vector*.

Pure monochromatic waves do not exist in nature. The reason we have been discussing monochromatic waves (and why we will continue to do so) is that they are mathematically quite simple and serve to illustrate many of the important properties of electromagnetic wave propagation. Waves that do actually occur are made up of waves with several frequencies, and form what is called a *wave group*. If the wave group is nearly monochromatic then the frequencies in the group are clustered in a narrow region around the main frequency. If a wave group is propagating in a dispersive medium each frequency component travels with a slightly different speed. The speed with which a component with angular frequency ω travels is given by Equation (11.22), and the wave number of the component wave by Equation (11.23). This speed is called the *wave velocity* or *phase velocity* of that component of the group which has angular frequency ω. The energy in the group of waves travels with the *group velocity*. This depends on the dispersion of the different frequencies in the group, i.e. on the variation of the relative permittivity of the medium with frequency*.

Sometimes the wave velocities of the different components of an electromagnetic wave are greater than the speed of light in free space. An example is in the ionosphere, which has a refractive index less than one for high frequency waves (see Problem 11.9) and so can totally reflect radio waves incident at a sufficiently oblique angle. There is no contradiction of the special theory of relativity since the energy in a wave travelling through the ionosphere moves with the group velocity, which is always less than the speed of light in free space.

There is no dispersion of a wave in free space. The wave velocities of all the different frequency components are the same, and equal to the speed of light in free space, c. The group velocity is also equal to c, and the energy associated with any frequency component in the wave moves in the direction the wave is moving with the same speed. We will calculate the rate of energy flow across unit area associated with a component of angular frequency ω. If the wave is moving in the z-direction the energy flow is in the z-direction and is the same at all points with the same value of the coordinate z. Since the fields are

---

* For a discussion of wave and group velocities see for example section 1.5 in Optics, by F. G. Smith and J. H. Thomson, J. Wiley, London, 1972.

oscillating in time the instantaneous energy flow oscillates also. However the instantaneous energy flow averaged over a whole period of oscillation of the fields is constant. Figure 11.7 shows a rectangular box of length c lying along the z-axis. The box has sides of lengths a and b. The energy contained in the box,

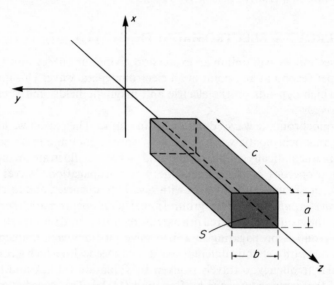

Figure 11.7. When a plane wave propagates in the z-direction in free space, the average energy crossing the surface S per second is equal to the average energy stored in the fields within a box of length equal to the speed of light.

averaged over a cycle of the fields, is the average energy crossing the area S per second, since the energy in the wave travels with the speed c.

In sections 2.4 and 6.3.1 we obtained expressions which together give the total energy U in a combination of static electric and magnetic fields.

$$U = \frac{1}{2} \int_V (\mathbf{E} \cdot \mathbf{D} + \mathbf{B} \cdot \mathbf{H}) \, d\tau, \tag{11.28}$$

where V is the volume over which the fields are non-zero. *We will assume that this equation also gives the energy when the fields change with time.* Furthermore we will assume that it correctly gives the total energy in an arbitrary volume V. Using Equation (11.28) and the relations $\mathbf{D} = \varepsilon_0 \mathbf{E}$, $\mathbf{B} = \mu_0 \mathbf{H}$, we find that at time t the energy density in the box drawn in the figure is equal to $\varepsilon_0 E^2/2 + \mu_0 H^2/2$ where the fields $\mathbf{E}$ and $\mathbf{H}$ are given by

$$\mathbf{E} = \mathbf{e}_x E_0 \exp\left[j(\omega t - kz)\right]$$

$$\mathbf{H} = \mathbf{e}_y H_0 \exp\left[j(\omega t - kz)\right].$$

In evaluating energy, which is a scalar, it is necessary to abandon the complex notation and remember that the fields are given by the real parts of the above expressions. With this change the total energy in the box at time t is given by

$$U = \frac{1}{2} \int_0^a \int_0^b \int_0^c [\varepsilon_0 E_0^2 \cos^2(\omega t - kz) + \mu_0 H_0^2 \cos^2(\omega t - kz)]\, dz\, dy\, dx$$

The average energy in the box is thus

$$\bar{U} = \frac{ab}{2T} \int_0^T \int_0^c [\varepsilon_0 E_0^2 \cos^2(\omega t - kz) + \mu_0 H_0^2 \cos^2(\omega t - kz)]\, dz\, dt$$

where T is the period of the oscillations and is equal to $2\pi/\omega$. Hence

$$\bar{U} = \frac{ab}{2} \int_0^c (\tfrac{1}{2}\varepsilon_0 E_0^2 + \tfrac{1}{2}\mu_0 H_0^2)\, dz$$

$$= \frac{ab}{4} c(\varepsilon_0 E_0^2 + \mu_0 H_0^2).$$

Using Equations (11.8) and (11.19) the above equation reduces to

$$\bar{U} = \frac{ab}{2} c\varepsilon_0 E_0^2 = \tfrac{1}{2} E_0 H_0 ab.$$

The average energy crossing unit area per second at any point in the plane wave is thus $E_0 H_0/2$, and its direction is the z-direction. Introducing the symbol $\mathbf{N}$ for the Poynting vector we thus have

$$\bar{\mathbf{N}} = \mathbf{e}_z \tfrac{1}{2} E_0 H_0.$$

This is the result we would obtain if the instantaneous value of the Poynting vector were

$$\mathbf{N} = \mathbf{e}_z EH,$$

or alternatively

$$\boxed{\mathbf{N} = \mathbf{E} \wedge \mathbf{H}.} \qquad (11.29)$$

This equation gives the Poynting vector for a plane wave in free space; however it can also be used to obtain the energy flow in *any* electromagnetic field. It can be shown that the interpretation of expression (11.29) as the energy crossing unit area per second in an electromagnetic wave is consistent with the conservation laws of physics and with Maxwell's equations. All experiments so far performed support this interpretation.

For a plane wave moving in the direction of the unit vector $\hat{\mathbf{k}}$ in an isotropic medium with the relative permittivity ε and relative permeability μ the Poynting vector is

$$\mathbf{N} = \mathbf{E} \wedge \mathbf{H}$$

$$= \mathbf{E} \wedge \frac{\mathbf{B}}{\mu\mu_0}$$

$$= \mathbf{E} \wedge (\hat{\mathbf{k}} \wedge \mathbf{E})\frac{\sqrt{\varepsilon\varepsilon_0\mu\mu_0}}{\mu\mu_0},$$

from Equations (11.22) and (11.25). This gives

$$\mathbf{N} = \hat{\mathbf{k}}E^2\sqrt{\frac{\varepsilon\varepsilon_0}{\mu\mu_0}}$$

or

$$\overline{\mathbf{N}} = \tfrac{1}{2}\hat{\mathbf{k}}E_0^2\sqrt{\frac{\varepsilon\varepsilon_0}{\mu\mu_0}}. \tag{11.30}$$

If the amplitude of the electric field in a plane wave is 0.1 V m^{-1} (a typical value a few km from a radio transmitter) the energy crossing unit area per second in air is $10^{-2}\sqrt{\varepsilon_0/\mu_0}/2 \sim 1.4 \times 10^{-5} \text{ W m}^{-2}$. If the radio wave were being propagated in water (which has a relative permittivity of about 80 at radio frequencies), and at some point the electric field had the amplitude 0.1 V m^{-1}, the Poynting vector would be increased to $\sim 1.2 \times 10^{-4} \text{ W m}^{-2}$.

We now outline a more general, but more mathematical approach to the Poynting vector in linear isotropic materials. The remainder of this section can be omitted without loss of continuity. We begin with Equation (11.28) which gives the energy in a volume V enclosed by a surface S. The material inside the surface S is taken to be uniform and isotropic and to have relative permittivity ε and relative permeability μ. The rate of change of energy inside the surface S is equal to

$$\frac{\partial U}{\partial t} = \frac{1}{2}\frac{\partial}{\partial t}\left\{\int_V (\mathbf{E}\cdot\mathbf{D} + \mathbf{B}\cdot\mathbf{H})\,d\tau\right\}$$

$$= \int_V \left(\frac{\partial\mathbf{D}}{\partial t}\cdot\mathbf{E} + \frac{\partial\mathbf{B}}{\partial t}\cdot\mathbf{H}\right)d\tau$$

But

$$\frac{\partial\mathbf{D}}{\partial t} = \text{curl }\mathbf{H} - \mathbf{j}$$

and

$$\frac{\partial\mathbf{B}}{\partial t} = -\text{curl }\mathbf{E},$$

from Maxwell's Equations (10.13) and (10.12). Hence

$$\frac{\partial U}{\partial t} = \int_V [(\text{curl } \mathbf{H} - \mathbf{j}) \cdot \mathbf{E} - \text{curl } \mathbf{E} \cdot \mathbf{H}] \, d\tau. \tag{11.31}$$

Using the vector identity (B.23),

$$\text{div} (\mathbf{E} \wedge \mathbf{H}) = \mathbf{H} \cdot \text{curl } \mathbf{E} - \mathbf{E} \cdot \text{curl } \mathbf{H},$$

Equation (11.31) can be rewritten as

$$\frac{\partial U}{\partial t} = - \int_V \mathbf{j} \cdot \mathbf{E} \, d\tau - \int_V \text{div} (\mathbf{E} \wedge \mathbf{H}) \, d\tau.$$

The last term on the right hand side of this equation can be transformed into an integral over the surface S by using the divergence theorem, and we have

$$\frac{\partial U}{\partial t} = - \int_V \mathbf{j} \cdot \mathbf{E} \, d\tau - \int_S (\mathbf{E} \wedge \mathbf{H}) \cdot d\mathbf{S}. \tag{11.32}$$

If the free charge density within the volume V is given by ρ, and if the average velocity of the charges is $\mathbf{v}$, the current density $\mathbf{j}$ is equal to $\rho \mathbf{v}$, and hence

$$- \int_V \mathbf{j} \cdot \mathbf{E} \, d\tau = - \int_V (\mathbf{E} \cdot \mathbf{v}) \rho \, d\tau.$$

The term on the right hand side of this equation is equal to the rate at which the electric field $\mathbf{E}$ does work on the free charges within the volume V. This work appears as heat, and the first term on the right hand side of Equation (11.32) thus represents the rate of Joule heating in the volume V. The energy in the electromagnetic field in the volume V decreases at a rate equal to the rate of Joule heating plus the energy flowing out of the surface S per unit time. Hence we are led to interpret the integral $\int_S (\mathbf{E} \wedge \mathbf{H}) \cdot d\mathbf{S}$ as the amount of energy flowing outwards over the surface S per second. This interpretation is consistent with the conservation of energy and the original assumption that the energy was given by Equation (11.28), i.e. has the same form as for static fields. We now go one step further and postulate that the vector $\mathbf{N} = \mathbf{E} \wedge \mathbf{H}$ represents the energy flow per unit area per second at every point.

11.4   THE ABSORPTION OF PLANE WAVES IN CONDUCTORS AND THE SKIN EFFECT

If a wave is propagating in a conducting medium the oscillating electric field in the wave sets up currents. Work must be done to drive the currents, and some of the energy in the wave is dissipated as heat in the medium. A plane wave progressing through a conducting medium must therefore be attenuated. In this section we determine the attenuation and discuss some of its practical consequences.

We will consider materials for which the current density $\mathbf{j}$ is proportional to the electric field $\mathbf{E}$, and write

$$\mathbf{j} = \sigma\mathbf{E}, \tag{11.33}$$

where the constant σ can be taken to be equal to the steady current electrical conductivity. This relationship is almost exact, up to the highest frequencies where classical theory is applicable, for most materials that obey Ohm's law with direct currents. For very rapidly changing currents, the relationship may no longer be an adequate approximation (see Problem (10.2)). We now use Maxwell's equations to derive an expression which represents an electromagnetic wave propagating in a conducting medium in which the relationship (11.33) is obeyed. In this case the last two Maxwell Equations (10.12) and (10.13) can be written

$$\operatorname{curl} \mathbf{E} = -\frac{\partial \mathbf{B}}{\partial t} \tag{11.34}$$

and

$$\frac{1}{\mu\mu_0} \operatorname{curl} \mathbf{B} = \varepsilon\varepsilon_0 \frac{\partial \mathbf{E}}{\partial t} + \sigma\mathbf{E}, \tag{11.35}$$

where ε and μ are the relative permittivity and permeability respectively. From these two equations we can derive an equation involving the field $\mathbf{E}$ alone, in a similar manner to the procedure used in section 11.1 for waves in free space. We again use the relation (B.20),

$$\operatorname{curl} \operatorname{curl} \mathbf{E} = -\nabla^2\mathbf{E} + \operatorname{grad} \operatorname{div} \mathbf{E}.$$

When a wave is travelling through a conducting medium the macroscopic charge density remains zero everywhere, and so we have $\operatorname{div} \mathbf{E} = 0$. The last equation under these conditions becomes

$$\operatorname{curl} \operatorname{curl} \mathbf{E} = -\nabla^2\mathbf{E}.$$

From Equation (11.34)

$$\operatorname{curl} \operatorname{curl} \mathbf{E} = -\operatorname{curl}\left(\frac{\partial \mathbf{B}}{\partial t}\right)$$

$$= -\frac{\partial}{\partial t}(\operatorname{curl} \mathbf{B}),$$

and hence

$$\nabla^2\mathbf{E} = \frac{\partial}{\partial t}(\operatorname{curl} \mathbf{B})$$

$$= \varepsilon\varepsilon_0\mu\mu_0 \frac{\partial^2 \mathbf{E}}{\partial t^2} + \mu\mu_0\sigma \frac{\partial \mathbf{E}}{\partial t}, \tag{11.36}$$

from Equation (11.35).

Let us look for a solution to Equation (11.36) which represents an attenuated plane wave moving in the z-direction, i.e. one in which the fields are constant everywhere in the x–y plane at a fixed value of the coordinate z, but whose amplitudes decrease exponentially as z increases. In such a wave

$$\nabla^2 \mathbf{E} = \frac{\partial^2 \mathbf{E}}{\partial z^2}.$$

If the wave has angular frequency ω,

$$\frac{\partial \mathbf{E}}{\partial t} = j\omega \mathbf{E},$$

and

$$\frac{\partial^2 \mathbf{E}}{\partial t^2} = j\omega \frac{\partial \mathbf{E}}{\partial t}.$$

Substituting these expressions into Equation (11.36) gives the equation

$$\frac{\partial^2 \mathbf{E}}{\partial z^2} = j\omega \varepsilon \varepsilon_0 \mu \mu_0 \frac{\partial \mathbf{E}}{\partial t} + \mu \mu_0 \sigma \frac{\partial \mathbf{E}}{\partial t}. \tag{11.37}$$

For almost all conducting media the product $\omega \varepsilon \varepsilon_0$ is very much less than the electrical conductivity σ. For example a weak electrolyte would have a conductivity of about $1\,(\Omega\,\text{m})^{-1}$. At a frequency of 1 MHz the value of $\omega \varepsilon \varepsilon_0$ is less than $10^{-2}\,(\Omega\,\text{m})^{-1}$. Thus it is usually a good approximation to neglect the first term on the right hand side of Equation (11.37). The differential equation obeyed by a plane wave in conducting media then becomes

$$\frac{\partial^2 \mathbf{E}}{\partial z^2} = \mu \mu_0 \sigma \frac{\partial \mathbf{E}}{\partial t}. \tag{11.38}$$

The attenuated plane wave solution we are looking for has the form

$$\mathbf{E} = \mathbf{E}_0 \exp\left[j(\omega t - \beta z)\right] \exp\left(-\alpha z\right), \tag{11.39}$$

where α and β are parameters to be determined. Substitution of the derivatives obtained from this equation into Equation (11.38) gives

$$\alpha = \beta = \sqrt{\frac{\mu \mu_0 \sigma \omega}{2}}. \tag{11.40}$$

The quantity

$$\delta = \frac{1}{\alpha} = \sqrt{\frac{2}{\mu \mu_0 \sigma \omega}} \tag{11.41}$$

measures how rapidly the wave is attenuated and is called the *skin depth*. Usually the relative permeability μ is very nearly unity and $\delta \sim \sqrt{2/\mu_0 \sigma \omega}$. For a frequency of 50 Hz, the skin depth in copper is

$$\delta = \sqrt{2/4\pi \times 10^{-7} \times 5.9 \times 10^{7} \times 100\pi} \sim 1 \text{ cm}.$$

As the frequency increases the skin depth decreases proportionally to $\omega^{-1/2}$ and at a frequency of 50 MHz the skin depth in copper has decreased to $\sim 10^{-3}$ cm.

The fact that the skin depth is small for good conductors at high frequencies has important consequences for the design of conductors to take high frequency currents. Suppose that an alternating current is passed along a long cylindrical conductor. The electric field is everywhere parallel to the axis of the cylinder, and by symmetry depends only on the distance z perpendicularly inwards from the surface. The exact determination of the fields requires the solution of Maxwell's equations fitted to the appropriate boundary conditions. However when examining the behaviour of the electric field near the surface of the conductor, it is a good approximation to consider the surface flat if the skin depth is small compared to the radius of the cylindrical wire. The field will be given in this approximation by Equations (11.39) and (11.40), and the amplitude of the current density $\mathbf{j} = \sigma\mathbf{E}$ in the rod decreases exponentially with distance inwards as shown in Figure 11.8.

At very high frequencies the current is all carried in a thin outer layer of *any* conductor. This phenomenon is known as the *skin effect*. At high frequencies

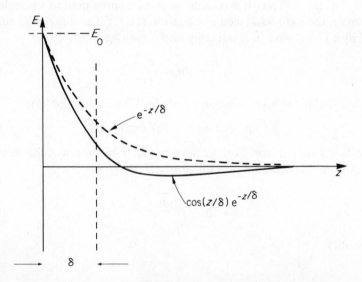

Figure 11.8. The variation of the electric field with distance inwards from the surface of a cylindrical conductor. The full curve is expression (11.39) for the field at time $t = 0$. The dotted curve is the locus of the electric field amplitude, i.e. a plot of the function $E_0 \exp(-z/\delta)$.

the resistance R of a conductor whose dimensions are much greater than the skin depth δ can be estimated by assuming that all the current is carried uniformly in a surface layer of thickness δ. If the conductor is a cylinder of length l and radius r, its resistance in this approximation is

$$R = \frac{l}{\sigma \cdot 2\pi r \delta}$$

$$= \frac{l}{2\pi r}\sqrt{\frac{\mu_0 \omega}{2\sigma}}. \tag{11.42}$$

If R_0 is the resistance of the rod to direct current

$$R = \frac{rR_0}{2\delta}.$$

The resistance to direct current of a copper rod 5 mm diameter is about 8×10^{-4} Ω per metre. At a frequency of 50 MHz it is about 0.1 Ω per metre. It would be an expensive waste of copper to use such a conductor at 50 MHz.

The absorption of plane waves in insulators

There can still be attenuation of a wave even though the conductivity of the medium through which the wave is travelling is zero. This is because the wave may lose energy in exciting various microscopic modes of excitation in the medium. For example, excited energy levels of the molecules of a gas, or levels associated with the regular lattice structure of a solid may absorb energy. This absorption of energy can be described by giving the medium a complex relative permittivity

$$\varepsilon = \varepsilon_1 + j\varepsilon_2.$$

A plane wave progressing in the medium can be represented by Equation (11.16),

$$\mathbf{E} = \mathbf{E}_0 \exp\left[j(\omega t - kz)\right],$$

or equivalently (using Equations (11.22) and (11.23)) by

$$\mathbf{E} = \mathbf{E}_0 \exp\left[j\omega(t - \sqrt{\varepsilon\varepsilon_0\mu\mu_0}z)\right]. \tag{11.43}$$

If ε is complex

$$\sqrt{\varepsilon\varepsilon_0\mu\mu_0} = \sqrt{(\varepsilon_1 + j\varepsilon_2)\varepsilon_0\mu\mu_0}.$$

This may be written in the form

$$\sqrt{\varepsilon\varepsilon_0\mu\mu_0} = (n - j\eta)\sqrt{\varepsilon_0\mu\mu_0}$$

where

$$n = \sqrt{\varepsilon_1}$$

and

$$\eta = -\frac{\varepsilon_2}{2\sqrt{\varepsilon_1}}.$$

We have assumed that ε_2 is small compared to ε_1, corresponding to a small absorption of energy. If we put the relative permeability μ equal to unity, Equation (11.43) can now be rewritten

$$\mathbf{E} = \mathbf{E}_0 \exp\left[j\omega\left(t - \frac{z}{v}\right)\right] \exp\left(-\frac{\omega\eta}{c}z\right), \qquad (11.44)$$

where

$$v = \frac{c}{\varepsilon_1}.$$

This expression represents a plane wave whose amplitude is decreasing exponentially with z. The amplitude falls to one half of its initial value over a distance equal to

$$\frac{c}{\omega\eta}\log_e 2 \sim \frac{0.7c}{\omega\eta}.$$

The quantities n and η vary with frequency. For radiation of wavelength 5×10^{-7} m (yellow visible light) in glass, η is very small. As the frequency is increased into the ultraviolet region of the spectrum, η increases due to the excitation of the electrons in the molecules of the glass. Glass is opaque to ultra-violet wavelengths shorter than about 2×10^{-7} m. As another example most of the infrared waves with wavelengths between about 10^{-3} and 10^{-5} m cannot penetrate the atmosphere because of absorption due to the excitation of rotations in the air molecules.

Classical theory, not surprisingly, has problems describing absorption due to atomic or molecular energy levels. Experimental data on, for example, the absorption of microwaves in the atmosphere are properly described by the quantum theory of the interaction of the electromagnetic field with the atmosphere's microscopic constituents.

11.5   THE REFLECTION AND TRANSMISSION OF ELECTROMAGNETIC WAVES

When plane waves are incident on a boundary between different media, some of the incident energy crosses the boundary and some is reflected. In this section we will determine the reflection and transmission coefficients in a few examples of practical importance. These coefficients depend on the relative permittivities and permeabilities of the media on either side of the boundary, i.e. on their refractive indices.

When static fields were being discussed we found that they had to obey certain conditions at the boundaries between different media. The solutions to the time independent Maxwell equations within regions of different permittivities or permeabilities had to be fitted to each other at the boundaries between the different regions, and this fitting procedure enables the fields everywhere to be uniquely determined. The same procedure has to be followed when the fields change with time, and we will have to digress a little in order to discuss the boundary conditions that the changing fields have to obey.

11.5.1   Boundary conditions on electric and magnetic fields

(i)   *The tangential component of the electric field* **E**

The integral form of Faraday's law is

$$\oint \mathbf{E} \cdot d\mathbf{l} = -\frac{d}{dt} \int_S \mathbf{B} \cdot d\mathbf{S}, \tag{11.45}$$

where S is a complete surface spanning the loop over which the line integral is taken, and the direction of the vector $d\mathbf{S}$ is given by the right hand rule. Let us apply Equation (11.45) to the small surface S shown in Figure 11.9 which cuts the interface between the two different media and is perpendicular to it. The sides AD and BC cutting the surface are of equal length Δt. The sides AB and DC, parallel to the interface just above and below it, and parallel to the tangential components of the fields $\mathbf{E}_1$ and $\mathbf{E}_2$, are of equal length Δl. Let us make the length Δt smaller and smaller, i.e. tend to zero. The area of the surface S then tends to zero. The magnetic field **B** and its derivatives are always finite in real situations involving finite charges and currents, so that the integral on the right hand side of Equation (11.45) tends to zero also. Hence

$$\mathbf{E}_1 \cdot \Delta \mathbf{l} - \mathbf{E}_2 \cdot \Delta \mathbf{l} = 0,$$

where $\Delta \mathbf{l}$ is a vector in the direction from A to B, and $\mathbf{E}_1$ and $\mathbf{E}_2$ are the fields along AB just above the interface, and along DC just below it. Hence the tangential components of the electric field are continuous at the boundary, and the condition

$$E_\parallel \quad \text{continuous,} \tag{11.46}$$

which we found earlier to be true for the electrostatic field, is true for the electric field in general.

(ii)   *The tangential component of the magnetic intensity* **H**

The integral form of the Maxwell equation relating the fields **H** and **D** is

$$\oint \mathbf{H} \cdot d\mathbf{l} = \int_S \left(\mathbf{j}_f + \frac{\partial \mathbf{D}}{\partial t} \right) \cdot d\mathbf{S}. \tag{11.47}$$

Let us continue to use Figure 11.9 where the sides AB and DC are now parallel to the projections in the surface of the fields $\mathbf{H}_1$ and $\mathbf{H}_2$ on either side, i.e. the

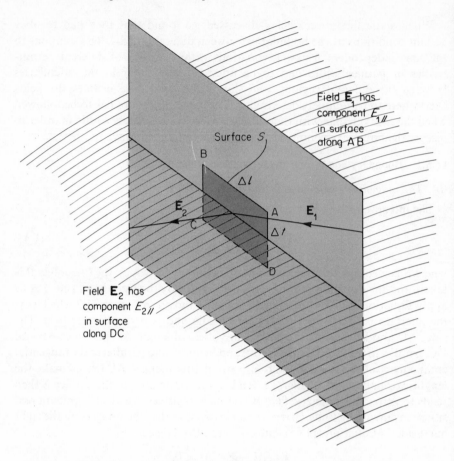

Field **E**₁ has component $E_{1//}$ in surface along AB

Surface S

Field **E**₂ has component $E_{2//}$ in surface along DC

Figure 11.9. Part of a surface separating two media. The component of the field **E**₁ in the surface is along AB. The component of the field **E**₂ in the surface is along DC.

vectors **H**₁ and **H**₂ lie in the same plane as the rectangle ABCD. Let the area S shrink as before. The second term on the right hand side of Equation (11.47) tends to zero as the length Δt tends to zero, since the field **D** and its derivatives are everywhere finite. The first term on the right hand side becomes

$$\text{Limit}_{\Delta t \to 0} \, (j_f \, \Delta l \, \Delta t), \tag{11.48}$$

which is the current through the surface S in the direction given by the right hand rule together with the sense of traversal of the line integral $\oint \mathbf{H} \cdot \mathbf{dl}$ around the rectangle. j_f is the magnitude of the free current density at S; the direction of the vector $\mathbf{j}_f$ is perpendicular to the plane of the rectangle ABCD since the sides AB and DC are parallel to the maximum components along the surface of the fields **H** either side of the interface. For media of finite electrical conductivity σ the limit (11.48) is always zero since the current density cannot be infinite in real situations involving finite charges. For finite conductivity Equation (11.47) thus reduces to

$$\mathbf{H}_1 \cdot \Delta\mathbf{l} - \mathbf{H}_2 \cdot \Delta\mathbf{l} = 0,$$

and the boundary condition (as worked out also in section 5.3.2) is that

$$H_\parallel \quad \text{continuous.} \tag{11.49}$$

The tangential components of the magnetic intensity are continuous at all points across the interface between two media.

In the treatment of certain problems involving good conductors it is often useful to assume that the conductivities of the conductors are infinite. This assumption greatly simplifies the mathematics, and causes little loss of understanding of the physical processes that occur. With this approximation the limit (11.48) becomes

$$\text{Limit}_{\Delta t \to 0} \, (j_f \, \Delta l \, \Delta t) = j_s \, \Delta l,$$

and Equation (11.47) applied to the loop ABCD can be written

$$\oint \mathbf{H} \cdot \mathbf{dl} = j_s \, \Delta l.$$

Here j_s is a *surface current per unit length* which flows in a vanishingly thin layer, on a macroscopic scale, at the interface between the two media. The current $j_s \, \Delta l$ is again to be taken in the direction given by the right hand screw rule and the sense of traversal of ABCD. There can be no electromagnetic field inside a perfect conductor since the skin depth is zero. There is thus a discontinuity in the field vector $\mathbf{H}$ at the surface. The discontinuity is related to the surface current. Figure 11.10 shows a surface current $\mathbf{j}_s$ flowing in the surface of a perfect conductor. $\mathbf{j}_s$ is perpendicular to the plane of the small rectangular surface ABCD cutting the interface since we have again chosen the sides AB and DC to lie parallel to the projection of the vector $\mathbf{H}_1$ on the surface. If we calculate the line integral $\oint \mathbf{H} \cdot \mathbf{dl}$ around the loop ABCD, traversing the loop from A to B and round to A, the only contribution to the integral is from the portion AB, and

$$\oint \mathbf{H} \cdot \mathbf{dl} = H_{1\parallel} \, \Delta l = j_s \, \Delta l.$$

Hence

$$H_{1\parallel} = j_s, \tag{11.50}$$

and at the interface between one medium and a perfect conductor the maximum component of the magnetic intensity along the surface is equal to the surface current per unit length, and their directions are perpendicular to each other.

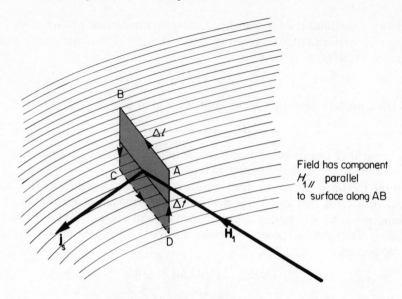

Field has component $H_{1//}$ parallel to surface along AB

Figure 11.10. Part of a surface separating two media. The lower medium has infinite electrical conductivity, and a surface current is induced in it in order to satisfy the boundary condition (11.50).

This boundary condition is useful for calculating surface currents induced in conductors by electromagnetic waves.

(*iii*)   *The normal component of the magnetic field* **B**

The equation div **B** = 0 is always true since the lines of the field **B** are continuous. The argument made in section 5.3.2 is true at all times and the boundary condition is

$$B_\perp \quad \text{continuous,} \tag{11.51}$$

where $B_\perp$ is the component of the field **B** normal to the surface.

(*iv*)   *The normal component of the field* **D**

The equation div **D** = ρ_f, where ρ_f is the volume density of free charge, leads to the same conclusion obtained in section 2.3.3 that

$$D_\perp \quad \text{continuous} \tag{11.52}$$

if there is no surface density of free charge at the interface, and when there is a surface charge density σ_f,

$$D_{1\perp} - D_{2\perp} = \sigma_f, \tag{11.53}$$

where $D_{1\perp}$ and $D_{2\perp}$ are the magnitudes of the perpendicular components of the field **D** either side of the interface.

11.5.2   Reflection at dielectric boundaries

We will now use the boundary conditions given in the previous section to determine what fraction of the energy in a plane wave incident on a dielectric boundary is reflected, and what fraction is transmitted. We will consider the simplest problem where a plane wave, of angular frequency ω, is incident normally on the boundary. All directions in the plane of the boundary are equivalent in this situation and so we need only consider a single plane polarized wave in order to obtain the answer for a wave of any state of polarization.

Figure 11.11 shows schematically the wave incident on the plane interface. Let the direction of propagation of the wave be the z-direction, and the electric

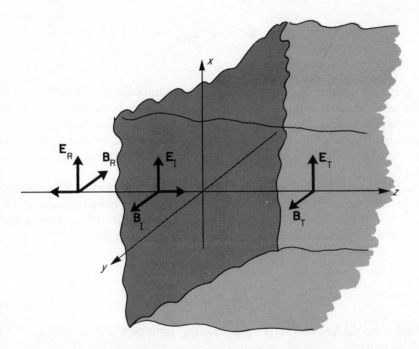

Figure 11.11. A plane polarized wave incident normally onto the plane boundary between a dielectric on the right and free space on the left. The arrows indicate the relative directions of the fields in the incident, reflected and transmitted waves.

vector lie along the x-axis, so that the magnetic field vectors lie along the y-axis. Let us take the medium on the left to be air or free space, and that on the right to have relative permittivity ε and relative permeability μ. On the left there is an incident wave, whose fields will be denoted by $\mathbf{E}_I$ and $\mathbf{B}_I$, and a reflected wave with fields $\mathbf{E}_R$ and $\mathbf{B}_R$. On the right there is a transmitted wave with fields $\mathbf{E}_T$ and $\mathbf{B}_T$. We can write down the plane wave expressions for these fields at once:

$$\mathbf{E}_I = \mathbf{e}_x E_{0I} \exp\left[j(\omega t - k_1 z)\right]$$

$$\mathbf{B}_I = \mathbf{e}_y B_{0I} \exp\left[j(\omega t - k_1 z)\right]$$

$$\mathbf{E}_R = \mathbf{e}_x E_{0R} \exp\left[j(\omega t + k_1 z)\right]$$

$$\mathbf{B}_R = -\mathbf{e}_y B_{0R} \exp\left[j(\omega t + k_1 z)\right]$$

$$\mathbf{E}_T = \mathbf{e}_x E_{0T} \exp\left[j(\omega t - k_2 z)\right]$$

$$\mathbf{B}_T = \mathbf{e}_y B_{0T} \exp\left[j(\omega t - k_2 z)\right],$$

where

$$k_1 = \omega\sqrt{\varepsilon_0 \mu_0}$$

$$k_2 = \omega\sqrt{\varepsilon \varepsilon_0 \mu \mu_0}.$$

In these expressions $E_{0I}, B_{0I}, \ldots$ etc., are the amplitudes of the fields, and we have used the result of Equation (11.25) for the relative directions of the electric and magnetic fields. The relationships between the amplitudes, and hence the transmission and reflection coefficients, are determined by making the total fields on either side of the boundary obey the conditions discussed in section 11.5.1.

Let us choose the boundary to be at $z = 0$ for convenience. The common factor $\exp(j\omega t)$ can be ignored; if we satisfy the boundary conditions at one time, say $t = 0$, they will be satisfied at all times. The fields in this problem are all tangential to the boundary, and condition (11.46) tells us that

$$\mathbf{E}_I + \mathbf{E}_R = \mathbf{E}_T,$$

or

$$E_{0I} + E_{0R} = E_{0T}. \tag{11.54}$$

There is no surface current at the dielectric boundary and condition (11.49) tells us that

$$\mathbf{H}_I + \mathbf{H}_R = \mathbf{H}_T,$$

or

$$\frac{B_{0I}}{\mu_0} - \frac{B_{0R}}{\mu_0} = \frac{B_{0T}}{\mu \mu_0}.$$

Using Equations (11.22) and (11.25) this last equality can be written

$$E_{0I}\sqrt{\frac{\varepsilon_0}{\mu_0}} - E_{0R}\sqrt{\frac{\varepsilon_0}{\mu_0}} = E_{0T}\sqrt{\frac{\varepsilon \varepsilon_0}{\mu \mu_0}}. \tag{11.55}$$

Replacing the relative permeability by unity (a good approximation for almost all dielectrics), and using the relation (11.27) between refractive index n and relative permittivity ε, Equation (11.55) becomes

$$E_{0I} - E_{0R} = nE_{0T}. \tag{11.56}$$

Combining Equations (11.54) and (11.56) we obtain

$$\frac{E_{0R}}{E_{0I}} = \frac{1-n}{1+n}, \tag{11.57}$$

and

$$\frac{E_{0T}}{E_{0I}} = \frac{2}{1+n}. \tag{11.58}$$

Equation (11.57) gives the ratio of the amplitudes of the reflected and incident waves. For reflection at a medium whose refractive index is greater than unity the ratio is negative. This corresponds to a phase change of π on reflection, i.e. the electric vector of the incident wave oscillates 180° out of phase with that in the reflected wave. This is the phase difference used in interference calculations in optics when one part of a split wave front is reflected at a medium of greater refractive index than the medium in which it is incident.

The reflection coefficient R is defined as the reflected energy divided by the incident energy. These quantities are given by the time average over a cycle of the Poynting vectors for the incident and reflected waves.

$$
R = \frac{\overline{\mathbf{E}_R \wedge \mathbf{H}_R}}{\overline{\mathbf{E}_I \wedge \mathbf{H}_I}}
$$

$$
= \frac{E_{0R}^2}{E_{0I}^2}
$$

$$
= \left(\frac{1-n}{1+n}\right)^2, \tag{11.59}
$$

from Equation (11.57).

The transmission coefficient T is defined as the energy transmitted divided by the energy incident.

$$
T = \frac{\overline{\mathbf{E}_T \wedge \mathbf{H}_T}}{\overline{\mathbf{E}_I \wedge \mathbf{H}_I}}
$$

$$
= \frac{\frac{1}{2}E_{0T}^2 \sqrt{\dfrac{\varepsilon\varepsilon_0}{\mu\mu_0}}}{\frac{1}{2}E_{0I}^2 \sqrt{\dfrac{\varepsilon_0}{\mu_0}}},
$$

from Equation (11.22) and (11.25). Again putting $\mu = 1$, we obtain from Equation (11.58)

$$T = \frac{4n}{(1 + n)^2}.$$ (11.60)

This could have been deduced also from Equation (11.59), since conservation of energy requires that

$$T + R = 1.$$

For glass, in the visible region of the electromagnetic spectrum, the refractive index n is about 1.5. The reflection and transmission coefficients, calculated from Equations (11.59) and (11.60), are about 4% and 96% respectively. The refractive index of water for waves of frequency 100 MHz is about 9. The reflection and transmission coefficients are about 65% and 35% respectively. It is seen that the larger the relative permittivity for a given frequency, the larger the fraction of energy reflected. The microscopic explanation of this is in terms of the waves radiated by the molecular charges as they oscillate due to their interaction with the incident wave. A large relative permittivity corresponds to a large oscillating dipole moment in each molecule, and these dipole moments give rise to a large radiated field. The forward wave in this field interferes destructively with the original wave within the dielectric, and gives rise to a small transmission.

11.5.3   Reflection at metallic boundaries

We will now consider the case of the normal reflection of a plane polarized wave at a metallic boundary. The wave is usually almost completely reflected, and for a perfect conductor with infinite conductivity the transmission is zero since there can be no electromagnetic field inside the conductor.

The procedure for determining the reflection and transmission coefficients of a metal of finite conductivity follows that of the previous section. The wave in the metal is now an attenuated wave in which the electric field is given by Equations (11.39) and (11.40);

$$\mathbf{E}_T = \mathbf{e}_x E_{0T} \exp\left[j(\omega t - \alpha z)\right] \exp\left(-\alpha z\right).$$ (11.61)

The corresponding magnetic field is

$$\mathbf{B}_T = \mathbf{e}_y B_{0T} \exp\left[j(\omega t - \alpha z)\right] \exp\left(-\alpha z\right). \qquad \cdot$$ (11.62)

Application of the boundary conditions (11.46) and (11.49) gives the equations

$$E_{0I} + E_{0R} = E_{0T}$$ (11.63)

and

$$\frac{B_{0I}}{\mu_0} - \frac{B_{0R}}{\mu_0} = \frac{B_{0T}}{\mu\mu_0}$$ (11.64)

as before. We may use Equation (11.25) again to give us the relation between E_{0I} and B_{0I}, and between E_{0R} and B_{0R}, but we have to return to Maxwell's equations to determine the relation between E_{0T} and B_{0T}.

The differential form of Faraday's law holds inside the metal, and everywhere

$$\text{curl } \mathbf{E}_T = -\frac{\partial \mathbf{B}_T}{\partial t}.$$

From Equation (11.61),

$$\text{curl } \mathbf{E}_T = -\mathbf{e}_y E_{0T} \exp\left[j(\omega t - \alpha z)\right] \exp(-\alpha z)\alpha(1 + j).$$

Integration of this equation with respect to time, gives, from Faraday's law,

$$\mathbf{B}_T = \mathbf{e}_y E_{0T} \frac{\alpha}{\omega}\left(\frac{1 + j}{j}\right) \exp\left[j(\omega t - \alpha z)\right] \exp(-\alpha z).$$

We have put the constant of integration equal to zero in the absence of a steady field. Comparison of the above equation with Equation (11.62) shows that

$$B_{0T} = E_{0T}\frac{\alpha}{\omega}(1 - j),$$

or

$$B_{0T} = E_{0T}\sqrt{\frac{\mu\mu_0\sigma}{2\omega}}\,(1 - j),$$

From Equation (11.40).

The two Equations (11.63) and (11.64) derived from the boundary conditions can now be expressed in terms of the amplitudes of the electric fields alone. We obtain

$$E_{0I} + E_{0R} = E_{0T}$$

$$E_{0I} - E_{0R} = E_{0T}\sqrt{\frac{\sigma}{2\omega\mu\varepsilon_0}}(1 - j).$$

Eliminating E_{0T} from these equations, and putting the relative permeability μ equal to unity, gives

$$\frac{E_{0R}}{E_{0I}} = -\frac{\sqrt{\dfrac{\sigma}{2\omega\varepsilon_0}}\,(1 - j) - 1}{\sqrt{\dfrac{\sigma}{2\omega\varepsilon_0}}\,(1 - j) + 1}. \tag{11.65}$$

The reflection coefficient R is given by

$$R = \frac{E_{0R}^2}{E_{0I}^2}$$

$$= \left| \frac{E_{0R}}{E_{0I}} \right|^2. \tag{11.66}$$

Substitution of the value of E_{0R}/E_{0I} from Equation (11.65) gives

$$R \simeq 1 - 2\sqrt{\frac{2\omega\varepsilon_0}{\sigma}}, \tag{11.67}$$

if higher powers of $\sqrt{2\omega\varepsilon_0/\sigma}$ are neglected.

For copper $\sigma \sim 6 \times 10^7$ (ohm m)$^{-1}$, and at infrared frequencies $\sqrt{2\omega\varepsilon_0/\sigma}$ is equal to about 0.01. The hot filament of an electric heater at 1000 K has an emission spectrum with a peak around a wavelength of 2.5×10^{-6} m. A metal reflector behind the filament will reflect about 97% of these infrared rays. The other 3% passes across the metal air boundary but is rapidly absorbed in the metal due to Joule heating. At lower frequencies the reflection is greater and clean metal plates reflect all but a few parts in a million of radio waves incident upon them. At optical frequencies classical theory cannot describe metallic reflection; atomic transitions take place, and these account for the colours of metals.

★ **11.5.4   Polarization by reflection**

In this section we will consider one way in which a plane polarized wave can be obtained from a wave that is initially randomly polarized, i.e. has its electric field vector changing randomly in the wave front. A randomly polarized plane wave can always be expressed as the sum of two waves which have their electric fields oscillating at right angles to each other. If we can find a process that affects the field in one of these directions differently to the way in which it affects the field in the other we may be able to produce a wave with some degree of plane polarization. If we could remove the field in one direction altogether we would produce a completely polarized plane wave. When radiation strikes a dielectric surface obliquely the reflection coefficient depends on the state of polarization of the incident wave. For plane waves polarized with their electric vectors lying in the reflection plane we will show that for one particular angle of incidence, called the Brewster angle, the reflection coefficient is zero. Hence, if randomly polarized light, such as that from a sodium vapour lamp, is reflected at the Brewster angle, plane polarized light will be generated with the electric field vector perpendicular to the reflection plane.

Consider Figure 11.12, in which is shown a view looking down on the boundary between air or free space and a medium which has relative permittivity

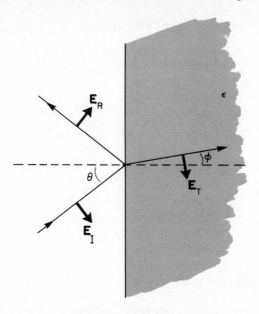

Figure 11.12. The reflection of a plane wave incident obliquely on the plane boundary between a dielectric and free space. The wave is linearly polarized, with the electric fields in the plane of incidence.

ε, relative permeability equal to one, and refractive index n. The plane wave is incident at an angle θ to the normal, and the transmitted wave is at an angle ϕ to the normal, where

$$\frac{\sin \theta}{\sin \phi} = n.$$

Let us resolve the electric field vector into two components, one in the plane of the boundary, the other in the plane of incidence, i.e. the plane of the paper in Figure 11.12. Consider the component in the latter plane. We will determine the reflection coefficient for this component of the wave by matching the fields either side of the boundary in the usual way. If $E_{0I}, B_{0I}, \dots$ etc. denote the amplitudes of the various fields (using the same nomenclature as before) then the boundary conditions (11.46) and (11.52) (in the absence of free charge on the surface) can be written as

$$E_{0I} \cos \theta - E_{0R} \cos \theta = E_{0T} \cos \phi$$

$$E_{0I} \sin \theta + E_{0R} \sin \theta = \varepsilon E_{0T} \sin \phi.$$

The above two equations reduce to the equations

$$E_{OI} - E_{OR} = E_{OT} \frac{\cos \phi}{\cos \theta}$$

$$E_{OI} + E_{OR} = \varepsilon E_{OT} \frac{\sin \phi}{\sin \theta} = n E_{OT},$$

since

$$n^2 = \varepsilon.$$

It is a straightforward matter to eliminate E_{OT} from these two coupled equations to obtain

$$\frac{E_{OR}}{E_{OI}} = \frac{n \cos \theta - \cos \phi}{n \cos \theta + \cos \phi}.$$

The reflection coefficient R is thus given by

$$R = \left(\frac{n \cos \theta - \cos \phi}{n \cos \theta + \cos \phi} \right)^2. \tag{11.68}$$

It can be seen at once that there is no reflected wave whose electric vector lies in the plane of incidence if

$$n \cos \theta = \cos \phi.$$

After some algebraic manipulation this condition reduces to

$$\tan \theta = n. \tag{11.69}$$

This equation gives the *Brewster angle*, for which the reflected wave is zero. An analysis similar to the above for that component of the original wave with the electric vector parallel to the boundary plane shows that there is no angle of reflection for which the amplitude of the reflected wave is zero. Hence if one reflects a randomly polarized wave at a dielectric with the angle of incidence equal to the Brewster angle the result is a plane polarized wave whose electric vector lies in a direction parallel to the boundary.

For yellow light on glass, the refractive index is about 1.5, and the Brewster angle is $\sim 56°$. Sunlight reflected off the sea or ground is partially polarized and a great deal of the glare is cut out by using polaroid sunglasses, which will not transmit light polarized in the Earth's horizontal plane.

Reflection at oblique incidence can also be used as an analyser of the degree of polarization of a plane wave. The small value of the reflection coefficient, given by Equation (11.68), makes this method less convenient to use than some other methods.

11.6   ELECTROMAGNETIC WAVES AND PHOTONS

In section 11.2.2 we described classically how a wave of fixed frequency travels with a different speed according to whether it is moving in free space or in a medium. If the relative permittivity of the medium is greater than unity the wave slows down. What is the microscopic description of the slowing down process? In asking this question we have already overstepped the bounds of continuum theory, which considers the medium to be continuous. The continuum theory breaks down in any event when the wavelength of the radiation is comparable with the spacing of the molecules. This spacing is a few Ångstroms in solids. At high frequencies, and for a microscopic description of many of the phenomena discussed in this chapter, it is convenient to look upon the electromagnetic field in a different way.

The classical theory considers the energy in the field to be distributed continuously, with an energy density $(\mathbf{E} \cdot \mathbf{D} + \mathbf{B} \cdot \mathbf{H})/2$, and to flow continuously at a rate given by the Poynting vector $(\mathbf{E} \wedge \mathbf{H})$. Quantum theory considers the energy in an electromagnetic field to exist in discrete light quanta, called photons. Each photon moves with the speed of light in free space and has zero rest mass. The energy E of the photons depends on the frequency v of the radiation and is given by

$$E = hv, \tag{11.70}$$

where h is Planck's constant. A suitable microscopic picture of the slowing down of a wave in a block of dielectric is in terms of a cloud of photons entering the block and being exchanged back and forth between the molecules before emerging. This increases the time it takes the photons to traverse the block and corresponds to a slower speed of travel in the medium than in free space.

We may use the relativistic relation between the total energy E and momentum p of a body of rest mass m_0 to calculate the momentum of a photon:

$$E^2 = c^2 p^2 + m_0^2 c^4.$$

The rest mass of photons is zero, and the momentum of a photon of energy E is given by

$$p = \frac{E}{c} = \frac{hv}{c}. \tag{11.71}$$

If we consider the energy in a wave to be carried by photons it can readily be seen that a plane wave incident on a metallic surface and reflected off it will exert a pressure on the surface. If the average value of the Poynting vector in the wave is $\overline{\mathbf{N}}$, the average momentum crossing unit area per second is given by Equation (11.71) to be $\overline{N}/c$. If all the wave is reflected at a perfect conductor the total change of momentum per unit area per second is equal to $2\overline{N}/c$ and this is the pressure P that the wave exerts on the conducting surface. If the amplitude of the electric field in the wave is E_0 and the wave is incident in air or free space we may write

$$\overline{N} = \tfrac{1}{2}E_0^2\sqrt{\frac{\varepsilon_0}{\mu_0}},$$

from Equation (11.30). Hence

$$P = \frac{E_0^2}{c}\sqrt{\frac{\varepsilon_0}{\mu_0}}$$

$$= \varepsilon_0 E_0^2. \tag{11.72}$$

We may make a rough estimate of the radiation pressure exerted by sunlight on the Earth if we make the assumption that all of the $1300\ \text{W m}^{-2}$ falling normally on the Earth arrives in a monochromatic wave. If E_0 is the amplitude of the electric field in this wave, we have

$$\tfrac{1}{2}E_0^2\sqrt{\frac{\varepsilon_0}{\mu_0}} = 1300,$$

which gives a value for E_0 of $\sim 1.4 \times 10^3\ \text{V m}^{-1}$. Assuming that all of the radiation falling on the Earth is absorbed, the radiation pressure will be one half of the value given by Equation (11.72). This works out at about $10^{-5}\ \text{N m}^{-2}$.

The radiation pressure on a conducting medium can be described classically by calculating the force on the conduction electrons (moving under the influence of the electric field of the wave) due to the magnetic field of the wave. (See Problem 11.13.) Some other electromagnetic phenomena can only be simply described if one uses the photon picture. One such phenomenon is the emission of electrons from certain materials when radiation of sufficiently short wavelength is shone on their surfaces. This effect is called the photoelectric effect. It is found that the number of photoelectrons emitted increases with the intensity of the wave shone on the surface of the material, but that the maximum energy of the emitted electrons is constant for a fixed frequency. Quantum mechanically an increase in the intensity of the radiation corresponds to an increase in the number of photons in the wave, the energy of the photons being constant at fixed frequency. Thus from the standpoint of the light quantum hypothesis both the results on the maximum energy of the photoelectrons and on their number can be readily understood.

PROBLEMS 11

11.1 The root mean square of the displacement current density in a linearly-polarized monochromatic plane wave in free space is $10^{-5}\ \text{A m}^{-2}$. The frequency is 10^8 Hz. Obtain values for the amplitudes of the electric and magnetic fields in the wave.

11.2 Show that the time average of the energy density in a monochromatic linearly-polarized plane wave moving in an isotropic non-conducting medium is distributed equally between the electric and magnetic fields.

Show that in a conducting medium the average energy in the magnetic field is greater than in the electric field.

11.3 A laser is a device that emits a parallel beam of monochromatic light. The intensity may be assumed constant across the beam. If the power is 1 watt and the beam has a diameter of 1 mm calculate the maximum amplitude of the magnetic field $\mathbf{B}$ in the beam in free space.

11.4 The centres of two close circular metal plates forming a parallel plate capacitor, initially charged, are connected internally by a fine wire of negligible self-inductance. Obtain an expression for the Poynting vector at the surface of the wire at any subsequent instant and hence evaluate the total flow of energy into the wire.

11.5 Deduce the conductivity of a medium of relative permittivity equal to five if the magnitude of the conduction and displacement current densities in it are equal when a monochromatic plane wave of frequency 10^8 Hz is propagated.

What is the attenuation per metre of the amplitude of a plane wave of frequency 10^6 Hz when propagated in such a medium? The relative permeability is unity.

11.6 A capacitor, formed by two flat parallel plates of area 113.1 cm^2, spaced 1 mm apart in air, is connected in series with a coil, wound with a single copper wire 40 m long and 0.2 cm in diameter, and an alternating voltage generator. If the inductance of the coil at resonance is 1 mH what is the Q of the circuit if the resistivity of copper is $1.72 \ \mu\Omega$ cm?

11.7 The skin depth of 1 MHz radio waves in pure copper is approximately 10^{-4} m at room temperature. Calculate the skin depth of 1 MHz radio waves at 15 K when the conductivity has increased by a factor of 10^4.

11.8 The electrical conductivity of sea water is about $4 \ (\Omega \, \text{m})^{-1}$. What is the skin depth for low frequency radio waves of wavelength 3000 m?

11.9 At heights greater than about 50 km above the surface of the Earth, in the ionosphere, the gas molecules in the atmosphere are ionized by ultra-violet rays from the sun. If the density of free electrons is n_0 per unit volume, determine the relative permittivity of the ionosphere for waves of frequency v. (There are magnetic fields in the ionosphere due to the motions of the charges, but these can be neglected when calculating the relative permittivity).

11.10 A monochromatic plane wave in free space is incident normally on the plane surface of a medium of refractive index equal to two. If the amplitude of the electric field in the incident wave is $10 \ \text{V m}^{-1}$, what is its value inside the medium?

11.11 Explain why the relative permittivity of water vapour begins to decrease at a lower frequency than the frequency at which that of nitrogen gas begins to decrease.

11.12 Show that if a material has a relative permeability greater than unity, at a certain angle of incidence there will be no reflection of a plane wave incident with its electric field vector parallel to the boundary.

11.13 Calculate classically the radiation pressure exerted on a conducting surface by a perpendicularly incident plane electromagnetic wave, and in the limit of infinite conductivity show that the answer reduces to Equation (11.72).

CHAPTER

Waveguides

The voltage and current waves along a coaxial cable or a pair of parallel metal plates were discussed in Chapter 9. The starting point of the discussion was that of electrical circuits, proceeding from the time delays introduced by successive sections of a ladder network to the smooth propagation of signals on transmission lines. It turned out that the velocity of propagation is always equal to the velocity of electromagnetic waves in the medium in which the line is embedded.

Now that we have considered the boundary conditions that electromagnetic waves have to obey at the surface of conductors, we can think of transmission lines in a different way, regarding them as systems that guide electromagnetic waves along certain paths. The conclusions of Chapter 9 could have been obtained by fitting solutions of Maxwell's equations to the boundary conditions appropriate to the conductors of the line.

A transmission line of a given shape may not be a very efficient way of transmitting power or signals from one point to another. For example two parallel wires radiate energy and there are heating losses in the wires: a coaxial cable with the outer cylinder earthed produces no fields outside, but there remain the energy losses due to resistive heating of the conductors, chiefly the inner one where the fields are largest. The most efficient way of transmitting energy over short distances at frequencies greater than about 1000 MHz is by using a waveguide. This is a hollow metal conductor whose cross-section is of a simple and *constant* shape. The walls reflect the waves to and fro, and there exists the possibility of combining the fields in the reflected waves in such a way as to add up to a wave that is propagated down the length of the pipe with very little attenuation. If the metallic walls were perfectly conducting there would be no attenuation at all since all of the incident energy would be reflected every time. This would be so if the metal used for the surfaces of the guide walls were superconducting. The technical problems connected with superconducting waveguides are presently being tackled. For waveguides normally used there are small energy losses at each reflection due to the currents induced in the walls by the transmitted fields. The energy losses in a length δx of the guide at a distance x from its beginning are proportional to the length δx and to the energy in the guide at that point. The mean energy in the wave in a waveguide thus falls off exponentially with distance travelled. In contrast, when radiation is emitted from antennas (this topic is discussed in the next chapter), the energy crossing unit area per second in the radiation fields falls off inversely as the square of the distance from the antenna. Hence if one is comparing antennas and waveguides from the point of view of energy losses it is usually best to use antennas to transmit waves over distances greater than a few kilometres. This is because at large distances the exponential rate of energy decay in a waveguide always results in a smaller energy than the inverse square law fall off in free space. Waveguides, however, are much better over shorter distances and are widely used in practice for the transmission of microwaves of wavelengths in the range from a few millimetres to about one metre. For example, 3 cm radar waves are usually generated in an area near the deck of a ship where they can be controlled, and are carried to the antenna on the top of the ship's mast with waveguides.

Waveguides of rectangular cross-section are the most commonly used and we shall restrict our attention to these alone.

12.1   THE PROPAGATION OF WAVES BETWEEN CONDUCTING PLANES

The main features of the propagation of waves in waveguides can be understood by considering the simpler problem of the propagation of waves between two parallel conducting planes.

Figure 12.1 shows parts of two parallel planes separated by a distance b. We will suppose that the planes are of infinite extent and furthermore that they are perfectly conducting. This last assumption is made in order to simplify the mathematics. The problem now is to determine the characteristics of the electromagnetic fields which can be propagated down the length of the plates. The fields we are looking for have to obey Maxwell's equations in the free space between the plates. They also have to obey certain boundary conditions at all points on the conducting walls. These conditions, as discussed in section 11.5.1, are that the tangential component of the electric field and the normal component of the magnetic field must be everywhere zero over the walls. The normal component of the electric field need not be zero, since there can be charges on the conducting surfaces, and the tangential component of the magnetic field need not be zero, since there can be surface currents in the

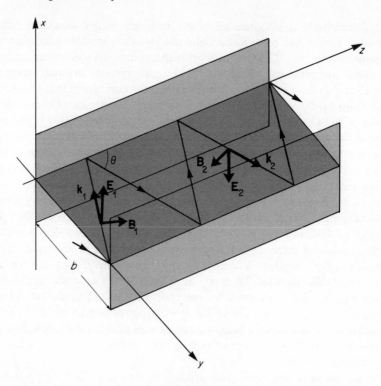

Figure 12.1. Parts of two parallel, infinite, perfectly conducting plates. One plate is the plane $y = 0$, the other in the plane $y = b$. A wave propagating in the z-direction between the plates is made up of a sum of waves travelling in the directions of the vectors $\mathbf{k}_1$ and $\mathbf{k}_2$.

perfectly conducting walls. The condition that the fields obey the last two of Maxwell's equations in the space between the planes means, just as in section 11.1, that they must satisfy the wave equations (11.5) and (11.6). The simplest type of solution to such wave equations is a plane wave, and it is immediately clear that the linearly polarized plane wave in which the fields are

$$\mathbf{E} = \mathbf{e}_y E_0 \exp\left[j(\omega t - kz)\right] \qquad (12.1)$$

and

$$\mathbf{B} = -\mathbf{e}_x \frac{E_0}{c} \exp\left[j(\omega t - kz)\right], \qquad (12.2)$$

satisfies all the boundary conditions at the walls. Waves of this form at any frequency can thus be propagated down the plates. The wave is called a *transverse electric and magnetic* or *TEM* wave because both the electric and magnetic fields are transverse to the direction of propagation. TEM waves, different to the plane wave above*, can be propagated along any pair of conductors of constant shapes, and they travel with the speed of light in the medium in which the conductors are embedded. These are the waves travelling on transmission lines which were discussed from a different point of view in Chapter 9. Inside a hollow conductor like a waveguide, however, TEM waves cannot be propagated. We will now try to find more general types of wave which satisfy the boundary conditions at the parallel plates: this will lead us to a solution of the waveguide problem.

Any superposition of plane waves between the plates is a solution to the wave equations, and we will try to find a combination of waves which satisfies the boundary conditions on the fields at the walls. Let us choose axes of coordinates as shown on Figure 12.1 and suppose that a linearly polarized plane wave is moving between the plates in a direction given by the vector $\mathbf{k}_1$. This vector lies perpendicular to the x-axis and makes an angle θ with the z-axis as shown. We will take the electric field vector in the wave to oscillate along the x-direction with amplitude E_0. The expression for the electric field in the plane wave is given by Equation (11.18)

$$\mathbf{E}_1 = \mathbf{e}_x E_0 \exp\left[\mathrm{j}(\omega t - \mathbf{k}_1 \cdot \mathbf{r})\right],$$

where ω and k, the modulus of $\mathbf{k}_1$, are the angular frequency and wavenumber of the wave *in free space*. The vector $\mathbf{k}_1$ is given by

$$\mathbf{k}_1 = -\mathbf{e}_y k \sin\theta + \mathbf{e}_z k \cos\theta.$$

Hence

$$\mathbf{k}_1 \cdot \mathbf{r} = -ky \sin\theta + kz \cos\theta,$$

and the wave may be written

$$\mathbf{E}_1 = \mathbf{e}_x E_0 \exp\left[\mathrm{j}(\omega t - kz \cos\theta + ky \sin\theta)\right].$$

When this wave strikes the plane $y = 0$ it is totally reflected, with a phase change of 180°, in the direction of the vector $\mathbf{k}_2$, given by

$$\mathbf{k}_2 = \mathbf{e}_y k \sin\theta + \mathbf{e}_z k \cos\theta.$$

The magnitude of the vector $\mathbf{k}_2$ is also equal to k, and the electric field in the reflected wave is

$$\mathbf{E}_2 = -\mathbf{e}_x E_0 \exp\left[\mathrm{j}(\omega t - kz \cos\theta - ky \sin\theta)\right].$$

Let us now see if the superposition of two fields described by the functions $\mathbf{E}_1$ and $\mathbf{E}_2$ can be made to fit the boundary conditions. The resultant field $\mathbf{E}$

---

* For example the TEM waves inside a coaxial cable transmission line have the electric field radially out from the centre wire, whilst the magnetic field lines are circular around the centre conductor.

is given by

$$\mathbf{E} = \mathbf{e}_x E_0 \exp\left[j(\omega t - kz \cos \theta)\right]$$

$$\times \left[\exp[jky \sin \theta) - \exp(-jky \sin \theta)\right]$$

$$= \mathbf{e}_x 2jE_0 \sin(ky \sin \theta) \exp\left[j(\omega t - kz \cos \theta)\right]. \tag{12.3}$$

The boundary condition that $E_x = 0$ at $y = 0$ is automatically satisfied. We can make the tangential field zero over the other plane by choosing the angle θ such that $E_x = 0$ at $y = b$, i.e. such that

$$kb \sin \theta = n\pi, \tag{12.4}$$

where n is an integer. There may be several different acceptable fields, corresponding to waves propagating in the z-direction, depending on the different possible values of the integer n. Since $\sin \theta$ cannot exceed unity there is a limitation on the number of different fields. For a given frequency of the radiation the maximum value of n is given by the condition

$$\frac{n\pi}{kb} \leqslant 1. \tag{12.5}$$

If $\pi/kb \geqslant 1$, i.e. $\lambda/2b \geqslant 1$, no wave at all of the type we have been considering is possible. There is thus a cut-off frequency below which such waves cannot be propagated. This cut-off frequency v_1 can be obtained by putting n equal to unity and taking the equality in the expression (12.5):

$$v_1 = \frac{c}{2b}. \tag{12.6}$$

The different waves that correspond to the different allowed values of the integer n are called *modes*. Each different mode has its own cut-off frequency v_n which can be determined from Equation (12.5) to be

$$v_n = \frac{nc}{2b}. \tag{12.7}$$

The cut-off frequencies of each mode increase as the order of the mode increases, i.e. as the value of the integer n increases. For a plate separation of 1.5 cm, radiation of free space wavelength 2 cm (frequency 1.5×10^{10} Hz) can only be propagated in the lowest $n = 1$ mode.

Substitution of the expressions for $\sin \theta$ and $\cos \theta$ obtained from Equation (12.4) into Equation (12.3) gives the electric fields in the waves in terms of the integer n.

$$\mathbf{E} = \mathbf{e}_x 2jE_0 \sin\left(\frac{n\pi y}{b}\right) \exp\left[j(\omega t - k_g z)\right], \tag{12.8}$$

where

$$k_g = \left(k^2 - \frac{n^2\pi^2}{b^2} \right)^{1/2}. \tag{12.9}$$

k_g is called the *guide wavenumber* and is equal to $2\pi/\lambda_g$, where λ_g is the *guide wavelength*. The electric fields in the waves described by Equation (12.8) have no components in the z-direction, the direction of propagation. The waves are thus called *transverse electric waves*, and the wave of order n is designated TE_n. The magnetic fields in the TE_n waves can easily be obtained from the relation

$$\text{curl } \mathbf{E} = -\frac{\partial \mathbf{B}}{\partial t} = -j\omega\mathbf{B}.$$

The expressions for the different components of the electric and magnetic fields are given below:

$$E_x = 2jE_0 \sin\left(\frac{n\pi y}{b}\right) \exp\left[j(\omega t - k_g z)\right]$$

$$E_y = E_z = 0$$

$$B_x = 0$$

$$B_y = 2jE_0 \frac{k_g}{\omega} \sin\left(\frac{n\pi y}{b}\right) \exp\left[j(\omega t - k_g z)\right]$$

$$B_z = 2E_0 \frac{(k^2 - k_g^2)^{1/2}}{\omega} \cos\left(\frac{n\pi y}{b}\right) \exp\left[j(\omega t - k_g z)\right]. \tag{12.10}$$

We may make several observations on these expressions. (i) They are independent of the coordinate x, as required by the symmetry of the problem. (ii) The fields are made up of a superposition of plane waves and therefore satisfy Maxwell's equations. (iii) The component B_y satisfies the boundary condition that the normal component of $\mathbf{B}$ be zero at the walls, i.e. at $y = 0$, and $y = b$. (iv) The wave propagates in the z-direction, and so the Poynting vector must be in this direction. Hence if E_x is the only non-zero component of the electric field, the field $\mathbf{B}$ must have a y-component, and this component must be in phase with E_x. If the field $\mathbf{B}$ has a z-component this must be out of phase with E_x so that the energy moving in the y-direction merely flows into and out of the walls, averaging to zero.

The variation of the electric field in the TE_1 mode with the coordinate y at time $t = 0$ and position $z = 0$ is shown on Figure 12.2. The field at any point $(x, y, 0)$ oscillates with time at the angular frequency ω, with the maximum amplitude of the oscillations in the middle of the plates at $y = b/2$. The variation of the electric field in the z-direction at time $t = 0$ in the middle of the plates

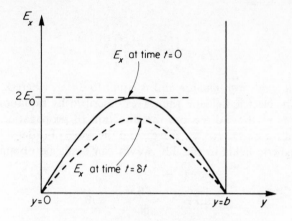

Figure 12.2. The electric field at position $z = 0$ and time $t = 0$ in the TE_1 mode propagating between parallel plates. The field is independent of the co-ordinate x, and at any point oscillates in time with angular frequency ω.

is shown on Figure 12.3. As time progresses the wave is propagated in the $+z$ direction with wave velocity

$$v_p = \left(\frac{\partial z}{\partial t}\right)_{\text{constant phase}}$$

$$= \frac{\omega}{k_g}$$

$$= c\frac{k}{k_g}. \tag{12.11}$$

Substituting into this the value of k_g from Equation (12.9) we obtain

$$v_p = c\left(1 - \frac{n^2\pi^2}{k^2b^2}\right)^{-1/2}. \tag{12.12}$$

From this equation it can be seen that the wave velocity exceeds the speed of light in free space. However, the energy of a group of waves travelling between the planes moves with the group velocity v_g, which is always less than the speed of light. v_g is given by*

$$v_g = \frac{d\omega}{dk_g}. \tag{12.13}$$

* See for example *Optics*, by F. G. Smith and J. H. Thomson, J. Wiley and Sons, Ltd., London 1972, section 1.5.

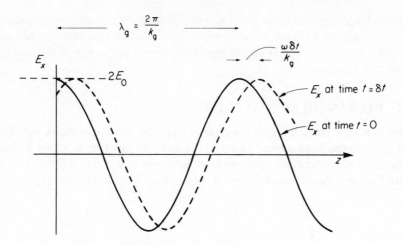

Figure 12.3. The variation of the electric field in the TE_1 wave with distance z at time $t = 0$. The field drawn is that in the centre of the plates where $y = b/2$. The electric field in the TE_1 mode in a rectangular waveguide is similar to the fields shown in this figure and on Figure 12.2.

This can be worked out from equation (12.9) to give

$$v_g = c\frac{k_g}{k}. \tag{12.14}$$

The product of the wave velocity and the group velocity is thus equal to the square of the speed of light,

$$v_g v_p = c^2. \tag{12.15}$$

The expressions (12.10), describing acceptable waves which propagate between the plates, are obtained by adding two waves in both of which the electric field vector is perpendicular to the z-direction. The electric field in the resultant wave is thus perpendicular to the z-axis, whilst the magnetic field does have a component in the z-direction. We can obtain another set of solutions which satisfy the boundary conditions and the wave equation and correspond to waves travelling in the z-direction by adding two waves in which the *magnetic* field has an x-component only. The resulting waves have no component of the magnetic field in the direction of propagation and are called *transverse magnetic waves,* or *TM waves.* Like the TE waves they are characterized by integers n which give the variation of the magnetic field in the y-direction. The different TM_n modes have cut-off frequencies at the same values as the TE_n modes, given by equation (12.7). The TE and TM modes, and the TEM wave discussed earlier, are the only solutions to the wave equation that obey the boundary conditions and correspond to waves travelling in the z-direction. For radiation of a given angular frequency the most general wave propagating in the z-direction is a mixture of all the different possible modes. The relative amounts and phases of the different modes depend on the precise nature of the source of the radiation between the plates.

12.2   RECTANGULAR WAVEGUIDES

We can now consider a rectangular waveguide, which we make by adding two sides to the conducting planes of the last section. This gives us a hollow metal pipe of rectangular cross-section, with width a in the x-direction, and width b in the y-direction, as shown on Figure 12.4. The direction of propagation

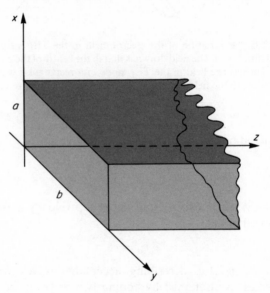

Figure 12.4. The coordinate system used in the discussion of rectangular waveguides.

of the waves is the z-direction. We will continue to suppose that the metal has infinite electrical conductivity, and will determine the characteristics of the waves that can travel down the waveguide.

There are two possible sets of waves which can propagate without attenuation. In one set, called TE waves, the electric field is perpendicular to the direction of propagation; in the other set, called TM waves, the magnetic field is perpendicular to the direction of propagation. The TEM plane wave which was a possible wave between the parallel plates no longer satisfies the boundary conditions over the new walls, and so is not an acceptable wave within the waveguide.

We will restrict our attention at first to the transverse electric TE waves. It can be seen at once that the TE_n waves of the infinite parallel plate problem are still solutions to Maxwell's equations which satisfy all the boundary conditions at the guide walls. They do not violate the boundary conditions on the new walls parallel to the y–z plane since the electric fields in the TE_n waves are everywhere perpendicular to the new walls and appropriate charges are induced on the walls; the magnetic fields are everywhere parallel to the new walls and appropriate surface currents are induced in them.

For all the TE_n waves the electric fields inside the guide are independent of the coordinate x. Now that the guide is not infinite in the x-direction we expect new modes to appear in which the field does vary in the x- direction. These modes are designated TE_{mn}, the integer index m playing a similar role for the variation with x as the index n does for the variation with y. Thus we relabel the TE_n modes of the parallel plate problem as TE_{0n} modes in the wave-guides. From our experience with the parallel plates we may guess that for a TE_{mn} mode, the x-component of the electric field will be of the form

$$E_x = Cf(x) \sin\left(\frac{n\pi y}{b}\right) \exp[j(\omega t - k_g z)], \qquad (12.16)$$

where C is a constant, and the function $f(x)$ gives the variation of the field component in the x-direction. k_g is again the wavenumber of the wave in the guide. If this expression is indeed an acceptable wave it must satisfy the wave equation (11.6), satisfy the boundary conditions at the walls, and also be the x-component of an electric field which satisfies the Maxwell equation

$$\text{div } \mathbf{E} \equiv \frac{\partial E_x}{\partial x} + \frac{\partial E_y}{\partial y} + \frac{\partial E_z}{\partial z} = 0.$$

For the TE waves the function $\partial E_z/\partial z$ is zero by definition. Hence

$$\frac{\partial E_y}{\partial y} = -\frac{\partial E_x}{\partial x},$$

and the field must have a y-component given by

$$\frac{\partial E_y}{\partial y} = -C\frac{\partial f(x)}{\partial x} \sin\left(\frac{n\pi y}{b}\right) \exp[j(\omega t - k_g z)].$$

Integration with respect to y, putting the constant of integration zero, gives

$$E_y = \frac{Cb}{n\pi}\frac{\partial f(x)}{\partial x} \cos\left(\frac{n\pi y}{b}\right) \exp[j(\omega t - k_g z)]. \qquad (12.17)$$

We can now determine the function $f(x)$ by making the component E_y satisfy the boundary condition that it be zero at the walls at $x = 0$ and $x = a$, and satisfy the wave equation. The function which does this, as can readily be verified, is

$$f(x) = -\left(\frac{n\pi}{b}\right) \cos\left(\frac{m\pi x}{a}\right),$$

where m is zero or any integer, and the parameters m, n and k_g must obey the equation

$$\frac{m^2\pi^2}{a^2} + \frac{n^2\pi^2}{b^2} = k^2 - k_g^2. \tag{12.18}$$

This last condition is obtained by substitution of E_y in the wave equation. It is called the *waveguide equation*, and gives the guide wavenumber k_g of the allowed waves in terms of the integers m and n, the dimensions a and b of the waveguide, and the wavenumber k of the wave in free space.

The expressions finally obtained for the electric field components in a TE$_{mn}$ wave are

$$E_x = -\left(\frac{Cn\pi}{b}\right) \cos\left(\frac{m\pi x}{a}\right) \sin\left(\frac{n\pi y}{b}\right) \exp\left[j(\omega t - k_g z)\right] \tag{12.19}$$

$$E_y = \left(\frac{Cm\pi}{a}\right) \sin\left(\frac{m\pi x}{a}\right) \cos\left(\frac{n\pi y}{b}\right) \exp\left[j(\omega t - k_g z)\right] \tag{12.20}$$

$$E_z = 0.$$

The magnetic fields can be determined in the usual way by using one of Maxwell's equations relating the fields **E** and **B**. We obtain

$$B_x = -\frac{k_g C m\pi}{\omega a} \sin\left(\frac{m\pi x}{a}\right) \cos\left(\frac{n\pi y}{b}\right) \exp\left[j(\omega t - k_g z)\right] \tag{12.21}$$

$$B_y = -\frac{k_g C n\pi}{\omega b} \cos\left(\frac{m\pi x}{a}\right) \sin\left(\frac{n\pi y}{b}\right) \exp\left[j(\omega t - k_g z)\right] \tag{12.22}$$

$$B_z = \frac{j(k^2 - k_g^2)C}{\omega} \cos\left(\frac{m\pi x}{a}\right) \cos\left(\frac{n\pi y}{b}\right) \exp\left[j(\omega t - k_g z)\right]. \tag{12.23}$$

The above fields correspond to waves propagated in the z-direction, which is therefore the direction of the Poynting vector. Now

$$N_z = E_x B_y - E_y B_x,$$

and together with the component E_x there goes an in-phase component B_y, and likewise for E_y and B_x. The component B_z is out of phase with E_x and E_y, and the energy in the x- and y-directions flows into and out of the walls, averaging to zero.

From expressions (12.21) to (12.23) for the magnetic field, and expressions (12.19) and (12.20) for the electric field, it can be seen that for the wave to progress without attenuation in the z-direction, the guide wavenumber must be real. Otherwise the solutions obtained represent fields decaying exponentially with distance z. The waveguide equation (12.18) tells us that k_g is real only if

$$k^2 \geqslant \frac{m^2\pi^2}{a^2} + \frac{n^2\pi^2}{b^2}.$$

The different TE_{mn} modes thus have different cut-off frequencies v_{mn} below which a TE_{mn} wave cannot be propagated in a guide of given size. The cut-off frequency of the TE_{0n} modes are the same as for waves between parallel plates. For the TE_{mn} mode the cut-off frequency is given by

$$v_{mn} = c\sqrt{\left(\frac{m}{2a}\right)^2 + \left(\frac{n}{2b}\right)^2}. \tag{12.24}$$

So far we have talked only of TE waves. There are also TM waves in which the magnetic field has no component in the z-direction, but in which the electric field has a non-zero z-component. The TM waves are also designated by integers m and n which characterize their behaviour in the x- and y-directions. They obey the identical wave guide equation to Equation (12.18) and have identical cut-off frequencies, given by Equation (12.24). There is one important difference however. There are no TM waves with either n or m equal to zero, i.e. no TM_{0n} or TM_{m0} waves. Let us see why this must be so. Figure 12.5 shows the lines of

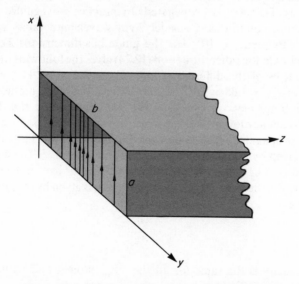

Figure 12.5. The lines of the electric field in the TE_{01} mode in a waveguide at time $t = 0$ over the plane $z = 0$.

the electric field in a TE_{01} mode in a waveguide at time $t = 0$ over the plane $z = 0$. The lines go from one wall straight across to the other, their density being greatest in the middle of the guide, as given by Equation (12.19). On the walls at $x = 0$ and $x = a$ there are induced appropriate surface charges on which the lines of the field $\mathbf{E}$ begin and end. If a TM_{01} mode existed in which the magnetic field were independent of the x-coordinate the field $\mathbf{B}$ would also have to go straight across from one wall to the other. However there are no magnetic charges and the field $\mathbf{B}$ has to vary with the x-coordinate in order to make the lines of the field continuous. The lowest TM mode is thus the TM_{11} mode.

The TE and TM modes are the only solutions to Maxwell's equations inside the guide that obey the boundary conditions. For radiation of a given frequency the general mathematical form of a wave travelling without attenuation in the z-direction is thus a mixture of all the possible TE and TM modes. The relative contributions to the total wave from the different modes is determined by the coupling of the waveguide to the source of radiation. In general the amplitudes of modes other than the mode with the lowest cut-off frequency are very small. Higher modes are also more strongly attenuated in a real guide than is the lowest mode.

The lowest cut-off frequency is given by the smaller of the quantities $c/2b$ or $c/2a$ and corresponds to the field varying along the longest side of the guide. The TE_{01} mode shown in Figure 12.5 has the lowest cut-off frequency $c/2b$. The vanishing of the TM_{01} and TM_{10} modes has the important consequence that for a given frequency one can always choose a size of waveguide such that only the lowest TE wave is propagated. In practice waveguides are usually used under those conditions. Consider 3 cm wavelength radar waves as an example; their frequency is 10^{10} Hz. If a guide has dimensions 2 cm one way ($b = 2$ cm) and 1 cm the other, Equation (12.24) gives the following values for the cut-off frequencies of the different modes: $\nu_{01} = 7.5 \times 10^9$ Hz; $\nu_{10} = \nu_{02} = 1.5 \times 10^{10}$ Hz; $\nu_{11} = 1.68 \times 10^{10}$ Hz; etc. The lowest TM mode, the TM_{11}, has a cut-off frequency $\nu_{11} = 1.68 \times 10^{10}$ Hz. Since only the TE_{01} mode has a cut-off frequency lower than the frequency of the radiation, only the TE_{01} mode can be propagated. The guide wavenumber k_g of the TE_{01} mode in the waveguide is given by Equation (12.18) to be about $1.4\ \text{cm}^{-1}$. The wave velocity v_p of the wave is ω/k_g which has the value $\sim 4.5 \times 10^8\ \text{m s}^{-1}$. The speed with which the energy in the wave travels down the guide is given by the group velocity of the TE_{01} wave

$$v_g = \frac{d\omega}{dk_g}.$$

The group velocity is the same for all the TE_{mn} modes, and its magnitude is found from equation (12.18) to be

$$v_g = c\frac{k_g}{k}. \tag{12.25}$$

It is thus generally true for a TE_{mn} wave that the product of the wave and group velocities is equal to the square of the speed of light in free space,

$$v_p v_g = c^2.$$

The energy in the TE_{01} mode in our example thus flows down the guide at a speed of about $2 \times 10^8 \text{ m s}^{-1}$.

The TE_{01} wave is the most important and most commonly encountered wave in a rectangular guide. The electric field in this wave is given by substituting $m = 0$, $n = 1$ in Equations (12.19) and (12.20), and we obtain

$$E_x = -\left(\frac{C\pi}{b}\right) \sin\left(\frac{\pi y}{b}\right) \exp\left[j(\omega t - k_g z)\right] \tag{12.26}$$

$$E_y = 0.$$

This field has already been shown on Figures 12.2 and 12.3. The magnetic field in the TE_{01} wave can be obtained by substituting $m = 0$, $n = 1$ in equations (12.21) to (12.23), to give

$$B_x = 0$$

$$B_y = -\frac{k_g C\pi}{\omega b} \sin\left(\frac{\pi y}{b}\right) \exp\left[j(\omega t - k_g z)\right] \tag{12.27}$$

$$B_z = \frac{j(k^2 - k_g^2)}{\omega} C \cos\left(\frac{\pi y}{b}\right) \exp\left[j(\omega t - k_g z)\right]. \tag{12.28}$$

The magnetic field described by these equations at time $t = 0$ is shown in Figure 12.6. The field lines have been drawn for the plane $x = a/2$ only. Since the field is independent of the coordinate x, the field lines look the same in any plane parallel to the one drawn.

The electric field in a waveguide can be very large. A ship's radar set typically works at about 0.5 MW power. (Of course the average power delivered by the source of waves is very much less than this, since the source is on for only a small fraction of every second). Thus the time average of the Poynting vector integrated over the cross-sectional area of the guide must equal 0.5 MW during operation. The Poynting vector for the fields in the waveguide is in the z-direction, and for the TE_{01} mode

$$\overline{N}_z = \overline{E_x H_y},$$

where the bars over the symbols indicate averages over a period of the oscillation. From Equations (12.26) and (12.27) we see that the y-component of the magnetic intensity is

$$H_y = \frac{k_g}{\mu_0 \omega} E_x.$$

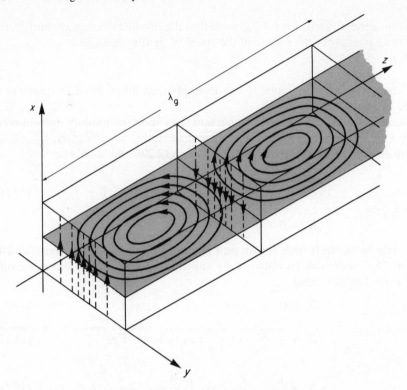

Figure 12.6. The magnetic field lines in the TE_{01} mode in a rectangular waveguide. The field lines are drawn at a time when the electric field, shown dotted,
is a maximum in the $+x$-direction over the plane $z = 0$. The magnetic field
lines look the same in any plane parallel to the plane $x = a/2$ drawn.

Hence if E_{0x} is the maximum amplitude of the component E_x in the guide

$$\overline{N_z} = \frac{1}{2} \frac{k_g}{\mu_0 \omega} E_{0x}^2 \sin^2\left(\frac{\pi y}{b}\right).$$

Integrating this expression over the cross-sectional area of the guide and
equating the result to the total energy flow per second, we have

$$0.5 \times 10^6 = \frac{1}{2} \frac{k_g}{\mu_0 \omega} E_{0x}^2 \cdot \frac{ab}{2}.$$

On substituting into this expression the typical values $a = 0.01$ m, $b = 0.02\,\eta$,
$\omega = 2\pi \times 10^{10}\,\text{s}^{-1}$ and $k_g = 1.4 \times 10^2\,\text{m}^{-1}$, we obtain the value of ~ 2.4 MV
m^{-1} for the maximum amplitude E_{0x}.

Waveguides commonly in use do not have superconducting walls, and so
there are losses, as mentioned in the introduction to this chapter. The energy

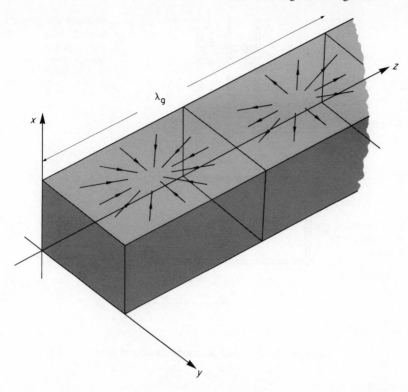

Figure 12.7. At times when the fields in the waveguide are as shown on
Figure 12.6, the surface currents over the top face are as shown above. The
currents at any time are flowing towards or away from the centre line, and a
slot cut down the guide in the middle will have little effect on the distribution
of surface currents and hence fields in the guide.

in a wave decreases as the wave propagates down the guide due to heating of
the guide walls. The heating arises because of the currents induced in the walls
by the wave, which penetrates to a depth of the order of the skin depth δ (discussed in section 11.4). The average heating loss in a small length dz of the guide
can be estimated by assuming that the currents in the walls flow only in a thin
layer of thickness δ. The currents can be determined by using Ampère's integral
law for the field **H** (Equation (5.29)), and hence the heating losses estimated in
terms of the conductivity σ and skin depth δ of the walls. Comparison of the
heating loss in the length dz with the energy flowing down the guide at the
coordinate z, then gives an estimate of how rapidly the energy is attenuated
(see Problem 12.6). For a typical brass rectangular waveguide, with sides 2 cm
and 1 cm wide, the energy of a TE_{01} wave of frequency 10^{10} Hz falls to one half
over a distance of about 16 m.

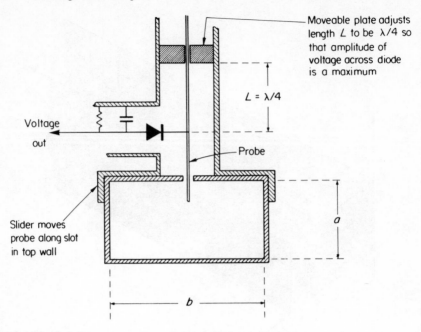

Figure 12.8. A crystal detector in a slotted guide.

There is always a wave reflected back from the end of a waveguide. If the guide is coupled to an antenna designed to radiate the wave into free space, the wave reflected off the specially designed endpiece is very small. If the waveguide is terminated by a metal sheet the reflected wave has an amplitude very nearly equal to the amplitude of the incident wave. The reflected wave combines with part of the forward going wave to produce a standing wave pattern, analogous to standing waves on a transmission line. The standing wave pattern in a wave-guide can be used to measure the amplitude of the reflected wave, and may be investigated by inserting a probe into the interior of the guide through a slot in a wall. (See Problem 12.9). The slot should be placed where it will not interfere with the surface currents in the wall, so that the wave in the guide suffers minimum interference. The surface currents in the walls are perpendicular to the magnetic field lines. For the TE_{01} mode the magnetic field lines are drawn in Figure 12.6 and the magnetic field gives rise to the surface current distribution over the top broad wall shown on Figure 12.7. A narrow slot along the z-axis in the middle of the wall will thus have minimum influence on the wave propagation. Figure 12.8 shows an arrangement often used to measure the standing wave ratio. A crystal diode rectifies the voltage produced between a point on the probe and earth. The diode gives a steady voltage out which is roughly proportioned to the square of the amplitude of the electric field in the middle of the guide.

12.3   CAVITIES

The oscillating electric currents that flow in antennas and give rise to radio waves, discussed in the next chapter, are produced in electrical circuits which use vacuum tubes and tuned lumped circuits or transmission lines. Vacuum tubes and the tuned circuits used at radio frequencies (up to 10^8 Hz) cannot be used at microwave frequencies (up to 10^{11} Hz). The devices usually used to cause the oscillations of charge which produce microwaves are *magnetrons* or *klystrons*. These generators use resonating cavities in a somewhat analogous fashion to the way in which a low frequency tube oscillator uses a tuned LC circuit. The cavity stores large amounts of energy at the resonant frequency with very little energy loss per cycle of the fields. The Q of the cavity (see section 7.6) is thus very high,

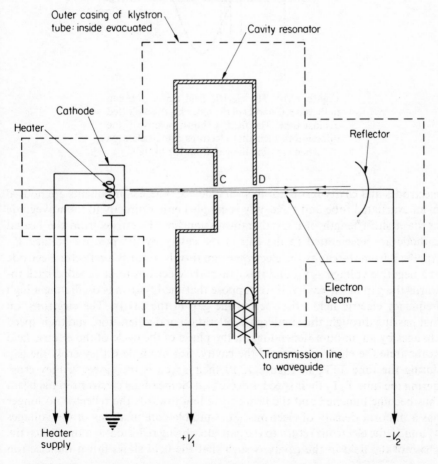

Figure 12.9. The basic structure of a reflex klystron. The electric field lines in the cavity go from side C to side D, and the field oscillates with time so that sometimes electrons are accelerated as they cross the gap, sometimes decelerated.

and the stored energy falls rapidly as the frequency goes off the resonant frequency. The generators thus oscillate at a resonant frequency of the cavity.

The simplest klystron, one frequently used in the teaching laboratory to generate waves of free space wavelengths a few cm at a power of a few hundred mW, is the reflex klystron*. Figure 12.9 shows schematically the arrangement of

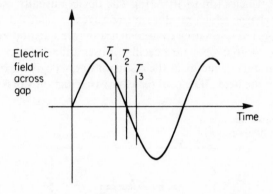

Figure 12.10. The electric field across the gap in the sides C and D of the reflex klystron plotted against time. The field is shown positive if the direction is such that electrons are accelerated when crossing the gap from left to right.

electrodes and cavity resonator in such a device. These components are housed in an evacuated tube and the cavity is coupled either directly to the waveguide or via a short length of coaxial transmission line. Electrons from the heated cathode are accelerated to the gap in the cavity by the positive voltage V_1. After exit from the cavity the electron beam travels towards a reflector electrode at a negative voltage $-V_2$, and this causes the electrons to be repelled back towards the gap in the cavity. If we suppose that the klystron is oscillating a high frequency electric field exists across the gap in the cavity. The electrons, on first passing through, thus undergo acceleration or deceleration, and their speed changes by an amount depending on the phase of the cycle of the electric field at the time the electrons entered the cavity. For example if they cross the gap during the time T_1T_2 in Figure 12.10 their speed is increased; if they cross during the time T_2T_3 their speed is decreased. Some of the electrons in the beam thus become bunched and the beam travelling towards the reflector no longer has a uniform density of electrons. For suitable combinations of the voltages V_1 and V_2 the electrons return to the gap, after being reflected, at a time when the phase of the field in the cavity is such that the field slows down the electron

* High power generation of radar waves is usually done by magnetrons. For a discussion of how a magnetron works see for example, *Electricity and Magnetism*; by B. I. Bleaney and B. Bleaney, Oxford University Press, 1965.

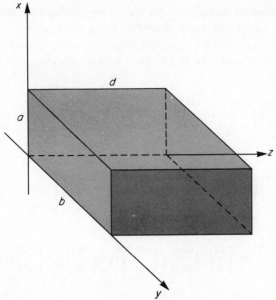

Figure 12.11. A rectangular cavity with sides of length
a, b and d.

bunch. There is thus a net delivery of energy from the electrons to the cavity, which keeps the oscillations going. The frequency is tunable over a narrow range of about 10% by flexing the walls of the cavity resonator, since the resonant frequency of the cavity depends on its dimensions.

The simplest cavity whose resonant frequencies can be calculated is a rectangular cavity. Consider the closed box shown in Figure 12.11 formed by adding two walls to the waveguide in Figure 12.4. The length of the cavity in the z-direction is d. The waves in the cavity are not progressive waves, they are standing waves. They may be written in the form

$$\mathbf{E} = \mathbf{E}_0 f(x, y, z) \exp(j\omega t),$$

where ω is the angular frequency, and the function $f(x, y, z)$ gives the spatial dependence of the amplitude of the wave. The form of this function can be guessed using our experience with the fields in waveguides. The field components E_x and E_y will now vary with the coordinate z, the variation being characterized by a third integer l. The electric field given by

$$E_x = -\left(\frac{Cn\pi}{b}\right) \cos\left(\frac{m\pi x}{a}\right) \sin\left(\frac{n\pi y}{b}\right) \sin\left(\frac{l\pi z}{d}\right) \exp(j\omega t) \tag{12.29}$$

$$E_y = \left(\frac{Cm\pi}{a}\right) \sin\left(\frac{m\pi x}{a}\right) \cos\left(\frac{n\pi y}{b}\right) \sin\left(\frac{l\pi z}{d}\right) \exp(j\omega t) \tag{12.30}$$

$$E_z = 0,$$

satisfies the boundary conditions on the cavity walls, and also satisfies the equation div $\mathbf{E} = 0$. It can easily be shown using the Maxwell equation curl $\mathbf{E} = -\partial \mathbf{B}/\partial t$, or by extending the earlier solutions in an obvious way, that the magnetic field associated with this electric field is given by

$$B_x = -j\left(\frac{l\pi}{d}\right)\left(\frac{m\pi}{a}\right)\frac{C}{\omega}\sin\left(\frac{m\pi x}{a}\right)\cos\left(\frac{n\pi y}{b}\right)\cos\left(\frac{l\pi z}{d}\right)\exp(j\omega t) \tag{12.31}$$

$$B_y = -j\left(\frac{l\pi}{d}\right)\left(\frac{n\pi}{b}\right)\frac{C}{\omega}\cos\left(\frac{m\pi x}{a}\right)\sin\left(\frac{n\pi y}{b}\right)\cos\left(\frac{l\pi z}{d}\right)\exp(j\omega t) \tag{12.32}$$

$$B_z = j\left\{\left(\frac{m\pi}{a}\right)^2 + \left(\frac{n\pi}{b}\right)^2\right\}\frac{C}{\omega}\cos\left(\frac{m\pi x}{a}\right)\cos\left(\frac{n\pi y}{b}\right)\sin\left(\frac{l\pi z}{d}\right)\exp(j\omega t). \tag{12.33}$$

In order for these electric and magnetic fields to be acceptable solutions to Maxwell's equations inside the cavity, they must also satisfy the wave equations (11.5) and (11.6). Substitution of any component into the appropriate wave equation shows that the fields are acceptable if the following equation is obeyed by the parameters m, n, l and k, the free space wavenumber of the radiation of angular frequency ω:

$$\frac{\pi^2 m^2}{a^2} + \frac{\pi^2 n^2}{b^2} + \frac{\pi^2 l^2}{d^2} = k^2. \tag{12.34}$$

There is thus an infinite number of resonant frequencies of the cavity corresponding to different values of the integers m, n and l. The cavity is usually used so that a low mode is excited, e.g. the lowest frequency corresponds to two of the integers equal to one, and the third integer, that associated with the shortest side of the cavity, equal to zero. Since the Q values of the cavity at each resonance are very high, typically a few thousand, when the lower resonances are used there is no overlapping.

The $m = 0$, $n = 1$, $l = 1$ resonance in which the fields are obtained by the appropriate substitutions in Equations (12.29) to (12.33) is designated the TE_{011} resonance because the electric field has no z-component. There is one other possible set of solutions for the electromagnetic fields in the cavity. These are TM_{mnl} standing waves in which the magnetic field has no z-component, but the electric field has a non-zero z-component. These waves occur at the same frequencies as the TE_{mnl} waves (for the same cavity), and so at a particular resonance the standing wave is the sum of two resonating waves, the TE mode and the TM mode*.

Cavities are often used as frequency meters as well as in microwave generators. A cavity will readily absorb energy from a waveguide (usually via a small hole in one wall) when the frequency of the wave is near a resonant frequency of the cavity. To measure frequency, a plunger is used to vary the length of the longest side of the cavity, which is operated near the $TE_{011}(TM_{011})$ resonance. A large absorption dip in the energy in the guide past the cavity indicates the resonance, and the length of the cavity side then gives the frequency from Equation (12.34). The resonant frequencies of cylindrical cavities are also amenable to calculation, and since cylindrical cavities can be made more accurately than rectangular cavities they are more often used in very accurate frequency measurements. Cavities are also used in applications where high microwave field levels are required. For example in electron spin resonance experiments†, in which a sample is placed in a high frequency magnetic field of large amplitude, the centre of a rectangular cavity resonating in the TE_{021} mode is often used.

PROBLEMS 12

12.1 Show that the field component

$$E_x = E_0 \sin\left(\frac{\pi y}{b}\right) \exp\left[j(\omega t - k_g z)\right]$$

corresponds to one of the TE waves propagated in the z-direction in a rectangular waveguide whose width in the y-direction is b. Show that the component E_y in the wave is zero. Determine the cut-off frequency of the wave.

12.2 Determine the maximum and minimum widths of a waveguide of square cross-section if it is to transmit waves of free-space wavelength λ in the TE_{01} mode only.

12.3 What is the maximum average power which an air-filled guide can transmit in the TE_{01} mode if the dimensions of the guide are $a = 1$ cm, $b = 2$ cm, and the frequency is 10^{10} Hz? The breakdown electric field in air is $30\,kV\,cm^{-1}$.

12.4 A source maintains fields of constant amplitude in the TE_{01} mode in a rectangular waveguide. Discuss how the power received at the far end of the guide varies as the frequency of the radiation is reduced. If the guide is terminated with an impedance such that there is no reflected wave what is the energy density in the waveguide?

12.5 Calculate how the wave and group velocities of the TE_{01} wave in a rectangular guide with $a = 1$ cm, $b = 2$ cm vary with frequency.

12.6 Using the method outlined on page 415, estimate the attenuation of the TE_{01} mode in a brass rectangular waveguide with $a = 1$ cm, $b = 2$ cm at a frequency of 10^{10} Hz. The conductivity of brass is $1.6 \times 10^{+7}\,(\Omega\,m)^{-1}$.

12.7 A rectangular cavity has dimensions $a = 2$ cm, $b = 3$ cm, $d = 4$ cm. How many resonances are there within the frequency range $v = 5 \times 10^9$ Hz to $v = 10^{10}$ Hz?

---

* In the theory of black body radiation (see for example M. Born, *Atomic Physics*, Blackie and Son Ltd., London, 1957, section 8.3) one needs to calculate the number of independent resonating modes of a cavity within a given frequency interval. At each resonance there are two independent waves, the TE and the TM, and this gives rise to a factor of two in the density of states function.

† See for example H. E. Hall, *Solid State Physics*. J. Wiley, London, 1974.

12.8 Show that the assumption that a TEM wave can be propagated within the interior of a hollow metal pipe leads to contradictions with electromagnetic theory. This constitutes a proof that there can be no TEM waves within a waveguide.

12.9 A crystal diode is used in a slotted waveguide to measure the power radiated by a horn aerial at the end of the guide. The maximum and minimum values of the voltage output from the diode detector as it is moved along the slotted guide are in the ratio 9 to 1. What fraction of the power in the wave travelling down the guide is radiated by the horn?

13

The generation of electromagnetic waves

In this chapter we discuss the electromagnetic fields produced by specified changing charge and current distributions. Radio waves are produced by currents oscillating at radio frequencies in aerials; microwaves with wavelengths of a few centimetres can be produced by making electrons oscillate to and fro at frequencies of about 10^{10} Hz in a tuned cavity; visible light arises from very rapid readjustments of the charge and current distributions within the electron clouds of atoms. The emission of visible light can only be fully described by quantum mechanics. However the classical radiation theory we will outline is a necessary guide to the quantum theory of radiation.

Any distribution of changing charge and current acts as a source of electromagnetic radiation. Far away from a source the electric and magnetic fields fall off inversely proportionally to the distance from the source. Nearer to the source there are other components of the fields which fall off more rapidly with distance, but these components make no contribution to the power continuously radiated. The part of the field inversely proportional to distance is called the *radiation field*, and the region far away from the source where the radiation field is the only important one is called the *radiation zone*. In Chapter 11 we considered the plane wave approximation to radiation fields. At sufficiently great distances from any source the radiation field can be adequately represented for most purposes by plane waves. In this chapter we derive expressions which enable us to describe more completely the fields of a given system of changing currents or charges. We then use the expressions to determine the total power radiated and the distribution of the radiated energy over space in a few simple examples.

13.1 THE RETARDED POTENTIALS

In the discussions of Chapters 11 and 12 on electromagnetic waves in free space, or in media where there were no free charges or currents present (other than those induced by the wave itself), there was no need to introduce the potentials A and ϕ. We were able to solve the problems quite simply without them. However if we wish to relate the fields to the changing charge and current distributions which produce them it is far easier to work with the potentials than with the fields directly.

In this section we will outline the steps taken to obtain the potentials in terms of given charge and current distributions. Once the potentials have been determined the fields can be obtained from the relations already given in Equations (4.42) and (6.18), viz.

$$\mathbf{B} = \text{curl } \mathbf{A} \tag{13.1}$$

and

$$\mathbf{E} = -\frac{\partial \mathbf{A}}{\partial t} - \text{grad } \phi. \tag{13.2}$$

We will first obtain differential equations obeyed by the potentials A and ϕ separately. We will only consider changing charge and current distributions in free space (or air), so that Maxwell's equations become:

$$\text{div } \mathbf{E} = \rho/\varepsilon_0 \tag{13.3}$$

$$\text{div } \mathbf{B} = 0 \tag{13.4}$$

$$\text{curl } \mathbf{E} = -\frac{\partial \mathbf{B}}{\partial t} \tag{13.5}$$

$$\frac{1}{\mu_0}\text{curl } \mathbf{B} = \mathbf{j} + \varepsilon_0 \frac{\partial \mathbf{E}}{\partial t}. \tag{13.6}$$

Substitution of the expression (13.2) for the field $\mathbf{E}$ into Equation (13.3) gives

$$-\nabla^2\phi - \frac{\partial}{\partial t}(\text{div } \mathbf{A}) = \rho/\varepsilon_0. \tag{13.7}$$

The potentials A and ϕ are not uniquely specified by equations (13.1) and (13.2). We can add to the vector A any function whose curl is zero and the same magnetic field results: we can add to the potential ϕ any function whose gradient is zero and the same electric field results. We are at liberty to impose any condition on the functions A and ϕ that does not contradict Maxwell's equations. In the discussion of steady magnetic fields we added the condition

$$\text{div } \mathbf{A} = 0.$$

In dealing with time-dependent fields it is convenient to generalize this to

$$\operatorname{div} \mathbf{A} = -\varepsilon_0\mu_0\frac{\partial \phi}{\partial t}. \tag{13.8}$$

The restriction on the potentials imposed by Equation (13.8) is known as the *Lorentz condition*. It involves no contradictions or difficulties and introduces a symmetry into the equations for the potentials that is very useful and elegant. It also enables electromagnetism to be brought into line with the special theory of relativity as discussed in the next chapter.

If we substitute Equation (13.8) into Equation (13.7) we obtain

$$-\nabla^2\phi + \varepsilon_0\mu_0\frac{\partial^2\phi}{\partial t^2} = \frac{\rho}{\varepsilon_0}, \tag{13.9}$$

an equation involving ϕ alone. It may be noted that in those regions of space where the charge density ρ is zero, the potential ϕ obeys the same wave equation as the fields $\mathbf{E}$ and $\mathbf{B}$. The equation for the vector potential $\mathbf{A}$ can be obtained by substituting expressions (13.1) and (13.2) for the electric and magnetic fields into Equation (13.6). The use of the Lorentz condition leads to an equation similar in form to Equation (13.9):

$$-\nabla^2\mathbf{A} + \varepsilon_0\mu_0\frac{\partial^2\mathbf{A}}{\partial t^2} = \mu_0\mathbf{j}. \tag{13.10}$$

We have now obtained differential equations for the potentials and it remains to solve them to give expressions for $\mathbf{A}$ and ϕ in terms of the charge and current distributions. We will not solve the equations formally* but will obtain the solutions by analogy with the solutions to the time independent problems discussed in electrostatics and magnetostatics. The solution to an electrostatic problem is obtained by solving the equation

$$-\nabla^2\phi = \rho/\varepsilon_0.$$

The solution has already been given in section 3.3, and is

$$\phi(\mathbf{r}) = \frac{1}{4\pi\varepsilon_0}\int_V \frac{\rho(\mathbf{r}')}{|\mathbf{r} - \mathbf{r}'|}\,\mathrm{d}\tau',$$

where V is the volume occupied by the charge distribution ρ. Now consider the situation where ρ changes with time. Let the point P on Figure 13.1 have position vector $\mathbf{r}$, and the point Q have position vector $\mathbf{r}'$. If the charge distribution near Q changes with time, the information that it has changed can only be appreciated

* See for example *Classical Electricity and Magnetism* by W. K. H. Panofsky and M. Phillips, Addison-Wesley, Reading, Mass., 1962; Chapter 14.

at the point P a time $|\mathbf{r} - \mathbf{r}'|/c$ after the change. This is the time it takes electromagnetic radiation to travel the distance $|\mathbf{r} - \mathbf{r}'|$. Hence it is plausible that at a time t the contribution to the potential $\phi(\mathbf{r}, t)$ at P from the charge within the volume element $d\tau'$ at Q depends not on what the charge is at time t but on what it *was* at the time $t - |\mathbf{r} - \mathbf{r}'|/c$. This is true for all the volume elements within the volume V. (Note that the time lapse $|\mathbf{r} - \mathbf{r}'|/c$ depends on the position

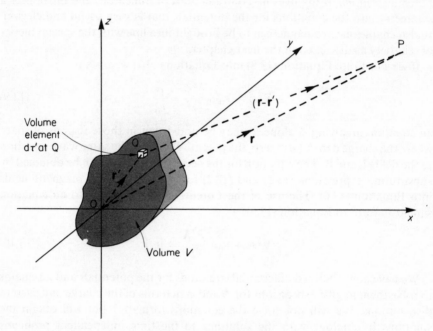

Figure 13.1. The coordinates used in the discussion of retarded potentials.

vector $\mathbf{r}'$ and so varies over the volume V.) The required solution to Equation (13.9) is thus obtained by integrating the contributions over the volume V,

$$\phi(\mathbf{r}, t) = \frac{1}{4\pi\varepsilon_0} \int_V \frac{\rho(\mathbf{r}', t - |\mathbf{r} - \mathbf{r}'|/c)}{|\mathbf{r} - \mathbf{r}'|} \, d\tau'. \tag{13.11}$$

The solution to Equation (13.10) is obtained in a similar manner by beginning with an equation which gives the magnetic vector potential due to a steady distribution of current*. If the current distribution within a volume V is changing then the potential $\mathbf{A}$ is made up of contributions which depend on the magnitude of the current density at appropriate times before the instant of observation.

---

* See Appendix C.

The solution to Equation (13.10) is

$$\mathbf{A}(\mathbf{r}, t) = \frac{\mu_0}{4\pi} \int_V \frac{\mathbf{j}(\mathbf{r}', t - |\mathbf{r} - \mathbf{r}'|/c)}{|\mathbf{r} - \mathbf{r}'|} \, d\tau'. \tag{13.12}$$

The potentials $\mathbf{A}$ and ϕ given by Equations (13.12) and (13.11) are known as the *retarded potentials*. The solution to the time dependent problem is just like the solution to the static problem, except that one sees the different parts of the source as they were at earlier times, the times taken by light to travel the appropriate distances.

It is usually no easy matter to obtain the retarded potentials from given charge and current distributions, or to calculate the fields $\mathbf{E}$ and $\mathbf{B}$ using Equations (13.1) and (13.2) if the potentials have been found. Complications arise due to the fact that the retarded times are different for different parts of the source. In certain simple situations, and with sensible approximations, the fields can be determined. An example is given in the next section.

13.2 THE HERTZIAN DIPOLE

In this section we calculate the radiation fields in the simple but important example of an oscillating electric dipole. Suppose that charge is transferred periodically from a very small sphere at one end of a short wire to another small sphere at the other end. If the charge on one sphere at time t is q, the charge on the other sphere is $-q$, and if the variation of the charge with time is sinusoidal q is given by

$$q = q_0 \sin \omega t,$$

where q_0 is the amplitude of the oscillating charge and ω is the angular frequency of the oscillations. If the length of the wire is very short compared to the wavelength of the radiation produced, the current I in the wire can be taken to be the same at all points along its length, and

$$I = \frac{dq}{dt} = I_0 \cos \omega t,$$

where

$$I_0 = \omega q_0. \tag{13.13}$$

The charge oscillating between the two spheres is equivalent to an oscillating dipole moment. If the distance between the spheres is l the instantaneous dipole moment p is given by

$$p = ql = p_0 \sin \omega t,$$

where

$$p_0 = q_0 l. \tag{13.14}$$

The small oscillating dipole is called a *Hertzian dipole*. The radiation field of such a dipole is important because many practical radiating systems can be considered to be made up by putting together a very large number of small Hertzian dipoles. The radiation fields of these larger aerials can then be calculated from a knowledge of that of the Hertzian dipole. In addition the very small dipole contains many features useful in the quantum theory of the emission of radiation by atoms, molecules and nuclei. A readjustment of the charge distribution within the microscopic system plays the role of the oscillating dipole, although in individual atoms or molecules the change in the charge distribution is not continuous, and is associated with a single photon rather than a steady emission of energy.

We will now determine the total power radiated by a Hertzian dipole, and the way in which the radiation field is distributed throughout space. We follow the procedure outlined in section 13.1: first the retarded potentials are calculated, making assumptions appropriate to the small dipole; secondly the electric and magnetic fields are calculated from the potentials; finally the Poynting vector gives the energy radiated. The dipole is shown in Figure 13.2. We have chosen

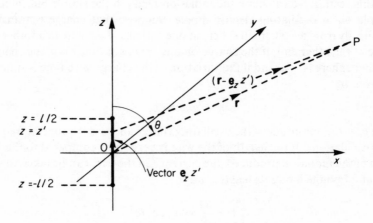

Figure 13.2. The coordinates used in the discussion of the Hertzian dipole.

the z-axis to lie along the wire, and the origin of coordinates to be at the centre of the wire. We will first calculate the vector potential $\mathbf{A}$. Let us suppose that the wire is very thin, so that when the potential is calculated from Equation (13.12) the integral over the volume occupied by the current becomes a line integral over the length of the wire. The vector $\mathbf{j}(\mathbf{r}', t - |\mathbf{r} - \mathbf{r}'|/c) \, d\tau'$ can then be replaced by the vector $I(\mathbf{r}', t - |\mathbf{r} - \mathbf{r}'|/c) \, d\mathbf{l}'$, where $d\mathbf{l}'$ is a vector in the direction of the wire at the point with position vector $\mathbf{r}'$. In the present example the direction of the current is always along the z-axis, and the vector $I(\mathbf{r}', t - |\mathbf{r} - \mathbf{r}'|/c)\,d\mathbf{l}'$ becomes a vector of magnitude $I(z', t - |\mathbf{r} - \mathbf{e}_z z'|/c)\,dz'$ with direction along the z-axis. Hence the magnetic vector potential is everywhere parallel to the z-axis, the components A_x and A_y always being zero. The component A_z is given by Equation (13.12),

$$A_z(\mathbf{r}, t) = \frac{\mu_0}{4\pi} \int_{-l/2}^{l/2} \frac{I(z', t - |\mathbf{r} - \mathbf{e}_z z'|/c)}{|\mathbf{r} - \mathbf{e}_z z'|}\,dz'. \qquad (13.15)$$

Let us consider the fields at points like the point P, far away from the dipole so that $r \gg l$. Under these circumstances the denominator in the integrand in Equation (13.15) can be put equal to r. The times $(t - |\mathbf{r} - \mathbf{e}_z z'|/c)$ at which the current density in the wire is to be determined can all be put equal to the same time $(t - r/c)$ if the biggest error in time Δt introduced by this approximation is very small compared with the period T of the oscillations. The biggest error in time is given by

$$\Delta t = \frac{r}{c} - \frac{|\mathbf{r} - \mathbf{e}_z l/2|}{c},$$

which can be expanded to first order in l/r to give

$$\Delta t \approx \frac{l \cos \theta}{2c}, \qquad (13.16)$$

where θ is the angle shown in Figure 13.2. The length l of the dipole is very much smaller than the wavelength of the radiation, i.e. $l \ll 2\pi c/\omega$. The period T is equal to $2\pi/\omega$, hence $l/c \ll T$, which from Equation (13.16), implies that the time Δt is indeed very small compared with the period T. Hence for points far away from the dipole Equation (13.15) can be written

$$A_z(\mathbf{r}, t) = \frac{\mu_0}{4\pi} \int_{-l/2}^{l/2} \frac{I(z', t - r/c)}{r}\,dz'.$$

Since the current is the same at all points along the short wire, this integral can easily be evaluated to give

$$A_z(\mathbf{r}, t) = \frac{\mu_0 l}{4\pi} \left\{ \frac{I(t - r/c)}{r} \right\}. \qquad (13.17)$$

Since A_x and A_y are both zero, we now know the vector potential everywhere. The potential ϕ can most easily be obtained from the Lorentz condition

$$\text{div } \mathbf{A} = -\varepsilon_0 \mu_0 \frac{\partial \phi}{\partial t}. \qquad (13.18)$$

Since only the z-component of the vector $\mathbf{A}$ is non-zero

$$\text{div } \mathbf{A} = \frac{\partial A_z}{\partial z},$$

and by combining Equations (13.17) and (13.18) we obtain

$$-\varepsilon_0\mu_0\frac{\partial\phi}{\partial t} = \frac{\mu_0 l}{4\pi}\frac{\partial}{\partial z}\left\{\frac{I(t-r/c)}{r}\right\}$$

$$= \frac{\mu_0 l}{4\pi}\left\{-\frac{I(t-r/c)z}{r^3} - \frac{\partial I(t-r/c)}{\partial(t-r/c)}\cdot\frac{z}{cr^2}\right\}.$$

This reduces to the equation

$$\frac{\partial\phi}{\partial t} = \frac{l}{4\pi\varepsilon_0}\left\{\frac{z}{r^3}I(t-r/c) + \frac{z}{cr^2}\frac{\partial I(t-r/c)}{\partial(t-r/c)}\right\},$$

and on integration we obtain

$$\phi = \frac{l}{4\pi\varepsilon_0}\left\{\frac{z}{r^3}q(t-r/c) + \frac{z}{cr^2}I(t-r/c)\right\}. \tag{13.19}$$

The forms of the potentials given by Equations (13.17) and (13.19) show that when the fields are calculated from them with the aid of Equations (13.1) and (13.2) there is a field component which falls off with distance proportionally to $1/r$, plus other components which decrease more rapidly. The first component is the radiation field, the other terms are important for the fields near to the dipole. (Close to the dipole, at distances such that $r \lesssim l$ the potentials given above are not sufficiently accurate to determine the fields.)

Let us restrict attention to the radiation field far away from the dipole, where the other field components are negligibly small, and calculate the total power radiated and the manner in which the radiated energy is distributed in space. The fields **E** and **B** can be calculated from the potentials by writing

$$q(t-r/c) = q_0\sin\{\omega(t-r/c)\}$$

and

$$I(t-r/c) = \omega q_0\cos\{\omega(t-r/c)\}.$$

The first term on the right hand side of equation (13.19) can be ignored in calculating the radiation field, since it gives an electric field which decreases faster than $1/r$. The obvious coordinate system to use in solving the problem is a spherical polar coordinate system. The expressions for the operators grad and curl in these co-ordinates are given in Appendix B. We will omit the mathematics of the derivation and quote the results. They are

$$E_r = 0$$

$$E_\phi = 0$$

$$E_\theta = -\frac{\omega l I_0\sin\theta}{4\pi\varepsilon_0 c^2}\cdot\frac{\sin\{\omega(t-r/c)\}}{r}, \tag{13.20}$$

$$B_r = 0$$

$$B_\theta = 0$$

$$B_\phi = -\frac{\mu_0 \omega l I_0 \sin\theta}{4\pi c} \cdot \frac{\sin\{\omega(t - r/c)\}}{r}. \tag{13.21}$$

The directions of the fields are shown in Figure 13.3. They are mutually perpendicular and the energy flow at all points is in the direction of the radius vector. The fields are outgoing spherical waves and can be written (in a format similar to that used in Chapter 11 for plane waves) as the real parts of the expressions

$$E_\theta = \frac{j\omega l I_0 \sin\theta}{4\pi\varepsilon_0 c^2} \cdot \frac{\exp\{j(\omega t - kr)\}}{r}$$

$$B_\phi = \frac{j\omega l I_0 \sin\theta}{4\pi\varepsilon_0 c^2} \cdot \frac{\exp\{j(\omega t - kr)\}}{cr}.$$

The fields are symmetric about the z-axis, i.e. they are independent of the azimuthal angle ϕ. This must be so since the dipole looks the same from all

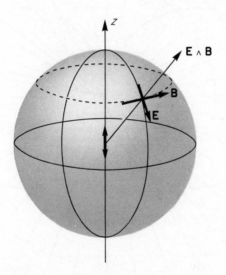

Figure 13.3. The relative directions of the radiation fields of a Hertzian dipole. The fields are tangential to the surface of a sphere centred on the dipole. The electric field lies along the direction of the unit vector $\mathbf{e}_\theta$, and the magnetic field lies along the direction of the unit vector $\mathbf{e}_\phi$.

points with the same values of the coordinates r and θ. The fields are zero at all points on the axis of the dipole, and rise to maximum values on a disc that passes through the middle of the dipole and cuts it perpendicularly, i.e. at $\theta = \pi/2$. The spatial distribution of the fields can be shown on a *polar diagram*, such as Figure 13.4, which gives the relative values of the fields at different positions on the surface of a sphere centred on the dipole. For example the ratio of the electric fields at two points on the sphere with polar angles θ_1 and θ_2 is given by the ratio of the lengths of the lines joining the origin O to the points P_1 and P_2 on the polar diagram.

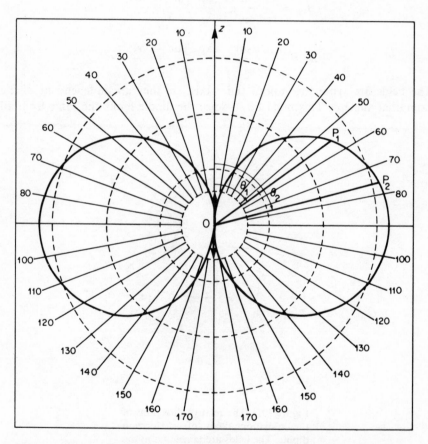

Figure 13.4. The polar diagram for a Hertzian dipole. The solid curved lines give the relative strengths of the radiation field at different points on the surface of a sphere centred on the dipole. The dipole lies along the z-axis and the field is independent of the azimuthal angle ϕ, depending only on the polar angles θ_1 and θ_2 of the points on the sphere.

The instantaneous total power W crossing the surface S of any sphere of radius r in the radiation zone is given by integrating the Poynting vector $\mathbf{N}$ over the surface.

$$W = \int_S \mathbf{N} \cdot d\mathbf{S}$$

$$= \frac{1}{\mu_0} \int_0^{2\pi} \int_0^{\pi} B_\phi E_\theta r^2 \sin\theta \, d\theta \, d\phi.$$

Hence

$$W = \frac{\omega^2 I_0^2 l^2}{16\pi^2 \varepsilon_0 c^3} \int_0^{2\pi} d\phi \int_0^{\pi} \sin^3\theta \, d\theta \cdot \sin^2\{\omega(t - r/c)\},$$

from Equations (13.20) and (13.21). The average power radiated $\overline{W}$ is given by averaging this expression over a cycle of the oscillations:

$$\overline{W} = \frac{\omega^2 I_0^2 l^2}{12\pi \varepsilon_0 c^3}. \tag{13.22}$$

This can be rewritten in terms of the amplitude p_0 of the oscillating dipole moment, using Equations (13.13) and (13.14):

$$\overline{W} = \frac{p_0^2 \omega^4}{12\pi \varepsilon_0 c^3}. \tag{13.23}$$

As mentioned earlier in this section, the results given above for the Hertzian dipole enable us to calculate the radiation fields of more practical systems, and are also useful in the quantum theory of the radiation from microscopic particles. An atom in an excited state readjusts its charge distribution at some instant and may emit a packet of energy in the form of electromagnetic radiation as the atom reverts to its stable ground state. This packet of energy takes away the original excitation energy E of the atom, and is called a photon. The excited state of the atom has a fixed probability per unit time λ_1 that it will decay by emission of a photon of energy E and leave the atom afterwards in its ground state. If there are N excited atoms the number of decays per second to the ground state is thus $N\lambda_1$. Hence the energy emitted per second $\overline{W}$ in the decays to the ground state is equal to $N\lambda_1 E$. The frequency ν of the electromagnetic wave associated with photons of energy E is given by the formula

$$h\nu = E,$$

where h is Planck's constant. The power radiated in the ground state decays of the N atoms is thus

$$\overline{W} = N\lambda_1 h\nu = N\lambda_1 h\omega/2\pi,$$

where ω is the angular frequency of the radiation. We may deduce an expression for the decay constant λ_1, by making an analogy between the change in the atomic charge distribution and a classical Hertzian dipole. The dipole moment p_0 in Equation (13.23) is replaced for the atomic radiators by a factor M which can be calculated using quantum mechanics, and depends on the structure of the two atomic states involved in the transition. Equation (13.23) then predicts that the power $\overline{W}$ radiated by the N atoms making transitions to the ground state is given by

$$\overline{W} = \frac{NM^2\omega^4}{12\pi\varepsilon_0 c^3}.$$

Equating this to the earlier expression for $\overline{W}$ gives

$$\lambda_1 = \frac{M^2\omega^3}{6h\varepsilon_0 c^3}.$$

This expression shows that the decay probability λ_1 is proportional to the cube of the energy of the photon released in the decay. This result is also obtained in a more rigorous quantum mechanical treatment of atomic electric dipole transitions.

13.3   ANTENNAS

The calculation of the last section for the small dipole is not applicable to aerial systems used for transmitting radio waves. Usually these aerials have lengths which are not short compared to the wavelength λ of the radiation they transmit, and the variation of the amplitude of the oscillating current along the aerial (antenna) has to be taken into account.

A simple antenna normally consists of a straight wire or metal tube an integral number of half-wavelengths long. The antenna has capacitance and inductance, and can carry current. It behaves like a length of open-ended transmission line: standing waves are set up, with the currents satisfying the condition that they be zero at each end of the wire. The current is usually fed into the aerial via a coaxial cable transmission line. Figure 13.5 shows schematically the arrangement of a transmitter which has a half-wave dipole antenna. The current I in this antenna at a point distance z' from the centre is given to a good approximation by the expression

$$I = I_0 \cos\frac{2\pi z'}{\lambda} \cos \omega t,$$

where ω is the angular frequency of the oscillations.

The radiation field from such an antenna can be determined by splitting it into infinitesimally small elements, each of which can be considered to be a Hertzian dipole. The contribution dE_θ to the electric field at a far off point P from the small dipole of length dz' at a point Q on the aerial with coordinate z'

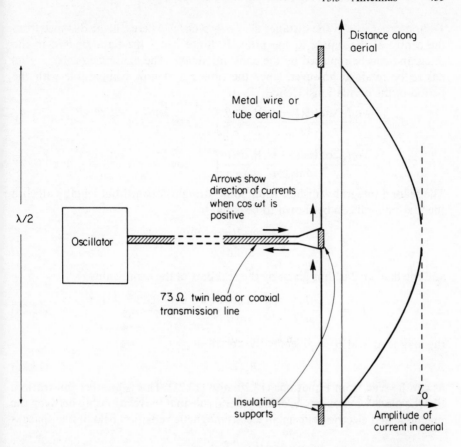

Figure 13.5. A schematic arrangement of a transmission line coupling an oscillator to a half-wave dipole aerial. The current amplitude varies approximately sinusoidally along the aerial.

is given by Equation (13.20) as

$$dE_\theta = -\frac{\omega \, dz' I_0 \cos (2\pi z'/\lambda) \sin \theta'}{4\pi\varepsilon_0 c^2} \cdot \frac{\sin \{\omega(t - r'/c)\}}{r'}.$$

In this equation r' is the distance from the point Q to the point P and θ' is the angle the line joining Q to P makes with the z-axis. The z-axis is taken to be along the aerial and we assume $r' \gg \lambda$. The total electric field E_θ at P is obtained by integrating over the length of the aerial, and we have

$$E_\theta = -\frac{\omega I_0}{4\pi\varepsilon_0 c^2} \int_{-\lambda/4}^{\lambda/4} \frac{\cos (2\pi z'/\lambda) \sin \theta'}{r'} \sin \{\omega(t - r'/c)\} \, dz'.$$

To first order in z'/r, the distance $r' = r - z' \cos \theta$, where r is the distance from the centre of the dipole to the point P. Since $r \gg \lambda$ the term $\sin \theta'/r'$ in the integrand can be replaced by the constant $\sin \theta / r$. The phase factor $\omega(t - r'/c)$ has to be retained however, since the time z'/c is now comparable with the period of the oscillations. Hence

$$E_\theta = -\frac{\omega I_0 \sin \{\omega(t - r/c)\} \sin \theta}{4\pi\varepsilon_0 c^2 r} \int_{-\lambda/4}^{\lambda/4} \cos \frac{2\pi z'}{\lambda} \cos \frac{\omega z' \cos \theta}{c} dz'$$

$$-\frac{\omega I_0 \cos \{\omega(t - r/c)\} \sin \theta}{4\pi\varepsilon_0 c^2 r} \int_{-\lambda/4}^{\lambda/4} \cos \frac{2\pi z'}{\lambda} \sin \frac{\omega z' \cos \theta}{c} dz'.$$

The second integral vanishes. After some straight-forward but lengthy algebra the first integral can be shown to be equal to

$$\frac{-2c \cos \{(\pi \cos \theta)/2\}}{\omega \sin^2 \theta},$$

and the field at P is thus given by the real part of the expression

$$E_\theta = -\frac{jI_0}{2\pi\varepsilon_0 c} \cdot \frac{\cos \{(\pi \cos \theta)/2\}}{\sin \theta} \cdot \frac{\exp \{j(\omega t - kr)\}}{r}. \tag{13.24}$$

the magnetic field B_ϕ is given by the equation

$$B_\phi = E_\theta/c, \tag{13.25}$$

as can be seen from Equations (13.20) and (13.21). This is another illustration, like the plane wave example, of the general rule that the relationship between the magnetic and electric fields in an electromagnetic radiation field in free space is

$$\mathbf{B} = \frac{1}{c}\hat{\mathbf{k}} \wedge \mathbf{E},$$

where $\hat{\mathbf{k}}$ is a unit vector in the direction of propagation of the wave.

For a full-wave aerial, one wavelength long, the current at a point with z-coordinate z' is

$$I = I_0 \sin \frac{2\pi z'}{\lambda} \cos \omega t.$$

The z-axis is taken to be along the aerial with the origin at its centre as before. A similar procedure to that followed with the half-wave aerial gives the following expressions for the fields:

$$E_\theta = -\frac{jI_0}{2\pi\varepsilon_0 c} \cdot \frac{\sin (\pi \cos \theta)}{\sin \theta} \cdot \frac{\exp \{j(\omega t - kr)\}}{r}, \tag{13.26}$$

and

$$B_\phi = E_\theta/c. \tag{13.27}$$

The polar diagrams for the fields from half-wave and full-wave aerials are shown in Figures 13.6 and 13.7.

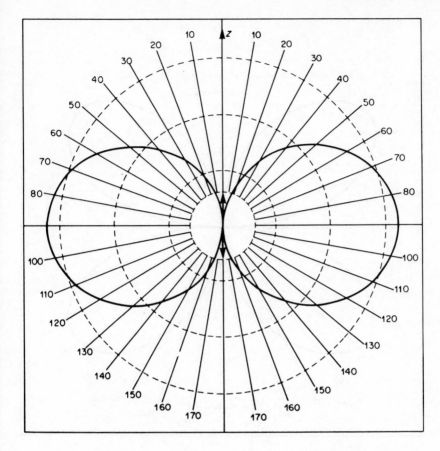

Figure 13.6. The polar diagram for a half-wave dipole aerial.

The average total power $\overline{W}$ radiated by a half-wave aerial can be obtained by integrating the Poynting vector over the surface of a complete sphere.

$$\overline{W} = \frac{I_0^2}{4\pi\varepsilon_0 c}\int_0^\pi \frac{\cos^2\left(\pi\cos\theta/2\right)}{\sin\theta}\,\mathrm{d}\theta.$$

The integral can be evaluated numerically to give

$$\overline{W} \simeq 73\frac{I_0^2}{2}.$$

In order to transmit 5 kW from a half-wave aerial the amplitude of the current oscillating in it at its centre must be about 12 A. If an oscillating current of amplitude I_0 were passing through a resistor of resistance 73 Ω the average power dissipated would be $73I_0^2/2$ watts. The *radiation resistance* of the half-

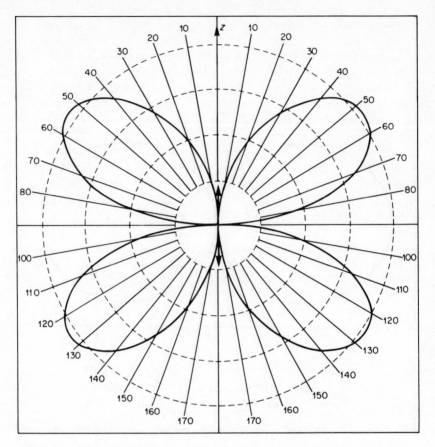

Figure 13.7. The polar diagram for a full-wave dipole aerial. Note the second zero at
$\theta = 90°$. This occurs because the full-wave aerial is similar to two half-wave aerials
placed end to end carrying out of phase currents.

wave antenna is thus 73 Ω, and is a convenient quantity used to represent the
aerial in an equivalent circuit.

The field patterns and other properties we have worked out above apply for
isolated aerials, those which are many wavelengths clear of the ground or other
bodies which will reflect electromagnetic waves. These patterns are almost impossible to obtain in practice, and at frequencies below about 3 MHz the height
of the antenna above ground is a major factor in determining its radiation pattern.
The optimum antenna system for a given requirement is usually determined
empirically, and the final radiation pattern measured experimentally. The
matching of the antenna to the transmission line feeding it is also usually done
empirically, using theory as a guideline. A half-wave dipole aerial looks like a
real impedance of about 73 Ω to the transmission line, since the ohmic losses in

the metal antenna are negligible compared to the energy radiated. The exact value of the equivalent impedance of the antenna again depends on several factors, among which are the height of the dipole above the ground, the point on the dipole to which the transmission line is joined, and the diameter of the dipole. A transmission line with characteristic impedance different from that of the aerial can be matched to the half-wave dipole in several ways, one of which is by using a length of transmission line of suitable characteristic impedance, as discussed in section 9.4.

An antenna system can be made more directional by using more than one dipole radiator. For example if two half-wave dipoles are placed parallel, side by side and separated by a short distance, the resultant field is the sum of the fields due to each separately. At some points the fields will be out of phase and will cancel each other. At other points the fields will interfere constructively. The aerial system will be directional, and the directional property can be varied by altering the relative phases of the currents fed to the two dipoles. For example consider two half-wave dipoles pointing vertically and separated in the horizontal planes by a distance equal to one half of a wave-length, as in Figure 13.8. If the same currents are fed to them in phase, the field at a particular point is given by the sum of two terms which will be the same except for a phase difference

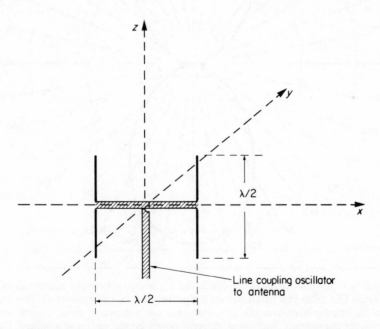

Figure 13.8. A simple aerial array consisting of two vertical half-wave dipoles separated by a distance equal to one half of a wavelength. The connections to the transmission line coupling the antenna to the oscillator are made so that the currents in both dipoles are in phase.

corresponding to the different path lengths from the dipoles to the point under consideration. Along the x-axis the fields everywhere cancel out, whilst they are always in phase along the y-axis. The relative fields in the $x-y$ plane are shown on Figure 13.9. The use of several dipoles instead of just two would enhance the directional properties of the antenna even further. The radiation pattern produced by a system consisting of a large number of dipoles fed with their currents all in phase is analogous to the diffraction pattern produced when a plane wave of monochromatic visible light is incident normally onto a diffraction grating*. Another way of obtaining a highly directional antenna is to situ-

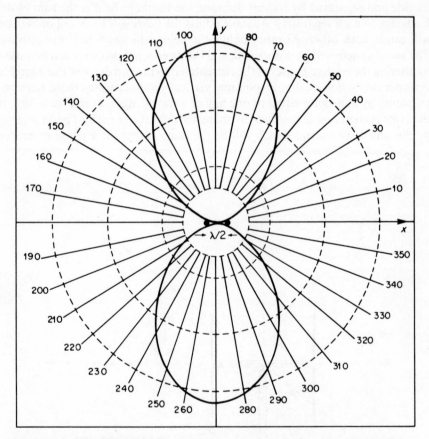

Figure 13.9. The relative field strengths in the $x-y$ plane far from the antenna shown in Figure 13.8. Note that this is not similar to the previous polar diagrams. This one gives the relative field strengths at points on a circle drawn in the $x-y$ plane and centred on the antenna, and not at different points on the surface of a sphere. The fields now depend on azimuthal angle ϕ, which is the angle shown on the scale, and a different diagram would have to be drawn to show the fields in another plane.

* See for example section 11.5 in *Optics* by F. G. Smith and J. Thompson, J. Wiley, London, 1972.

ate a half-wave dipole at the focus of a parabolic reflector. This produces a narrowly diverging beam whose direction can be simply varied by movement of the system as a whole. The angular half-width of the beam θ is that appropriate to the diffraction pattern of a circular aperture* of diameter d equal to the diameter of the parabolic reflector. θ is thus roughly equal to (λ/d) where λ is the wavelength of the radiation.

PROBLEMS 13

13.1 A half-wave dipole transmitter in the centre of a city is required to give signals to cars travelling in the city within a radius of 10 km. If the minimum field amplitude required is 0.02 V m^{-1} what must be the mean power of the transmitter? Why would the frequency chosen be likely to be around 100 MHz?

13.2 Give dimensions for a quarter wavelength length of coaxial line suitable for connecting a 50 Ω line to a 72 Ω aerial.

13.3 An antenna consists of four vertical half-wave dipoles separated from each other by one wavelength in the East–West direction. Calculate and sketch the horizontal polar diagram of the array when the dipoles are fed in phase with equal currents.

13.4 Why is an antenna much shorter than a wavelength impractical as a radiator of radio waves? If a commercial radio station transmitting at a wavelength of 500 m did not wish to string a cable 250 m long well above ground to form a half-wave dipole antenna, what might it do to make an efficient aerial system?

13.5 Electric quadrupole radiation is produced when the currents in two very small dipoles, placed side by side very close to each other, oscillate 180° out of phase. Show that the electric radiation field produced is proportional to the cube of the angular frequency. Hence show that the probability per unit time for the decay of an excited microscopic energy level via quadrupole radiation is proportional to the fifth power of the photon energy.

13.6 Determine the ratio of the e.m.f.s set up in a short wire of length l and in a closed loop of wire of area S when they are used as receiving aerials. Consider the cases where both aerials are orientated so as to produce maximum e.m.f.'s.

* *Ibid*: section 9.4.

CHAPTER

Electromagnetism and special relativity

14.1 INTRODUCTORY REMARKS

Maxwell's equations were used in section 11.1 to predict that light (or other electromagnetic waves outside the visible spectrum) travels in vacuo at the constant speed $c = 1/\sqrt{\varepsilon_0 \mu_0}$. The derivation of this result made no reference to the velocity of the source emitting the light. Experiment confirms that an observer always finds the value c for the speed of light, whatever may be the velocity of the light source relative to a frame of reference fixed in his laboratory (i.e. the velocity measured in a set of coordinates fixed in his laboratory). For example, the speed of the radiation emitted in the decay of π-mesons is measured to be c, even if the mesons themselves have been moving through the laboratory at speeds close to c*. Furthermore, there is no reason to suppose that the constancy of the speed of light is a peculiarity of our laboratory frame of reference—one would expect to find the same result in a space station or in a laboratory on another planet.

Questions about observations in different frames of reference are the concern of the theory of relativity. The special theory of relativity is restricted to statements about inertial frames of reference. Inertial frames are those in which the law of inertia holds, i.e. those in which all bodies move with a constant velocity unless they are subjected to external forces. All inertial frames move with a constant velocity relative to one another. The special theory accepts that there

---

* This experiment is described by F. J. M. Farley, J. Bailey and E. Picasso in *Nature*, **217**, 17 (1968). The radiation was emitted by mesons moving at a speed greater than $0.99975c$ and the speed of the radiation was measured to be $(2.9979 \pm 0.0004) \times 10^8$ m/sec.

is no way to pick out one of these frames as having a more fundamental importance than any of the others, and treats all inertial frames on an equal footing. If inertial frames are to be treated alike, then it must be possible to find a way of expressing physical laws which is common to all of them.

The equivalence of inertial frames and the constancy of the speed of light are the two ideas forming the basis of the special theory of relativity: they are the *axioms* of the theory. Written out in full, the axioms are

Axiom No. 1   *Physical laws must have the same form in all inertial frames of reference.*

Axiom No. 2   *All observers find the same value c for the speed of light.*

The results of the special theory of relativity follow from the requirement that the laws of physics should be in accordance with these two axioms. In mechanics, for example, the equivalence of mass and energy, and the increased mass of bodies moving at speeds close to c, follow from the relativistic laws proposed by Einstein. These effects are both well established experimentally, and constitute a powerful corroboration of the theory.

The purpose of this chapter is to test the laws of electromagnetism against the axioms of special relativity. We shall show that although observers in different frames of reference disagree about the magnitude of electric and magnetic fields, they all agree that the fields obey Maxwell's equations. In other words the laws of electromagnetism, as embodied in Maxwell's equations, are in accordance with the first axiom. Now from Maxwell's equations the speed of light is deduced to be c, and the second requirement of the theory is automatically fulfilled. The laws of electromagnetism therefore need no modification in moving frames of reference: they are already fully consistent with the special theory of relativity.

There is another way of looking at the matching of electromagnetism and special relativity. Not only is relativity a well-established theory, but its axioms are of such a general nature (axiom No. 1 seems almost intuitive) that we should like to place the theory very high in the hierarchy of physics. Other laws *must* be made to fit in with the special theory of relativity. In this spirit, we start in section 14.7 with the Coulomb field of a stationary point charge, and use relativistic arguments to acquire information about the fields generated by moving charges. In effect Maxwell's equations are derived from Coulomb's law, so that the whole of electromagnetism is seen to follow naturally from electrostatics.

14.2   THE LORENTZ TRANSFORMATION

Before discussing electromagnetic fields, we shall remind the reader of the *Lorentz transformation* relating space and time coordinates in different frames of reference. Consider two inertial frames S and S' having a common origin at $t = 0$, but with the origin of S' moving along the x-axis of S at a speed v (see Figure 14.1). An event occurring at time t in the frame S at the point with coordinates (x, y, z) is seen in S' as occurring at time t' at the point (x', y', z'). The relation between the two sets of coordinates is given by the Lorentz transformation, which follows from the postulates of special relativity. The relation is

$$x' = \gamma(x - vt), \qquad y' = y, \qquad z' = z,$$

$$t' = \gamma\left(t - \frac{vx}{c^2}\right),$$

(14.1)

where

$$\gamma = \frac{1}{\sqrt{1 - v^2/c^2}}.$$

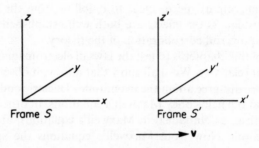

Frame S Frame S'

$\longrightarrow$ v

Figure 14.1. Two inertial frames of reference in
relative motion.

For an observer in S', the origin of S appears to be moving along the x'-axis at a speed $(-v)$. The two observers are on an equal footing, and the inverse Lorentz transformation, giving coordinates in S in terms of their values in S', is obtained simply by replacing v by $(-v)$ in Equation (14.1):

$$x = \gamma(x' + vt'), \qquad y = y', \qquad z = z'$$

$$t = \gamma\left(t' + \frac{vx'}{c^2}\right)$$

(14.2)

(The consistency of the two sets of equations can be checked by substituting (14.2) into (14.1), leading to $x' = x'$ etc.).

Two features of the Lorentz transformation will be particularly prominent when we go on to electric and magnetic fields. These are the FitzGerald contraction, and time dilatation, which refer to changes in the scales of length and time respectively, for observers in the frames S and S'. First, time dilatation. Suppose that a clock at a fixed point x' in the frame S' measures a time interval $(t'_2 - t'_1)$ between events occurring at x' at the times t'_1 and t'_2. For an observer in

S, these events appear to occur at times given by Equation (14.2) as

$$t_1 = \gamma\left(t_1' + \frac{vx'}{c^2}\right)$$

$$t_2 = \gamma\left(t_2' + \frac{vx'}{c^2}\right).$$

He thus considers the time interval to be $(t_2 - t_1) = \gamma(t_2' - t_1')$. This is larger than the time $(t_2' - t_1')$ measured by the clock in S', since $\gamma = 1/\sqrt{1 - v^2/c^2}$ is greater than unity. In other words, *the moving clock appears to run slow.*

We will now examine the analogous problem of the apparent length of a moving rigid rod. Imagine a rigid body, with ends which are fixed in S' at the points x_1' and x_2'. At a definite time t in the frame S, the ends are seen to be at x_1 and x_2, where from (14.1)

$$x_1' = \gamma(x_1 - vt)$$

$$x_2' = \gamma(x_2 - vt).$$

Now the length of the moving body as measured in S is the distance between its ends at the same time t, i.e.

$$x_1 - x_2 = \frac{1}{\gamma}(x_1' - x_2').$$

The moving body appears to be shorter, its length having contracted by the factor $1/\gamma$. This effect is called the FitzGerald contraction. Notice that the FitzGerald contraction only operates along the direction of motion: since $y' = y$ and $z' = z$, the apparent dimensions of the body are unaltered in directions perpendicular to its motion.

Although distances and time intervals are changed by the Lorentz transformation, observers in S and S' must still agree that light waves travel at a speed c. If a light signal leaves the origin at time zero and propagates as a spherical wave, an observer in S describes the locus of the spherical wave-front at time t by the equation

$$x^2 + y^2 + z^2 - c^2t^2 = 0.$$

Similarly, as seen by an observer in S' the locus at time t' is given by

$$x'^2 + y'^2 + z'^2 - c^2t'^2 = 0.$$

These two equations are consistent with the Lorentz transformation, since we find by substituting from equations (14.1) that

$$x'^2 + y'^2 + z'^2 - c^2t'^2 = x^2 + y^2 + z^2 - c^2t^2.$$

A quantity like $(x^2 + y^2 + z^2 - c^2t^2)$ which always has the same value for observers in S and S' is said to be *invariant under the Lorentz transformation*, or to be *Lorentz invariant*.

14.3 CHARGES AND FIELDS AS SEEN BY DIFFERENT OBSERVERS

Charge

An atom carries negative charges in its electron cloud, and positive charges on the protons in its nucleus. Both protons and electrons are moving about within the atom, yet their positive and negative charges exactly cancel so that the atom as a whole preserves strict electrical neutrality. This implies that the charge on each elementary particle appears to be $\pm e$ whether or not the particle is moving relative to the observer. In other words, the electronic charge is invariant under Lorentz transformations. On a charged body containing many atoms, different observers could in principle count up the number of neutral atoms and the number of spare elementary charges left over. Since the observers agree that each spare elementary charge has a magnitude $+e$ or $-e$, they would also agree about the net charge on the body. It follows that *the net charge carried by any macroscopic body is Lorentz invariant.*

Charge density of a line of equally spaced charges

Charge density, unlike total charge, is *not* Lorentz invariant. Imagine a series of positive charges each of magnitude $+e$, strung out along a line with equal spacings as shown in Figure 14.2. Suppose that an observer in a frame of reference S is at rest relative to the charges, and that in S the N charges labelled

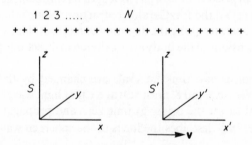

Figure 14.2. A line of charges which is stationary in S appears to be moving from right to left in S'.

1 to N span a distance of unit length. Then this observer sees a linear charge density Ne coulombs/m. The observer in S' moving with a speed v parallel to the line of charges agrees that each of the N charges carries a charge $+e$, but he does not agree that they cover a unit of length. Because of the Lorentz contraction, in his view the N charges have been squashed up into a distance $1/\gamma$, and the linear charge density is

$$\gamma Ne = \frac{Ne}{\sqrt{1 - v^2/c^2}} \text{ coulombs/m.}$$

Electric and magnetic forces

Let us now investigate the forces exerted by electric and magnetic fields, as viewed by observers in S and S'. Suppose that in the frame S, a wire which has no net charge carries a current in the x-direction. A charge $+q$ moving parallel to the wire at a speed v experiences a magnetic force pushing it towards the wire. But from the point of view of the observer in a frame S', in which this charge is at rest, it cannot be acted upon by any magnetic forces! Yet he must agree with the observer in S that there is a force on q, pointing towards the current-carrying wire. What does he consider to be the origin of this force? The answer is that the observer in S' does not think that the wire is electrically neutral; he says that there is an *electrostatic* field around the wire. As we shall see, the difference arises because of the FitzGerald contraction.

For an observer in S, the neutral wire contains equal numbers of positive and negative charges $\pm e$, each with a density N charges per unit length, say. Assume that all the negative charges are at rest, and that all the positive charges are moving at the same speed v (the calculations are made easier by this assumption, although the final result does not depend on it). The current flowing along the wire is

$$I = Nev,$$

and at a distance r from the wire the observer in S sees a magnetic field of magnitude (see Equation (4.32))

$$B = \frac{\mu_0 I}{2\pi r} = \frac{\mu_0 Nev}{2\pi r}.$$

The charge $+q$ moves at a speed v in the x-direction in S, perpendicularly to the magnetic field, and if its distance from the wire is r it experiences a force

$$F = qvB = \frac{\mu_0 Nev^2}{2\pi r}. \tag{14.3}$$

Now let us evaluate the force from the point of view of the observer in S'. The negative charges which were at rest in S are moving at a speed v relative to him. They are squashed up by the FitzGerald contraction, and he measures their linear charge density to be $(-\gamma Ne)$. On the other hand the positive charges are at rest in his frame; the contraction works the other way round, and he measures a diminished density Ne/γ. From his point of view the wire has a net linear charge density*

$$\lambda' = -\gamma Ne\left(1 - \frac{1}{\gamma^2}\right) = -\gamma Ne\frac{v^2}{c^2} = -\frac{\gamma Iv}{c^2}. \tag{14.4}$$

---

* It may seem paradoxical that the positive and negative charge densities on the wire transform differently, although the total charge on a body is invariant. But real steady currents always flow in closed circuits. The return half of the circuit (which may be too far away to make much difference to the field) has an opposite change in charge density, ensuring that the total charge on the complete circuit appears to be the same in both frames.

Distances perpendicular to the motion are not affected by the Lorentz transformation, and in S' the stationary charge $+q$ is still at a distance r from the wire. It experiences an electric field pointing towards the wire, of magnitude

$$E' = \frac{\lambda'}{2\pi\varepsilon_0 r} = \frac{\gamma N e v^2}{2\pi\varepsilon_0 r c^2},$$

or, replacing $\varepsilon_0 c^2$ by $1/\mu_0$,

$$E' = \frac{\gamma\mu_0 N e v^2}{2\pi r} = \gamma v B.$$

The electrostatic force on the charge is

$$F' = qE' = \gamma q v B = \gamma F, \tag{14.5}$$

by comparison with Equation (14.3).

The force which was purely magnetic in the frame S appears to be purely electrostatic in S'. Observers in the two frames disagree not only about the origin of the force but also about its magnitude, the electrostatic force in S' being larger by the factor γ. The factor γ arises because of time dilation, as we shall see by considering momentum changes in the frames S and S'. In S' the force F' gives q a component of momentum p'_r towards the wire such that $F' = dp'_r/dt'$. In a time dt' the charge thus acquires a momentum $dp'_r = F'\,dt'$. Now transverse momentum is not altered by a Lorentz transformation; during the time measured in S' as dt', an observer in S agrees that q acquires a transverse momentum $dp_r = dp'_r$. But the observer in S thinks that a clock which is at rest relative to q goes slow, and he reckons that the time taken to acquire the momentum dp_r is $dt = \gamma\,dt'$. Hence

$$F'\,dt' = dp'_r = dp_r = F\,dt = F\gamma\,dt',$$

and

$$F' = \gamma F,$$

in agreement with Equation (14.5)

For simplicity we have illustrated the interchangeability of electric and magnetic fields by imagining a current composed of a set of charges all travelling at the same speed as the moving charge q. But the argument applies more or less unaltered however the current is constituted. An observer moving at a speed v parallel to the direction of a current I in a wire always considers that the wire has a line charge density λ' given by Equation (14.4): $\lambda' = -\gamma I v/c^2$. In ordinary conductors the mean drift velocity $\bar{v}$ of the moving charge carriers is very small— about 10^{-4} m/sec for a current of one amp flowing in a wire of area 1 mm². If the line density of moving charge carriers in the wire is κ, then $I = \kappa\bar{v}$, and $\lambda' = -\kappa\gamma\bar{v}v/c^2$. The charge density of the wire as seen by the moving observer is thus extremely small compared to the density of moving charge carriers.

14.4   FOUR-VECTORS

According to the special theory of relativity there is no absolute frame of reference; all coordinate systems in uniform relative motion are equally accept-able, and physical laws must take the same form in all of them. How can one test whether or not a particular law is in accordance with this requirement? Let us illustrate the kind of argument used to make such a test by considering first of all two fixed coordinate systems which have the same origin but whose axes point in different directions. For example, if the frame of reference S in Figure 14.3 is rotated through 90° about the z-axis, it coincides with the frame S'. The coordinates of the point P are not the same in S and S'. Rotation of the axes mixes up the x- and y-coordinates so that if P is at (x, y, z) in S, its coordinates (x', y', z') in S' are given by $x' = y, y' = -x, z' = z$. Any rotation of the axes leads to a *linear* transformation of this kind, in which the new coordinates are linear combinations of the old ones.

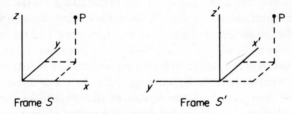

Frame S Frame S'

Figure 14.3. The frame S' is generated by rotating frame
S through 90° about the z-axis.

Because the orientation of the axes is of no consequence, it is often a good idea to use the vector notation. In specifying the point P by a position vector $\mathbf{r}$, no reference at all is made to the orientation of the axes; $\mathbf{r}$ represents the same point in space whatever set of axes is chosen. There is a further advantage in using the vector notation. Other quantities besides position are vectors, and when the axes are rotated, the components of these other vectors transform in exactly the same way as the components of a position vector. For example, looking again at the frames S and S' in figure 14.3, if a momentum vector $\mathbf{p}$ has components (p_x, p_y, p_z) in S, then its components in S' are $p'_x = p_y, p'_y = -p_x,$ $p'_z = p_z$. Because transformations of this kind are always linear, the sum of any two vectors will also transform in the same way when the axes are rotated. Now look at a vector equation such as the equation defining the force $\mathbf{F}$ acting on a charge q in fields $\mathbf{E}$ and $\mathbf{B}$ (Equation 4.16):

$$\mathbf{F} = q\mathbf{E} + q\mathbf{v} \wedge \mathbf{B}.$$

Each term on both sides of the equation is a vector, and the *whole equation* transforms in the same way as $\mathbf{r}$. If the equation is true for one set of axes, it must also be true for axes at any other orientation. There is no need to bother working out components for a number of different axes: just by writing the equation in terms of vectors we have ensured that the result is independent of the orientation of the axes.

The fact that vector equations are not affected by rotating the coordinate system may seem obvious, but it is worth spelling out in detail because we now want to apply the same idea to the less easily visualized transformations of relativity. Like the rotational transformation, the Lorentz transformation is a linear relation between different coordinate systems, but it mixes up time with the spatial coordinates. An *event* occurring at a particular time and place in a frame of reference S is *specified by the four coordinates* (x, y, z) *and* t. The Lorentz transformation tells us the coordinates (x', y', z') and t' of the same event in a frame S' moving relative to S. The event itself can be regarded as a four-dimensional vector, or *four-vector* for short.

The three dimensional vector $\mathbf{r}$ is characterized by its direction and length. The square of the length is found from the scalar product $\mathbf{r} \cdot \mathbf{r} = x^2 + y^2 + z^2$; $(x^2 + y^2 + z^2)$ has the same value for all orientations of the axes, i.e. it is *invariant* under rotations of the axes. A similar invariant scalar product can be formed from the components of a four-vector. In section 14.2 we have seen that the quantity $s^2 = x^2 + y^2 + z^2 - c^2t^2$ is *Lorentz invariant*, i.e. that s^2 has the same value when worked out in different inertial frames which are connected by a Lorentz transformation. By analogy with the three-dimensional scalar product r^2, we interpret s^2 as the square of the 'length' of the four-vector (x, y, z, t). Notice that s^2 is not always positive, because of the term $(-c^2t^2)$. To make s^2 look more like a familiar scalar product, it is convenient to replace the time coordinate by a new variable jct, which has the dimensions of a length. We shall write the four-vector representing an event as x_μ, with the suffix μ taking the values 1 to 4 for the four components $x_1 = x$, $x_2 = y$, $x_3 = z$, $x_4 = jct$. The invariant scalar product s^2 is given by

$$s^2 = x_\mu x_\mu = x_1^2 + x_2^2 + x_3^2 + x_4^2 = x^2 + y^2 + z^2 - c^2t^2. \qquad (14.6)$$

(The sum over μ has not been written explicitly. We use the convention that whenever a Greek suffix appears twice in a single term of a four-vector expression, this suffix is assumed to be summed from 1 to 4.)

In the new notation, the Lorentz transformation is

$$x'_1 = \gamma\left(x_1 + \frac{jv}{c}x_4\right),$$

$$x'_2 = x_2, \qquad x'_3 = x_3, \qquad (14.7)$$

$$x'_4 = \gamma\left(x_4 - \frac{jv}{c}x_1\right).$$

and the inverse transformation, obtained by substituting $(-v)$ for v, is

$$x_1 = \gamma\left(x_1' - \frac{jv}{c}x_4'\right),$$

$$x_2 = x_2', \qquad x_3 = x_3', \tag{14.8}$$

$$x_4 = \gamma\left(x_4' + \frac{jv}{c}x_1'\right).$$

Just as there are other three-dimensional vectors besides position vectors, so there are other four-vectors besides those whose space and time coordinates represent an event. For example, the momentum and energy of a particle together form a four-vector p_μ with components $(p_x, p_y, p_z, jE/c)$. These components transform according to the Lorentz transformation, with the same coefficients as occur in Equation (14.7) for the four-vector x_μ. The scalar product $p_\mu p_\mu$ is invariant, i.e.

$$p_\mu p_\mu = p_x^2 + p_y^2 + p_z^2 - E^2/c^2 = -m_0^2 c^2, \text{ a constant.}$$

All observers in inertial frames agree about the value of the *rest mass* m_0 of the particle, although they may differ about its momentum and energy.

Four-vectors may be combined together to form four-vector equations, analogous to the vector equations in three dimensions. A trivial example of a four-vector equation is found in the addition of the momenta and energies of two particles. The total momentum $\mathbf{P}$ of the pair of particles is the sum of their individual momenta $\mathbf{p}_1$ and $\mathbf{p}_2$, and similarly the total energy E is the sum of their energies E_1 and E_2:

$$\mathbf{P} = \mathbf{p}_1 + \mathbf{p}_2,$$

$$E = E_1 + E_2.$$

Because the Lorentz transformation is linear, $(\mathbf{P}, E)$ transforms in the same way as $(\mathbf{p}_1, E_1)$ and $(\mathbf{p}_2, E_2)$, or in other words $\mathbf{P}$ and E form the components of a new four-vector P_μ defined by the four-vector equation $P_\mu = p_{1\mu} + p_{2\mu}$.

In order to investigate the relativistic behaviour of Maxwell's equations, we shall need to know the transformation properties of differential operators. The space and time differential operators fit easily into the four-vector scheme, since together they form yet another four-vector. To see that this is so, we apply the rules of partial differentiation to write

$$\frac{\partial}{\partial x_1'} = \frac{\partial x_1}{\partial x_1'}\frac{\partial}{\partial x_1} + \frac{\partial x_2}{\partial x_1'}\frac{\partial}{\partial x_2} + \frac{\partial x_3}{\partial x_1'}\frac{\partial}{\partial x_3} + \frac{\partial x_4}{\partial x_1'}\frac{\partial}{\partial x_4}$$

$$= \gamma\left(\frac{\partial}{\partial x_1} + \frac{jv}{c}\frac{\partial}{\partial x_4}\right),$$

using Equations (14.8). This is the same transformation law as applies to the coordinate x_1 itself. Similar equations for the other components $\partial/\partial x_2, \partial/\partial x_3,$ $\partial/\partial x_4$ show that they too transform according to the Lorentz transformation. In other words, $\partial/\partial x_\mu$ is a four-vector.

From the four vector $\partial/\partial x_\mu$ we can form the invariant scalar product

$$\frac{\partial}{\partial x_\mu}\frac{\partial}{\partial x_\mu} = \frac{\partial^2}{\partial x_1^2} + \frac{\partial^2}{\partial x_2^2} + \frac{\partial^2}{\partial x_3^2} + \frac{\partial^2}{\partial x_4^2}$$

$$= \frac{\partial^2}{\partial x^2} + \frac{\partial^2}{\partial y^2} + \frac{\partial^2}{\partial z^2} - \frac{1}{c^2}\frac{\partial^2}{\partial t^2}.$$

This is the operator occurring in the wave equation for waves moving at a speed c: it must be an invariant, since all observers agree about the speed of light.

14.5 MAXWELL'S EQUATIONS IN FOUR-VECTOR FORM

In section 14.3 we found that electric and magnetic fields transform into one another in different inertial frames. Electric and magnetic fields are therefore not simply the space-like components of two different four-vectors. In the four-dimensional picture the behaviour of the fields is rather complicated—together they form some of the components of a quantity called a four-tensor. For our purposes there is fortunately no need to know anything about tensors, since it turns out that the vector and scalar potentials do make up a four-vector. By expressing Maxwell's equations in terms of the potentials, we shall be able to write them in a way which is obviously relativistically invariant. But before doing this, we must first show that current density and charge density also constitute a four-vector, and write the equation of continuity in an invariant form.

The equation of continuity

We have already seen that the charge on any body is invariant, so that it is assigned the same value whatever may be the velocity of the body relative to an observer measuring the charge. Charge is also conserved, since a charged elementary particle can only be created or annihilated in association with another particle carrying charge of exactly the same magnitude but opposite size: this must evidently be true in all inertial frames. Now the mathematical statement of charge conservation is contained in the equation of continuity (Equation (10.1)) and this equation must therefore have the same form in all inertial frames. In the frame S, the equation of continuity is

$$\text{div } \mathbf{j} + \frac{\partial \rho}{\partial t} = 0.$$

This equation takes on the required invariant form only if we introduce a four-vector j_μ made up from the current density and the charge density. The current density $\mathbf{j}$ is the space-like part of j_μ, and the fourth component is $j_4 = jc\rho$.

Rewriting the equation of continuity in terms of j_μ,

$$\text{div } \mathbf{j} + \frac{\partial \rho}{\partial t} = \frac{\partial j_1}{\partial x_1} + \frac{\partial j_2}{\partial x_2} + \frac{\partial j_3}{\partial x_3} + \frac{\partial j_4}{\partial x_4} = 0,$$

or

$$\frac{\partial j_\mu}{\partial x_\mu} = 0. \tag{14.9}$$

The equation of continuity is now in four-vector form, and manifestly invariant. As we are insisting from the start that the law of charge conservation must be invariant, Equation (14.9) constitutes a proof that $(\mathbf{j}, jc\rho)$ is indeed a four-vector. Its components transform in the same way as the components of $(\mathbf{r}, jct)$. Thus in the frame S', moving at a speed v along the x-axis of S,

$$\rho' = \gamma \left(\rho - \frac{v j_x}{c^2} \right).$$

In the example of the line charges discussed in section 14.3, the charges were at rest in frame S, and $j_x = 0$. As a result of the FitzGerald contraction, the transformation is then simply $\rho' = \gamma\rho$.

Maxwell's equations

To show that Maxwell's equations are Lorentz invariant, it is sufficient to express them in four-vector form. We have already pointed out that the electromagnetic fields are not components of four-vectors, and we can most easily test the invariance of Maxwell's equations by writing them in terms of the scalar and vector potentials. This has already been done in the previous chapter for the case when no dielectric or magnetic materials are present; we shall continue to make this restriction in order to avoid undue complication. The fields are derived from the potentials using Equations (13.1) and (13.2):

$$\mathbf{B} = \text{curl } \mathbf{A}$$

$$\mathbf{E} = -\text{grad } \phi - \frac{\partial \mathbf{A}}{\partial t}.$$

Because of the vector identities div curl $\mathbf{A} \equiv 0$ and curl grad $\phi \equiv 0$, the law that no free magnetic poles exist (div $\mathbf{B} = 0$) and Faraday's law (curl $\mathbf{E} = -\partial \mathbf{B}/\partial t$) are automatically satisfied. However, there is still some arbitrariness in the potentials, which can have a variety of different forms while still remaining consistent with Maxwell's equations. Most of the possible choices of potentials will not lead to a Lorentz invariant form for the equations describing how $\mathbf{A}$ and ϕ vary with $\mathbf{r}$ and t. To achieve invariance, we must require the potentials to satisfy the Lorentz condition (Equation (13.8))

$$\text{div } \mathbf{A} + \frac{1}{c^2} \frac{\partial \phi}{\partial t} = 0.$$

This restriction on the potentials was chosen in section 13.1, where it was shown to lead to the following forms for the remaining two of Maxwell's equations:

$$\text{Gauss' law becomes} \quad -\nabla^2\phi + \frac{1}{c^2}\frac{\partial^2\phi}{\partial t^2} = \frac{\rho}{\varepsilon_0},$$

$$\text{and Ampère's law becomes} \quad -\nabla^2\mathbf{A} + \frac{1}{c^2}\frac{\partial^2\mathbf{A}}{\partial t^2} = \mu_0\mathbf{j}.$$

Apart from constant factors, the right-hand sides of these equations are the components of the four-vector $j_\nu = (\mathbf{j}, jc\rho)$. On the left-hand sides the operator

$$-\nabla^2 + \frac{1}{c^2}\frac{\partial^2}{\partial t^2} = -\frac{\partial^2}{\partial x_\mu^2}.$$

operates on the scalar and vector potentials. This operator is Lorentz in-variant, and under Lorentz transformations both sides of the equations trans-form in the same way only if the quantity $A_\nu = (\mathbf{A}, j\phi/c)$ is a four-vector. By comparison with Ampère's law, the fourth component equation is found by making the substitution $\mathbf{A} \to j\phi/c$, $\mathbf{j} \to jc\rho$, i.e.

$$\frac{1}{c}\left\{-\nabla^2\phi + \frac{1}{c^2}\frac{\partial^2\phi}{\partial t^2}\right\} = \mu_0 c\rho,$$

or

$$-\nabla^2\phi + \frac{1}{c^2}\frac{\partial^2\phi}{\partial t^2} = \mu_0 c^2\rho = \rho/\varepsilon_0$$

which is indeed Gauss' law. Ampère's law and Gauss' law can then be written together in the single four-vector equation

$$-\frac{\partial^2}{\partial x_\mu^2}A_\nu = \mu_0 j_\nu. \tag{14.10}$$

Notice that the Lorentz condition itself is also invariant. It can now be written in terms of four-vectors, and is

$$\text{div }\mathbf{A} + \frac{1}{c^2}\frac{\partial\phi}{\partial t} = \frac{\partial A_\mu}{\partial x_\mu} = 0. \tag{14.11}$$

To sum up, we have shown that without any modifications at all, Maxwell's equations can be written in Lorentz invariant form, as required by the first axiom of the special theory of relativity. Maxwell's equations are also automati-cally in accordance with the second axiom that all observers measure the speed of light to be c, since in free space equation (14.10) reduces to a wave equation for electromagnetic waves travelling at the speed c. As well as finding that Maxwell's equations are consistent with the special theory of relativity, we have also learnt something new, namely that the quantities A_ν and j_ν are four-vectors. Knowing

A_v and j_v in one inertial frame, we can now work out their values in any other inertial frame by using the Lorentz transformation.

We have not yet completed the test of electromagnetism against relativity, because the equation defining the Lorentz force $\mathbf{F} = q(\mathbf{E} + \mathbf{v} \wedge \mathbf{B})$ has not so far been shown to be invariant. Since the Lorentz force is expressed in terms of the fields $\mathbf{E}$ and $\mathbf{B}$, we must first investigate the transformation of the fields, and then make sure that the force itself transforms in the proper manner. This is done in the next section.

14.6   TRANSFORMATION OF THE FIELDS

Electric and magnetic fields do not form the space-like parts of four-vectors. However, the fields can be derived by differentiating the vector and scalar potentials. Knowing how to transform the four-potential A_μ and the differential operator ∂/∂_μ, we can work out the transformation of the fields one component at a time. For example, let us evaluate the electric field component E_z' in the S' frame in terms of the fields $\mathbf{E}$ and $\mathbf{B}$ in the S frame.

Now

$$E_z' = -\frac{\partial \phi'}{\partial z'} - \frac{\partial A_z'}{\partial t'}.$$

Since $A_\mu = (\mathbf{A}, j\phi/c)$ and $\partial/\partial x_\mu = (\nabla, (j/c)\,\partial/\partial t)$ are both four-vectors, this equation may be rewritten as

$$E_z' = jc\left\{\frac{\partial A_4'}{\partial x_3'} - \frac{\partial A_3'}{\partial x_4'}\right\}. \tag{14.12}$$

Using the Lorentz transformation given in Equation (14.7),

$$A_4' = \gamma\left(A_4 - \frac{jv}{c}A_1\right), \qquad A_3' = A_3,$$

$$\frac{\partial}{\partial x_3'} = \frac{\partial}{\partial x_3} \quad \text{and} \quad \frac{\partial}{\partial x_4'} = \gamma\left(\frac{\partial}{\partial x_4} - \frac{jv}{c}\frac{\partial}{\partial x_1}\right).$$

Hence

$$E_z' = jc\gamma\left\{\frac{\partial}{\partial x_3}\left(A_4 - \frac{jv}{c}A_1\right) - \left(\frac{\partial}{\partial x_4} - \frac{jv}{c}\frac{\partial}{\partial x_1}\right)A_3\right\}$$

$$= jc\gamma\left(\frac{\partial A_4}{\partial x_3} - \frac{\partial A_3}{\partial x_4}\right) + v\gamma\left(\frac{\partial A_1}{\partial x_3} - \frac{\partial A_3}{\partial x_1}\right).$$

By comparison with Equation (14.12), the first term is seen to be simply γE_z. The second term looks unfamiliar in its four-vector disguise, but it contains only space-like quantities, and is

$$v\gamma\left(\frac{\partial A_x}{\partial z} - \frac{\partial A_z}{\partial x}\right) = v\gamma(\text{curl }\mathbf{A})_y = v\gamma B_y.$$

Finally we have $E_z' = \gamma(E_z + vB_y)$.

This is in agreement with the result derived earlier (see Equation 14.5) for the special case in which there is no electric field in the S frame.

Proceeding in the same way, all the other components of $\mathbf{E}'$ and $\mathbf{B}'$ can be evaluated in terms of the components in the S frame. The results are

$$E'_x = E_x,$$
$$E'_y = \gamma(E_y - vB_z), \tag{14.13}$$
$$E'_z = \gamma(E_z + vB_y),$$

and

$$B'_x = B_x,$$
$$B'_y = \gamma\left(B_y + \frac{v}{c^2}E_z\right), \tag{14.14}$$
$$B'_z = \gamma\left(B_z - \frac{v}{c^2}E_y\right).$$

The inverse transformations for $\mathbf{E}$ and $\mathbf{B}$ in terms of $\mathbf{E}'$ and $\mathbf{B}'$ have the same form except that v is replaced by $(-v)$.

For a macroscopic body moving at a speed small compared with c, there is only an insignificant difference between the electric and magnetic fields as measured by a stationary observer and by one moving with the body. Suppose, for example, that an object is travelling at the speed of sound (300/m sec) perpendicularly to an electric field of magnitude 10^6 volts/m—about as large a field as can be maintained over a distance of a few metres in vacuo. According to Equation (14.14), an observer moving with the object sees not only an electric field, but also a magnetic field of magnitude $\gamma vE/c^2 \simeq 3 \times 10^{-9}$ tesla. This is completely negligible even by comparison with the earth's rather weak field. On the atomic scale, on the other hand, speeds near to c are attained, and the role of external electric and magnetic fields can be very much modified. Consider a neutral hydrogen atom with an energy of 100 keV in the laboratory frame; a neutral atom can be given this energy by accelerating a proton through 100 kV and subsequently allowing it to capture an electron. The speed of such an atom is roughly 5×10^6 m/sec. If the atom now enters a magnetic field of 1 T, then in the frame S', at rest with respect to the atom, Equation (14.13) tells us that there is a huge electric field of magnitude 5×10^6 volts/m! The neutral atom, which has only recently captured its electron, is most likely to be in an excited state, and when subjected to such a strong electric field, it will very quickly ionize again*.

---

* The field is not strong enough to pull the electron straight away from the proton. But it is energetically favourable for the electron to separate from the proton in the direction opposite to that of the electric field. It does separate, because of the same quantum mechanical tunnelling effect which explains field emission (see Problem 1.11). Lorentz ionization provides us with another of the apparent paradoxes of relativity. Work must be done to ionize an atom, yet magnetic forces can do no work. However, although work is done by the electric field in the frame at rest relative to the atom, in the laboratory frame the kinetic energy alters; kinetic and potential energy are traded, but no external work is done.

This process is called Lorentz ionization. In the future Lorentz ionization may have technological importance, since it provides a possible means of injecting ions into the plasma of a fusion reactor.

Fields due to a moving point charge

The field transformations allow us to work out the fields generated by a point charge q moving with a constant velocity. Choosing the x-axis to be in the direction of motion of the charge, we can arrange that the charge is at rest at the origin of the S' frame. In this frame there is no magnetic field, and the electric field is simply

$$\mathbf{E}' = \frac{q}{4\pi\varepsilon_0} \frac{\mathbf{r}'}{r'^3},$$

with components

$$E'_x = \frac{qx'}{4\pi\varepsilon_0(x'^2 + y'^2 + z'^2)^{3/2}}, \qquad E'_y = \frac{qy'}{4\pi\varepsilon_0(x'^2 + y'^2 + z'^2)^{3/2}},$$

$$E'_z = \frac{qz'}{4\pi\varepsilon_0(x'^2 + y'^2 + z'^2)^{3/2}}.$$

The components of the fields in S now follow from the inverse of the transformations given in Equations (14.13) and (14.14) together with the Lorentz transformation as in Equation (14.1). We have

$$E_x = E'_x = \frac{qx'}{4\pi\varepsilon_0(x'^2 + y'^2 + z'^2)^{3/2}}$$

$$= \frac{q\gamma(x - vt)}{4\pi\varepsilon_0[\gamma^2(x - vt)^2 + y^2 + z^2]^{3/2}},$$

$$E_y = \gamma E'_y = \frac{q\gamma y}{4\pi\varepsilon_0[\gamma^2(x - vt)^2 + y^2 + z^2]^{3/2}}, \qquad (14.15)$$

$$E_z = \gamma E'_z = \frac{q\gamma z}{4\pi\varepsilon_0[\gamma^2(x - vt)^2 + y^2 + z^2]^{3/2}},$$

and

$$B_x = 0,$$

$$B_y = -\frac{\gamma v E'_z}{c^2} = -\frac{v E_z}{c^2}$$

$$B_z = \frac{\gamma v E'_y}{c^2} = \frac{v E_y}{c^2},$$

or equivalently

$$\mathbf{B} = \frac{1}{c^2}\mathbf{v} \wedge \mathbf{E}. \qquad (14.16)$$

The electric field lines are still straight lines radiating from the charge, but the field pattern has been squashed in the direction of motion. At speeds close to c, when γ is large, the field on the x-axis is directed along the x-axis and has the magnitude

$$E_x = \frac{q}{4\pi\varepsilon_0 \gamma^2 (x - vt)^2},$$

the same inverse square behaviour as for a stationary charge at $(x - vt)$, but diminished by the factor $1/\gamma^2$. On the other hand, the field radial to the charge at $x = vt$ is *increased* by the factor γ. Thus for very fast charged particles, such as the π-mesons in the experiment mentioned at the beginning of this chapter, for which $v/c > 0.99975$, the electric field is confined to a thin disc perpendicular to the direction of motion, as illustrated in Figure 14.4. Lines of **B** circulate around the moving charge, and equation (14.16) shows that they too are confined to the disc where the electric field is appreciable.

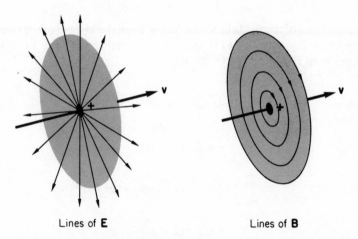

Lines of **E**   Lines of **B**

Figure 14.4. Around a charged particle moving at a speed near to c, the fields are confined to a thin disc which moves along with the particle.

Transformation of forces

Look again at the transformation of the electric field from S to S' in Equations (14.13). The right-hand side is reminiscent of the components of the Lorentz force on a charged particle moving at a speed v in the x-direction. This is no coincidence. In the S' frame such a particle is at rest, and the whole of the Lorentz force must be accounted for by the electrostatic force. From Equation

(14.13) the force on a charge q at rest in the S' frame has components

$$F'_x = qE'_x = qE_x,$$

$$F'_y = qE'_y = q\gamma(E_y - vB_z),$$

$$F'_z = qE'_z = q\gamma(E_z + vB_y).$$

By comparison, in the S frame the components of the Lorentz force $\mathbf{F} = q(\mathbf{E} + \mathbf{v} \wedge \mathbf{B})$ are

$$F_x = qE_x = F'_x,$$

$$F_y = q(E_y - vB_z) = \frac{F'_y}{\gamma}, \qquad (14.17)$$

$$F_z = q(E_z + vB_y) = \frac{F'_z}{\gamma}.$$

Is this the right way to transform forces between the frames S and S'? In order to be consistent, force must equal rate of change of momentum in both frames, i.e.

$$\mathbf{F} = \frac{d\mathbf{p}}{dt} \quad \text{and} \quad \mathbf{F}' = \frac{d\mathbf{p}'}{dt'},$$

where the time derivative applies to a point fixed relative to the particle. An observer in S thinks that a clock moving with the particle runs slow by the factor γ, and thus $dt = \gamma\, dt'$. The increment dp_y of the transverse momentum component is unaltered by the Lorentz transformation, so that $dp_y = dp'_y$. The y-component of the force therefore transforms as follows:

$$F_y = \frac{dp_y}{dt} = \frac{dp'_y}{dt'} = \frac{F'_y}{\gamma}.$$

Similarly $\qquad\qquad\qquad\qquad\qquad\qquad\qquad\qquad\qquad\qquad\qquad$ (14.18)

$$F_z = \frac{F'_z}{\gamma}.$$

The transformation of the x-component of the energy-momentum four-vector $(\mathbf{p}, jE/c)$ is given by Equation (14.8) to be

$$dp_x = dp_1 = \gamma\left(dp'_1 - \frac{jv}{c}\,dp'_4\right) = \gamma\left(dp'_x + \frac{v}{c^2}\,dE'\right).$$

Hence

$$F_x = \frac{dp_x}{dt} = \frac{1}{\gamma}\frac{dp_x}{dt'} = \frac{dp'_x}{dt'} + \frac{v}{c^2}\frac{dE'}{dt'}$$

$$= \frac{dp'_x}{dt'} = F'_x, \qquad (14.19)$$

since in the S' frame the particle is momentarily at rest, and $dE'/dt' = 0$. These results are consistent with the laws of transformation of force derived from the Lorentz force given in Equation (14.17).

This completes the demonstration that the classical laws of electromagnetism are consistent with the special theory of relativity. We have been able to express Maxwell's equations for the potentials as a single four-vector equation, so ensuring that this equation has the same form in all inertial frames. Finally, the Lorentz force exerted by electromagnetic fields on charged particles transforms in the appropriate way to agree with the relativistic relations between force and momentum. Different observers therefore have consistent descriptions of the motion of charged particles in electromagnetic fields.

14.7   MAGNETISM AS A RELATIVISTIC PHENOMENON

In this section the argument of the rest of the chapter is reversed. Up to now we have accepted the laws of electromagnetism, and tested them against the theory of relativity. Here we take the view that any acceptable theory must fall in with the requirements of relativity, since relativity is well-established in other fields of physics besides electromagnetism. Electric charge is a quantity which fits very easily into relativity theory: there is rather direct evidence that charge is invariant and that it is conserved for all observers in inertial frames. Accepting charge invariance and conservation, what else do we need to know? If we start with Coulomb's law for the force between stationary charges, is it possible, by using arguments based on relativity, to deduce the laws of magnetism? The answer to this question is that although a rigorous deduction is not possible, it can be made very plausible that the laws embodied in Maxwell's equations represent the simplest conceivable generalization of Coulomb's law which is consistent with relativity.

Let us see how far we can go from the starting-point of a distribution of stationary charges. As before, the charges are taken to be at rest in the frame S'. Applying Coulomb's law and the principle of superposition, the force on a test charge q at the origin of S' is given in terms of the charge density $\rho'(\mathbf{r}')$:

$$\mathbf{F}' = q\mathbf{E}' = \frac{-q}{4\pi\varepsilon_0} \int_{\text{all space}} \frac{\rho'(\mathbf{r}')\,\mathbf{r}'\,d\tau'}{r'^3}. \tag{14.20}$$

Here $\mathbf{r}'$ is the vector from the origin to the volume element $d\tau'$, as shown in Figure 14.5.

How is this situation described by an observer in a frame S, in which the charge is all moving at a speed v along the x-axis? From mechanics we know how to transform force (according to the transformation in Equations (14.18) and (14.19)) and since charge conservation requires that charge and current density form a four-vector $(\mathbf{j}, jc\rho)$, we know how to transform ρ'. The transformation properties of each of the quantities in Equation (14.20) are thus known. Nevertheless it is not possible to transform the whole equation for the most

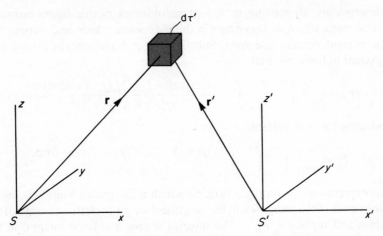

Figure 14.5. A charge element at rest in S' is moving along the x-axis at a speed v in S.

general charge distribution. The trouble arises because the charge density at a fixed point in S may not be constant, and the integral on the right-hand side of the transformed equation cannot be evaluated without making additional assumptions. However, if the charge density $\rho'(\mathbf{r}') = \rho'(y', z')$ as measured in S' does not vary with x', then $\rho(\mathbf{r})$ is independent of time in S. In this special case, we can complete the transformation and look for a way of expressing the result in which $\mathbf{j}$ and $jc\rho$ are treated on an equal footing (the search is made considerably easier by the fact that we know the answer beforehand!). Then we may be able to discover the four-vector analogue of Coulomb's law.

Because of the uniformity along the x-direction, the x-component of the force on the test charge is zero in both S and S' frames. In S' the y- and z-components are given by

$$F'_y = \frac{-q}{4\pi\varepsilon_0} \int_{-\infty}^{\infty} \int_{-\infty}^{\infty} \int_{-\infty}^{\infty} \frac{\rho'(y', z')y' \, dx' \, dy' \, dz'}{(x'^2 + y'^2 + z'^2)^{3/2}}$$

and   (14.21)

$$F'_z = \frac{-q}{4\pi\varepsilon_0} \int_{-\infty}^{\infty} \int_{-\infty}^{\infty} \int_{-\infty}^{\infty} \frac{\rho'(y', z')z' \, dx' \, dy' \, dz'}{(x'^2 + y'^2 + z'^2)^{3/2}}$$

For simplicity we shall choose to measure the force on q at the moment when the origins of S and S' coincide, i.e. at $t = 0$. The transformations needed for the y-component are then

$$F'_y = \gamma F_y \quad \text{(from Equation (14.18))}$$

$$y' = y, \qquad z' = z, \qquad x' = \gamma x$$

and

$$\rho'(y', z') = \gamma \left\{ \rho(y, z) - \frac{v}{c^2} j_x(y, z) \right\}$$

The charges are all moving in S, and an observer in this frame measures a current density which is everywhere in the x-direction; the y- and z-components of the current density are zero. Substituting in Equation (14.21) for the y-component of force, we find

$$\gamma F_y = \frac{-q}{4\pi\varepsilon_0} \int_{-\infty}^{\infty} \int_{-\infty}^{\infty} \int_{-\infty}^{\infty} \frac{\gamma^2[\rho(y,z) - (v/c^2)j_x(y,z)]y\,dx\,dy\,dz}{(\gamma^2 x^2 + y^2 + z^2)^{3/2}}.$$

Introducing the new variable $x = \gamma x$

$$F_y = \frac{-q}{4\pi\varepsilon_0} \int_{-\infty}^{\infty} \int_{-\infty}^{\infty} \int_{-\infty}^{\infty} \frac{[\rho(y,z) - (v/c^2)j_x(y,z)]y\,dx\,dy\,dz}{(x^2 + y^2 + z^2)^{3/2}}.$$

In this expression x is simply a variable which is integrated from $-\infty$ to $+\infty$, and the result of the integral will be the same if we go back to our original set of variables and replace x by x. The integral is now a volume integral in the S frame, which can be written

$$F_y = \frac{-q}{4\pi\varepsilon_0} \int_{\text{all space}} \frac{\rho y\,d\tau}{r^3} + \frac{qv}{4\pi\varepsilon_0} \int_{\text{all space}} \frac{j_x y\,d\tau}{r^3}.$$

Similarly

$$F_z = \frac{-q}{4\pi\varepsilon_0} \int_{\text{all space}} \frac{\rho z\,d\tau}{r^3} + \frac{qv}{4\pi\varepsilon_0 c^2} \int_{\text{all space}} \frac{j_x z\,d\tau}{r^3}.$$

Because of the uniformity in the x-direction,

$$\int \frac{\rho(y,z)x\,d\tau}{r^3} = 0, \quad \text{and} \quad F_x = 0.$$

Remembering that both $\mathbf{j}$ and $\mathbf{v}$ are pointing along the x-axis, the force can be expressed in vector notation as

$$\mathbf{F} = \frac{-q}{4\pi\varepsilon_0} \int_{\text{all space}} \frac{\rho \mathbf{r}\,d\tau}{r^3} - \frac{qv}{4\pi\varepsilon_0 c^2} \wedge \int_{\text{all space}} \frac{(\mathbf{j} \wedge \mathbf{r})\,d\tau}{r^3}$$

$$= q\mathbf{E} + q\mathbf{v} \wedge \mathbf{B}, \tag{14.22}$$

writing

$$\mathbf{B} = -\frac{1}{4\pi\varepsilon_0 c^2} \int \frac{\mathbf{j} \wedge \mathbf{r}\,d\tau}{r^3} = -\frac{\mu_0}{4\pi} \int \frac{\mathbf{j} \wedge \mathbf{r}\,d\tau}{r^3}. \tag{14.23}$$

This expression is simply the Biot–Savart law* for the steady field $\mathbf{B}$ at the origin.

---

* There is an inevitable confusion of notation because we are now using primes to distinguish moving frames of reference. The vector $\mathbf{r}$ of equation (14.23) is the same as the $\mathbf{r}'$ of Equation (4.43) which is

$$\mathbf{B}(\mathbf{r}) = \frac{\mu_0 I}{4\pi} \oint \frac{d\mathbf{l} \wedge (\mathbf{r} - \mathbf{r}')}{|\mathbf{r} - \mathbf{r}'|^3}$$

Putting $I\,d\mathbf{l}' = \mathbf{j}\,d\tau'$, and evaluating this expression at the origin, this expression takes the same form as Equation (14.23).

We have thus derived the Biot–Savart law from Coulomb's law in the special case when there is a stationary charge distribution in the S' frame, independent of x'. It is proved in Appendix C that the Biot–Savart law can be written in the equivalent form

$$\mathbf{A} = \frac{1}{4\pi\varepsilon_0 c^2} \int \frac{\mathbf{j}\,\mathrm{d}\tau}{r}, \tag{14.24}$$

where $\mathbf{B} = \operatorname{curl} \mathbf{A}$.

Compare this equation with Equation (1.24) for the electrostatic potential ϕ; we have not included any surface charges in the present discussion and the potential reduces to

$$\phi = \frac{1}{4\pi\varepsilon_0} \int \frac{\rho\,\mathrm{d}\tau}{r}.$$

We see that $\mathbf{A}$ is related to $\mathbf{j}/c^2$ in exactly the same way as ϕ is related to ρ, and we can construct a four-vector $(\mathbf{A}, j\phi/c)$ which transforms in the same way as $(\mathbf{j}, jc\rho)$. This four-vector is defined by Equations (14.24) and (1.24) for any distribution of charge and current density, and it is not restricted to charges which are all moving with the same velocity. However, after removing this restriction it is not possible to *prove* that the force on a charge q is still given by Equation (14.22). From the starting-point of Coulomb's law the Lorentz transformation cannot by itself tell us about the interaction of charges in relative motion. But let us now *assume* that equation (14.22) holds in general, with $\mathbf{B} = \operatorname{curl} \mathbf{A}$. Does the force then have the correct properties when it undergoes a Lorentz transformation? The answer to this question, which was discussed in section 14.6, is yes, provided that a time-dependent term is added to the electric field, which must become $\mathbf{E} = -\operatorname{grad} \phi - \partial\mathbf{A}/\partial t$. This is Faraday's law, since

$$\operatorname{curl} \mathbf{E} = -\operatorname{curl} \operatorname{grad} \phi - \frac{\partial}{\partial t}(\operatorname{curl} \mathbf{A})$$

$$= -\frac{\partial\mathbf{B}}{\partial t}.$$

To sum up, after making a Lorentz transformation of electrostatic forces, a simple generalization leads very persuasively to the other laws governing the forces between charged particles. There is a sense, then, in which Coulomb's law can be regarded as fundamental, with magnetic forces and induced e.m.f.'s appearing as relativistic side-effects. Yet one ought not to attach too much importance to the philosophical game of choosing the neatest set of initial postulates as the basis of electromagnetism. The real test of a physical theory comes in comparing its predictions with observation. On this criterion electro-magnetism is extraordinarily successful; as we pointed out at the beginning of this book, when quantum effects have been taken into account, no accurate observation has ever detected the slightest deviation from the theoretical predictions.

14.8 RETARDED POTENTIALS FROM THE RELATIVISTIC STANDPOINT

A general solution of Maxwell's equations in the absence of dielectric and magnetic media was given in Chapter 13 in terms of retarded scalar and vector potentials. The potentials are called 'retarded' because the contribution of each element of charge and current is delayed by the time taken for light to travel to the point where the potential is being calculated: information about the present value of charge and current propagates outwards at a speed c. In this section the retarded potentials are derived in an elegant way using relativistic arguments. We start by using the Lorentz transformation to find the potentials generated by a small charge when it is viewed from a moving frame of reference. Then the potentials caused by any distribution of charge and current can be found by superposition.

A charge $\rho'\,d\tau' = \rho'\,dx'\,dy'\,dz'$ is stationary in a frame S' and is located at the origin of S' as in Figure 14.6. What are the potentials seen at a time t in the S

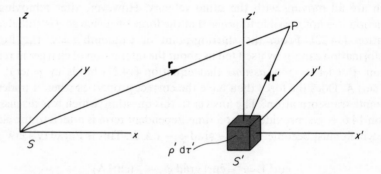

Figure 14.6. An element of charge at the origin of S' moves past the point P which is at rest in S.

frame, at the point P which is stationary in S? We shall choose the origin of S so that a light signal leaving the origin at $t = 0$ reaches P at the time t—this choice of origin simplifies the algebra. The arrival of the light signal at P is an event with coordinates $(\mathbf{r}, jct)$, where $t = r/c$. The origins of S and S' coincide at $t = 0$, and for an observer in S' the light signal arrives at P at the time $t' = r'/c$. In S', the contributions to the scalar and vector potentials at P due to the volume element at the origin are

$$d\phi' = \frac{\rho'\,d\tau'}{4\pi\varepsilon_0 r'}, \qquad dA' = 0.$$

Knowing that $(\mathbf{A}, j\phi/c)$ is a four-vector, the potentials at P as seen in S can now be obtained from the Lorentz transformation:

$$d\phi(\mathbf{r}, t) = \gamma \, d\phi' = \frac{\gamma\rho' \, d\tau'}{4\pi\varepsilon_0 r'}$$

$$dA_x(\mathbf{r}, t) = \frac{\gamma v \, d\phi'}{c^2} = \frac{\gamma v\rho' \, d\tau'}{4\pi\varepsilon_0 c^2 r'}, \tag{14.24}$$

$$dA_y = 0, \qquad dA_z = 0.$$

The right-hand sides of these equations have to be transformed back again to give the potentials in terms of the charge and current densities in S. The time in S is $t = r/c$, and we have

$$x' = \gamma(x - vt) = \gamma\left(x - \frac{vr}{c}\right). \tag{14.25}$$

Hence

$$
\begin{aligned}
r' &= \left\{\gamma^2\left(x - \frac{vr}{c}\right)^2 + y^2 + z^2\right\}^{1/2} \\
&= \gamma\left\{\left(x - \frac{vr}{c}\right)^2 + y^2\left(1 - \frac{v^2}{c^2}\right) + z^2\left(1 - \frac{v^2}{c^2}\right)\right\}^{1/2} \\
&= \gamma\left\{x^2 + y^2 + z^2 - \frac{2vrx}{c} + \frac{v^2 x^2}{c^2}\right\}^{1/2} \\
&= \gamma\left(r - \frac{vx}{c}\right). \tag{14.26}
\end{aligned}
$$

The transformation of $d\tau'$ is tricky, because the position of the origin of S varies as x' varies. Even across the distance dx' there is a difference in the time taken for a light signal to reach P, with the result that the length dx in S is stretched out by comparison with a frame having a fixed origin. This is a kind of shock-wave effect. As we have seen in section 14.6, the fields caused by a particle moving at a speed very close to c pile up into a thin disc travelling along with the particle, like the shock-front accompanying an aeroplane flying at a little less than the speed of sound. To account for this shock-wave effect properly we must remember that in equation (14.25), r is a variable. Differentiating this equation, we find

$$dx' = \gamma \, dx\left(1 - \frac{v}{c}\frac{dr}{dx}\right) = \gamma \, dx\left(1 - \frac{vx}{cr}\right) \tag{14.27}$$

As before $dy' = dy$ and $dz' = dz$.

Finally, the transformation of charge density is

$$\gamma\rho' = \rho. \tag{14.28}$$

Substituting Equations (14.26), (14.27) and (14.28) into (14.24),

$$d\phi(\mathbf{r}, t) = \frac{\rho \, d\tau}{4\pi\varepsilon_0 r},$$

$$dA_x(\mathbf{r}, t) = \frac{v\rho \, d\tau}{4\pi\varepsilon_0 c^2 r} \tag{14.29}$$

or, since $v\rho$ represents the current density $\mathbf{j}$, which is pointing in the x-direction,

$$d\mathbf{A}(\mathbf{r}, t) = \frac{\mathbf{j} \, d\tau}{4\pi\varepsilon_0 c^2 r} = \frac{\mu_0 \mathbf{j} \, d\tau}{4\pi r}. \tag{14.30}$$

The potentials are the same as those due to a stationary charge $\rho \, d\tau$ and a steady current element $\mathbf{j} \, d\tau$ situated at the origin at the earlier time $(t - r/c)$, and the volume element $d\tau$ can be taken as enclosing the origin, provided that ρ and $\mathbf{j}$ are evaluated at the time $(t - r/c)$. Equations (14.29) and (14.30) thus represent retarded potentials. All that remains to be done is to add together the contributions to ϕ and $\mathbf{A}$ from charge elements moving at different speeds in different directions, and to integrate over all of space. Returning to the notation of earlier chapters, we put $\mathbf{r}'$ for the position coordinate of a volume element with respect to a fixed origin and integrate Equations (14.29) and (14.30) to find

$$\phi(\mathbf{r}, t) = \frac{1}{4\pi\varepsilon_0} \int_{\text{all space}} \frac{\rho(\mathbf{r}', t - |\mathbf{r} - \mathbf{r}'|/c) \, d\tau'}{|\mathbf{r} - \mathbf{r}'|}$$

and

$$\mathbf{A}(\mathbf{r}, t) = \frac{\mu_0}{4\pi} \int_{\text{all space}} \frac{\mathbf{j}(\mathbf{r}', t - |\mathbf{r} - \mathbf{r}'|/c) \, d\tau'}{|\mathbf{r} - \mathbf{r}'|}. \tag{14.31}$$

This is the same as the solution of Maxwell's equations given in Equations (13.11) and (13.12).

PROBLEMS 14

14.1 In its rest frame, a rectangular loop of wire with sides of length a and b is seen to be uncharged, and to carry a current I. Viewed from a frame moving at speed v in a direction parallel to the side of length a, what is the distribution of line charges on the loop? What is its electric dipole moment in this frame, expressed in terms of the magnetic dipole moment in the rest frame?

14.2 A cosmic ray proton of energy 10^{15} eV approaches the earth, moving perpendicularly to a magnetic field of magnitude 10^{-5} T. What is the magnitude of the electric field experienced in the proton's rest frame?

14.3 Two point charges are moving in a frame S as shown in Figure 14.7. For an observer in S, the charge moving in the y-direction crosses the x-axis at the time when the other charge is at the origin. Using Equations (14.15) and (14.16) calculate the force he considers to be acting on the first charge. Is the force on the charge at the origin equal and opposite?

14.4 Generalize the result of Problem 14.3 to express the force between two circuits carrying steady current, and check that there is agreement with the Biot–Savart law.

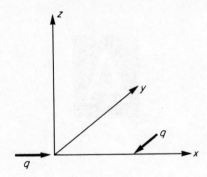

Figure 14.7.

APPENDIX

Units

Throughout this book the units used for all physical quantities are the units of the Système International—SI units for short. SI units, which are recommended by the International Union of Pure and Applied Physics, are based on the metre, the kilogramme and the second as the fundamental units of length, mass and time. The SI electrical units are the same as the practical units in general use—amps, volts, ohms, etc. Section 1 of this appendix explains how the SI units are defined, and then describes the observations needed to set up precise electrical standards.

Other systems of units besides the Système International are still in common use. In particular, the Gaussian system of units is frequently used, expecially in solid state physics. Gaussian units are defined in section A.2, and a summary is given of the change in form of the important equations of electromagnetism when they are expressed in Gaussian units. Finally, a table of conversion factors between the two sets of units is given in section A.3, together with a discussion of the dimensions of electrical units.

A.1 ELECTRICAL UNITS AND STANDARDS

A.1.1 The definition of the ampere

The constants ε_0 and μ_0 determine the magnitudes of SI electrical units in terms of the fundamental units of length, mass and time. The electrical units are directly linked to these fundamental units through the force laws for stationary charges and for steady currents. Coulomb's law (Equation 1.3) gives the force between two stationary charges as

$$\mathbf{F} = \frac{1}{4\pi\varepsilon_0} \frac{q_1 q_2 (\mathbf{r}_1 - \mathbf{r}_2)}{|\mathbf{r}_1 - \mathbf{r}_2|^3},$$

and the Biot–Savart law (Equation (4.43)) leads to the force between circuits made up of thin wires carrying steady currents as given in Equation (4.59):

$$\mathbf{F} = \frac{\mu_0}{4\pi} I_1 I_2 \int_{s_1} \int_{s_2} d\mathbf{l}_1 \wedge \left\{ \frac{d\mathbf{l}_2 \wedge (\mathbf{r}_1 - \mathbf{r}_2)}{|\mathbf{r}_1 - \mathbf{r}_2|^3} \right\}.$$

The proportionality constants $1/4\pi\varepsilon_0$ and $\mu_0/4\pi$ occurring in these equations are not independent of one another, since the units of charge and current are not independent: the current flowing in a conductor is just the rate at which charge crosses a surface cutting the conductor. Consequently we are free to choose an arbitrary value for only one of the constants ε_0 and μ_0. In the Système International the permeability of free space μ_0 is given the value

$$\boxed{\mu_0 = 4\pi \cdot 10^{-7} \text{ henries/metre.}}$$

With this choice the unit of current defined by the Biot–Savart law is the *ampere*, which is already established as the practical unit. The definition can be given in physical terms by choosing circuits for which the integrals occurring in the Biot–Savart law can be carried out exactly. For example, if two thin parallel wires of infinite length are at a distance r apart and each carrying a current I, then the force per unit length between them is

$$\frac{\mu_0 I^2}{2\pi r} = 2 \cdot 10^{-7} \frac{I^2}{r}.$$

This result leads to the following physical definition of the ampere in terms of the SI units of force and length:

> The ampere is the steady current which, if flowing in two parallel thin conductors of infinite length and one metre apart, would produce a force of 2×10^{-7} newtons per metre between them.

Once the ampere has been defined, the unit of charge, which is called the *coulomb*, must be such that one coulomb per second passes along a conductor if it is carrying a current of one ampere. Now that the unit of charge has been defined, the constant of proportionality in Coulomb's law is no longer arbitrary, and the permittivity of free space ε_0 must be determined by experiment. In principle ε_0 could be derived directly from Coulomb's law by measuring the force between stationary charges, but it is not possible to make such measurements with the required precision. However, the constant ε_0 occurs in many of the other equations of electromagnetism, and any of these may be used for an experimental determination of ε_0. By far the most accurate determination of ε_0 comes from the measurement of the speed of light. It is shown in Chapter 11 that electromagnetic waves of any frequency—light waves, radio waves, etc.— propagate in free space with the same speed

$$c = 1/\sqrt{\varepsilon_0\mu_0}.$$

Hence

$$\varepsilon_0 = \frac{1}{\mu_0 c^2} = \frac{10^7}{4\pi c^2}.$$

The value of c currently accepted as the best is

$$c = (2.9979250 \pm 0.0000010) \times 10^8 \text{ m/sec,}$$

leading to

$$\boxed{\varepsilon_0 = (8.854185 \pm 0.000006) \times 10^{-12} \text{ farads/metre.}}$$

Notice that although ε_0 is known with great precision, it is subject to some uncertainty, whereas μ_0 is defined to be *exactly* $4\pi \times 10^{-7}$ henries/m, independently of any observations.

A.1.2   Calibration of electrical standards

It is very difficult to make precise absolute measurements of electrical quantities in terms of the units of length, mass and time. Absolute measurements are normally carried out only at standards laboratories such as the National Physical Laboratory in Britain or the National Bureau of Standards in the U.S.A. These laboratories make absolute calibrations of standard resistors and standard sources of e.m.f. The standard resistors are made of manganin, an alloy with a very low temperature coefficient of resistance, and they are constructed so that their resistance is very close to one ohm. The standards of e.m.f. are Weston cells, which at 20 °C have an e.m.f. close to 1.018 volts. Both these standards are very stable, and retain the same value to within less than 1 part in 10^6 over a period of a year or more. By making electrical measurements alone, the primary standards can be quickly and easily compared with secondary standards which are made available for other users.

The standard cell is not calibrated directly, but its voltage is derived from Ohm's law in an absolute measurement of the current through a standard resistor, the current being controlled so that the voltage drop across the resistor exactly balances the e.m.f. of the cell. Once the standards of resistance and voltage are set up, the values of other electromagnetic quantities can be derived by purely electromagnetic measurements, together with measurements of length and frequency. For example, a reference magnetic induction field is established by passing a known current through a coil of accurately known dimensions.

Determination of the ampere

The ampere is calibrated by direct measurement of the force between current-carrying coils in a 'current balance'. The arrangement of coils in the balance at the National Physical Laboratory is sketched in Figure A.1. On each arm

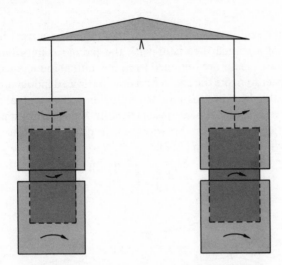

Figure A.1. A current balance. The arrows indicate the direction of current flow in the coils. The right-hand arm of the balance is pulled down, the left-hand arm pushed up.

of the balance a coil is hung at the centre of a larger coil placed on the same axis. The larger coils are split into two equal halves, with the current passing in opposite senses through the two halves. The coils are connected together so that the same current flows through all of them. Because parallel currents attract while opposite currents repel, each small coil experiences a net force pushing it towards the half of the larger coil surrounding it in which current flows in the same sense. The coils are arranged so that one of the arms of the balance is pushed up, the other down. The resulting couple is opposed by hanging masses on the arms until the balance shows no deflection. The force between the coils is thus known in terms of these masses and the gravitational acceleration g, and the current in the coils can be deduced from the Biot–Savart law.

For the current balance at the National Physical Laboratory, the lengths of the small and large coils are 15 cm and 25 cm respectively, and their diameters 20 cm and 32 cm. In spite of their considerable size, the forces on each small coil is less than 2 gm wt when the apparatus carries the current of 1.018 amps needed to balance a Weston cell against the potential drop across a 1 Ω standard resistor. However, the balancing mass of about 4 gm can be measured to within 1 part in 10^6. The main errors in the absolute determination of the ampere with the current balance arise from the difficulty in measuring the dimensions of the coils and from uncertainties in g. The overall accuracy in the calibration is about 4 parts in 10^6.

Determination of the ohm

The ohm can be calibrated absolutely with alternating current by comparing the resistance of a standard resistor with the inductive impedance of a coil. The most accurate results are obtained from the mutual inductance of two coils, because it is easier to make the correction due to the lead inductances extremely small for mutual inductance than for self-inductance. The mutual inductance is determined entirely by the dimensions of the coils. Apart from a small correction to allow for the thickness of the wires, the mutual inductance is given by Neumann's formula (Equation (4.58));

$$M_{12} = \frac{\mu_0}{4\pi} \oint_{s_1} \oint_{s_2} \frac{d\mathbf{l}_1 \cdot d\mathbf{l}_2}{r_{12}}.$$

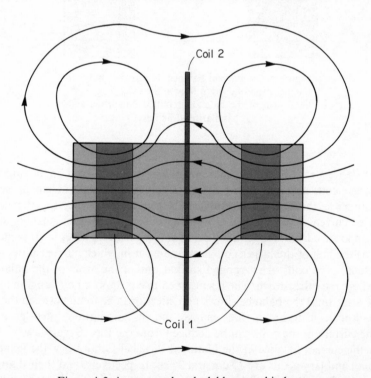

Figure A.2. An accurately calculable mutual inductor.

By clever design of the two coils their mutual inductance can be made insensitive to the dimensions of one of the coils: this coil can then have many turns, thus increasing the mutual inductance without incurring a penalty from dimensional inaccuracies. A cross-section through a calculable mutual inductor is shown in Figure A.2, with lines of the field **B** which is generated by a current through coil 1. Coil 1 is split so that the lines of **B** bulge out between the two halves. Where the lines linking the two halves of coil 1 approach close to the return lines passing outside the coil, there is a toroidal volume where the field generated by coil 1 is very small. If the turns of coil 2 are placed in this volume, the total flux from coil 1 linking these turns is rather insensitive to their exact position. The mutual inductance of the coils is almost completely determined by the dimensions of coil 1, which is made of a single layer of turns on a carefully machined former.

The impedance of the mutual inductance is compared with a standard resistor in an A.C. bridge. The bridge must be operated at a frequency low enough for skin-depth corrections to the resistance and capacitance between the coils to be negligible. The circuit of the A.C. bridge and its balance condition are discussed in Problem 8.7. In this method of calibrating the ohm an accuracy of about 1 part in 10^6 is achieved.

Even better accuracy can be obtained by comparing resistance with capacitative impedance. A cross-section of a standard capacitor is shown in Figure A.3.

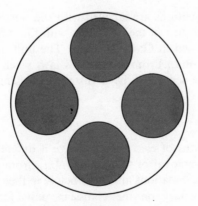

Figure A.3. Cross-section of a standard capacitor. The capacitance between opposite pairs of cylinders is measured.

The capacitor has four electrodes, and looks rather like the quadrupole lens described in section 3.9. Provided that the cross-section of such a four electrode capacitor is symmetrical, then apart from end corrections, the capacitance per unit length between opposite pairs of electrodes is exactly $\varepsilon_0 \ln 2/\pi$, no matter what shape or size the capacitor may be. The proof of this remarkable result is left as an exercise in Problem 3.11. Of course in any real laboratory capacitor the end corrections are large, but they can be eliminated by comparing the capacitance with the standard resistor twice, using different lengths of the capacitor. The errors in this measurement amount altogether to only 2 parts in 10^7. This is better than the present accuracy of the speed of light c, which is known to within 3 parts in 10^7; since $\varepsilon_0 = (\mu_0 c^2)^{-1}$, the error in the ohm arising from the uncertainty in c is 6 parts in 10^7.

A.2   GAUSSIAN UNITS

The most commonly used system of electrical units other than the Système International is the Gaussian system (sometimes called the c.g.s. system). This system is based on the centimetre, the gramme and the second as the units of length, mass and time. Hence the unit of force is the dyne, and the unit of energy the erg. *In the Gaussian system electrical units are defined by choosing the constant of proportionality in Coulomb's law to be unity.* Coulomb's law therefore takes the form

$$\mathbf{F} = \frac{q_1 q_2 \mathbf{r}_{21}}{r_{21}^3}. \tag{1.2G}$$

(In this section the equations are labelled with the same number as their first occurrence in the main text of the book, but with the suffix G to indicate that they have been expressed in Gaussian units. The unit of charge is called the e.s.u. or electrostatic unit.) From Coulomb's law, when two charges each of magnitude 1 e.s.u. are separated by a distance of 1 cm, each experiences a force of 1 dyne.

Electrostatics

Most of the equations of electrostatics have a different form in Gaussian units because of the removal of the factor $1/4\pi\varepsilon_0$ from Coulomb's law. The important electrostatic equations are listed below in their Gaussian form, with comments where necessary on the Gaussian units for various quantities. Gauss' law becomes

$$\operatorname{div} \mathbf{E} = 4\pi\rho \tag{1.16G}$$

The electric field $\mathbf{E}$ is measured in *statvolts/cm*, and is related to the potential (measured in *statvolts*) by the equation

$$\mathbf{E} = -\operatorname{grad} \phi. \tag{1.21G}$$

The potential energy of an assembly of point charges is

$$U = \tfrac{1}{2} \sum_i q_i \phi_i \text{ ergs}. \tag{1.31G}$$

The energy density of an electric field $\mathbf{E}$ is $E^2/8\pi$ ergs/c.c., leading to a total energy

$$U = \frac{1}{8\pi} \int E^2 \, d\tau. \tag{1.37G}$$

The dipole moment per unit volume $\mathbf{P}$ has the same dimensions as the field $\mathbf{E}$, and in an isotropic material of relative permittivity ε has the value

$$\mathbf{P} = \frac{1}{4\pi}(\varepsilon - 1)\mathbf{E}. \tag{2.6G}$$

The electric susceptibility is now defined by $\mathbf{P} = \chi_e \mathbf{E}$, so that

$$\chi_e = \frac{1}{4\pi}(\varepsilon - 1). \tag{2.7G}$$

(This is especially confusing: note that the *dimensionless* quantity χ_e has different values in the two systems of units.) The electric displacement

$$\mathbf{D} = \mathbf{E} + 4\pi\mathbf{P} \tag{2.20G}$$

satisfies Gauss's law for the free charge density;

$$\operatorname{div} \mathbf{D} = 4\pi\rho_f. \tag{2.19G}$$

Finally, Poisson's equation becomes

$$\nabla^2 \phi = -4\pi\rho/\varepsilon, \tag{3.2G}$$

for a material with uniform relative permittivity ε.

Currents and magnetic fields

In the Gaussian system of units the Lorentz force is written in such a way that the magnetic field $\mathbf{B}$ has the same dimensions as the electric field $\mathbf{E}$, namely force per unit charge. The force law for a charge q moving in an electromagnetic field is

$$\mathbf{F} = q\left(\mathbf{E} + \frac{1}{c}\mathbf{v} \wedge \mathbf{B}\right), \tag{4.16G}$$

where c is the speed of light. The speed of light also occurs in the law for the force between two circuits carrying currents I_1 and I_2:

$$\mathbf{F} = -\frac{I_1 I_2}{c^2} \oint_{s_1} \oint_{s_2} \frac{d\mathbf{l}_1 \cdot d\mathbf{l}_2}{r_{12}^3} \mathbf{r}_{12}. \tag{4.59G}$$

In magnetic material with a dipole moment per unit volume $\mathbf{M}$, the magnetic intensity $\mathbf{H}$ is defined as

$$\mathbf{H} = \mathbf{B} - 4\pi\mathbf{M}. \tag{5.27G}$$

The relative permeability μ is defined by

$$\mathbf{B} = \mu\mathbf{H}, \tag{5.32G}$$

and the magnetic susceptibility χ_H by

$$\mathbf{M} = \chi_H\mathbf{H}. \tag{5.31G}$$

Notice that as for the electric susceptibility, the magnetic susceptibilities in SI and Gaussian units differ by a factor 4π. Ampère's law for a steady density of free current $\mathbf{j}_f$ takes the form

$$\text{curl } \mathbf{H} = \frac{4\pi}{c}\mathbf{j}_f. \tag{5.30G}$$

The chief practical objection to the Gaussian system is that its units for the quantities required in circuit analysis do not coincide with the universally accepted units based on volts and amps. The statvolt is rather large (about 300 volts) and the e.s.u./sec is a very small unit of current (about 3.3×10^{-10} amps).

The units of impedance are based on the usual equations:

$$\text{Resistance } R: V = IR$$
$$\text{Capacitance } C: Q = CV$$
$$\text{Inductance } L: V = L\frac{\mathrm{d}I}{\mathrm{d}t}.$$

The units of R, C and L have no special names, and are referred to as 'Gaussian (or c.g.s.) units of resistance' etc.

Maxwell's equations

Maxwell's equations in Gaussian units take the form

$$\text{div } \mathbf{D} = 4\pi\rho_f, \tag{10.10G}$$

$$\text{div } \mathbf{B} = 0, \tag{10.11G}$$

$$\text{curl } \mathbf{E} = -\frac{1}{c}\frac{\partial\mathbf{B}}{\partial t}, \tag{10.12G}$$

$$\text{curl } \mathbf{H} = \frac{4\pi}{c}\mathbf{j}_f + \frac{1}{c}\frac{\partial\mathbf{D}}{\partial t}. \tag{10.13G}$$

A.3 CONVERSION BETWEEN SI AND GAUSSIAN UNITS

A table is given overleaf of the conversion factors between the two systems of units.

Dimensions

Charge has been given the dimension $[Q]$ in the table. Coulomb's law then requires that the permittivity of free space has the dimensions

$$[\varepsilon_0] = [\text{charge}]^2[\text{force}]^{-1}[\text{length}]^2 = [M^{-1}L^{-3}T^2Q^2].$$

Since $\varepsilon_0\mu_0 = 1/c^2$, the permeability of free space has the dimensions

$$[\mu_0] = [MLQ^{-2}].$$

Dimensional analysis is applied to the electrical quantities in the usual way. Thus the product CR has dimensions

$$[M^{-1}L^{-2}T^2Q^2][ML^2T^{-1}Q^{-2}] = [T].$$

Notice that the dimensions in the third column do not apply to Gaussian units, since the two sets of units are used in electromagnetic equations of a different form. In Gaussian units the fields **D** and **E** have the same dimensions, as do the fields **B** and **H**. However, since $\mathbf{B} = \mu\mathbf{H}$ in Gaussian units, a field of one gauss only corresponds to a field of one oersted in the absence of magnetic materials.

CONVERSION FACTORS BETWEEN SI AND GAUSSIAN UNITS

Quantity	SI unit	Dimensions of SI unit	Gaussian unit	No. of Gaussian units in one SI unit
Force	newton	$[MLT^{-2}]$	dyne	10^5 (i.e. 10^5 dyne = 1 newton)
Energy	joule	$[ML^2T^{-2}]$	erg	10^7
Power	watt	$[ML^2T^{-3}]$	erg/s.	10^7
Charge	coulomb	$[Q]$	e.s.u.	3×10^9
Current	ampere	$[T^{-1}Q]$	e.s.u./s.	3×10^9
Potential	volt	$[ML^2T^{-2}Q^{-1}]$	statvolt	$1/300$
Capacitance	farad	$[M^{-1}L^{-2}T^2Q^2]$		9×10^{11}
Resistance	ohm	$[ML^2T^{-1}Q^{-2}]$		$1/(9 \times 10^{11})$
Inductance	henry	$[ML^2Q^{-2}]$		$1/(9 \times 10^{11})$
E	volt/metre	$[MLT^{-2}Q^{-1}]$	statvolt/cm.	$1/(3 \times 10^4)$
D	coulomb/sq. metre	$[L^{-2}Q]$	statvolt/cm.	$12\pi \times 10^5$
B	tesla	$[MT^{-1}Q^{-1}]$	gauss	10^4
H	ampere/metre	$[L^{-1}T^{-1}Q]$	oersted	$4\pi \times 10^{-3}$

In this table c has been taken as 3×10^8 m/sec.

Vectors

B.1 FIELDS AND DIFFERENTIAL OPERATORS

A scalar or vector function of position is called a *field*. Fields may be operated on by *differential operators* which generate new fields. The results of the operations of div, grad, curl and ∇^2 are defined below.

Divergence

The divergence operator acts on a vector field, and the resultant is a scalar field. The divergence of an arbitrary vector field $\mathbf{F}(\mathbf{r})$, written div $\mathbf{F}$ or $\nabla \cdot \mathbf{F}$, represents the outward flux of $\mathbf{F}$ per unit volume at the point $\mathbf{r}$.

The flux of $\mathbf{F}$ out of a closed surface S enclosing a small volume $\delta\tau$ is $\int_S \mathbf{F} \cdot d\mathbf{S}$, taking $d\mathbf{S}$ to be the outward normal to the element of area dS. In the limit, as $\delta\tau$ tends to zero, the outward flux of $\mathbf{F}$ per unit volume is independent of the shape of S, provided that $\mathbf{F}$ is smoothly varying. Thus

$$\text{div } \mathbf{F} = \nabla \cdot \mathbf{F} = \lim_{\delta\tau \to 0} \frac{\int_S \mathbf{F} \cdot d\mathbf{S}}{\delta\tau}. \tag{B1}$$

As an example, consider the divergence of the current density $\mathbf{j}(\mathbf{r})$. Current density is the rate of flow of charge per unit area, and $\int_S \mathbf{j} \cdot d\mathbf{S}$ is therefore the net rate of removal of charge from the volume within S. It follows that div $\mathbf{j}$ is zero in any conductor which remains electrically neutral throughout its volume, whatever the pattern of currents flowing in it.

Gradient

The gradient operator acts on a scalar field, and the resultant is a vector field. The gradient of an arbitrary scalar field $\psi(r)$ is written as grad ψ or $\nabla\psi$. At any point, grad ψ lies along the direction in which ψ is most rapidly increasing, and its magnitude is equal to the slope of ψ in that direction.

If **e** is a unit vector (Figure B.1) in the direction of the line PQ joining two points P and Q with position vectors **r** and $(\mathbf{r} + \delta\mathbf{r})$, then at P the component

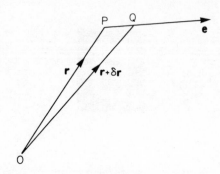

Figure B.1.

of grad ψ along **e** is

$$\text{grad } \psi \cdot \mathbf{e} = \lim_{\delta\mathbf{r} \to 0} \frac{\psi(\mathbf{r} + \delta\mathbf{r}) - \psi(\mathbf{r})}{\delta r}. \tag{B2}$$

For example, the electrostatic potential $\phi(\mathbf{r})$ is a scalar field, and the electrostatic field $\mathbf{E} = -\text{grad } \phi$ is the force per unit charge pushing positive charges towards regions of lower potential energy.

Curl

The curl operator acts on a vector field, and the resultant is another vector field. The curl of an arbitrary vector field $\mathbf{F}(\mathbf{r})$ is written as curl **F** or $\nabla \wedge \mathbf{F}$. The curl of a vector field is a measure of its vorticity. If the field lines of a vector field circulate around a central line like the streamlines of a whirlpool around its vortex, the curl of the vector lies in the direction of the central line.

If **e** is a unit vector perpendicular to a small surface of area δS located at **r**, then the component of curl **F** along **e** is

$$\text{curl } \mathbf{F} \cdot \mathbf{e} = \lim_{\delta S \to 0} \frac{\int_s \mathbf{F} \cdot d\mathbf{l}}{\delta S}, \tag{B3}$$

where the line integral is evaluated around the perimeter s of the area δS, in a right handed sense with respect to **e** as indicated in Figure B2. An example of a field which can have non-zero curl is the magnetic field **B**. The field lines of **B** close up around the currents which generate the field. But in regions where there is no free current or magnetization current, closed loops cannot be found among the field lines, and in these regions curl **B** is zero.

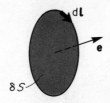

Figure B.2.

The Laplacian operator ∇^2

By operating on a scalar field $\psi(\mathbf{r})$ with the gradient and divergence operators in turn, another scalar field div grad ψ is formed. Since each of the three components of a vector field $\mathbf{F}$ are independent, the Laplacian operator ∇^2 may be allowed to operate separately on the three components, to form the components of a vector field $\nabla^2\mathbf{F}$.

The fields resulting from operating with differential operators must be worked out in a coordinate system suited to each particular problem. Formulae for the fields are given below in Cartesian, spherical polar and cylindrical polar coordinates.

Cartesian coordinates

In Cartesian coordinates, a vector field $\mathbf{F}(x, y, z)$ is resolved at any point into its components F_x, F_y and F_z along the directions of the x-, y- and z-axes. The divergence of $\mathbf{F}$ is given in terms of these components as

$$\text{div } \mathbf{F} = \frac{\partial F_x}{\partial x} + \frac{\partial F_y}{\partial y} + \frac{\partial F_z}{\partial z}. \tag{B4}$$

The components of curl $\mathbf{F}$ are

$$(\text{curl } \mathbf{F})_x = \frac{\partial F_z}{\partial y} - \frac{\partial F_y}{\partial z}, \quad (\text{curl } \mathbf{F})_y = \frac{\partial F_x}{\partial z} - \frac{\partial F_z}{\partial x}, \quad (\text{curl } \mathbf{F})_z = \frac{\partial F_y}{\partial x} - \frac{\partial F_x}{\partial y}. \tag{B5}$$

For a scalar field $\psi(x, y, z)$ the components of grad ψ are

$$(\text{grad } \psi)_x = \frac{\partial \psi}{\partial x}, \quad (\text{grad } \psi)_y = \frac{\partial \psi}{\partial y}, \quad (\text{grad } \psi)_z = \frac{\partial \psi}{\partial z}. \tag{B6}$$

The Laplacian operator is

$$\nabla^2 = \frac{\partial^2}{\partial x^2} + \frac{\partial^2}{\partial y^2} + \frac{\partial^2}{\partial z^2}. \tag{B7}$$

Spherical polar coordinates

A point P has spherical polar coordinates (r, θ, ϕ), where r is the length OP, as indicated in Figure B.3; θ the angle between OP and the z-axis; and ϕ the

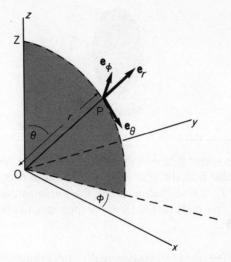

Figure B.3.

angle between the x-axis and the plane OPZ containing P and the z-axis. In this coordinate system a vector field is resolved into components along directions which depend on where the field is being evaluated. At P a vector field $\mathbf{F}$ is resolved into components F_r, F_θ and F_ϕ along the directions of the orthogonal unit vectors $\mathbf{e}_r$, $\mathbf{e}_\theta$ and $\mathbf{e}_\phi$. The vector $\mathbf{e}_r$ points in the direction OP; $\mathbf{e}_\theta$ lies in the plane OPZ perpendicular to $\mathbf{e}_r$ and pointing in the direction of increasing θ; and $\mathbf{e}_\phi$ is normal to the plane OPZ, pointing in the direction of increasing ϕ.

Expressions are given below for the results of operating with differential operators on arbitrary fields $\mathbf{F}(r, \theta, \phi)$ and $\psi(r, \theta, \phi)$.

The divergence of $\mathbf{F}$ is

$$\operatorname{div} \mathbf{F} = \frac{\partial F_r}{\partial r} + \frac{2}{r}F_r + \frac{1}{r}\frac{\partial F_\theta}{\partial \theta} + \frac{\cot \theta}{r}F_\theta + \frac{1}{r \sin \theta}\frac{\partial F_\phi}{\partial \phi}. \tag{B8}$$

The components of curl $\mathbf{F}$ are

$$(\operatorname{curl} \mathbf{F})_r = \frac{1}{r}\frac{\partial F_\phi}{\partial \theta} + \frac{\cot \theta}{r}F_\phi - \frac{1}{r \sin \theta}\frac{\partial F_\theta}{\partial \phi}.$$

$$(\operatorname{curl} \mathbf{F})_\theta = \frac{1}{r \sin \theta}\frac{\partial F_r}{\partial \phi} - \frac{\partial F_\phi}{\partial r} - \frac{1}{r}F_\phi.$$

$$(\operatorname{curl} \mathbf{F})_\phi = \frac{F_\theta}{r} + \frac{\partial F_\theta}{\partial r} - \frac{1}{r}\frac{\partial F_r}{\partial \theta}. \tag{B9}$$

The components of grad ψ are

$$(\operatorname{grad} \psi)_r = \frac{\partial \psi}{\partial r}, \qquad (\operatorname{grad} \psi)_\theta = \frac{1}{r}\frac{\partial \psi}{\partial \theta}, \qquad (\operatorname{grad} \psi)_\phi = \frac{1}{r \sin \theta}\frac{\partial \psi}{\partial \phi}. \tag{B10}$$

The Laplacian operator is

$$\nabla^2 = \frac{\partial^2}{\partial r^2} + \frac{2}{r}\frac{\partial}{\partial r} + \frac{1}{r^2}\frac{\partial^2}{\partial \theta^2} + \frac{\cot\theta}{r^2}\frac{\partial}{\partial\theta} + \frac{1}{r^2\sin^2\theta}\frac{\partial^2}{\partial\phi^2}. \tag{B11}$$

Cylindrical polar coordinates

A point P has cylindrical polar coordinates (r, ϕ, z) as indicated in Figure B.4. Here r is the perpendicular distance from P to the z-axis, ϕ the angle between the x-axis and the plane OPZ, and z the same as the Cartesian coordinate z.

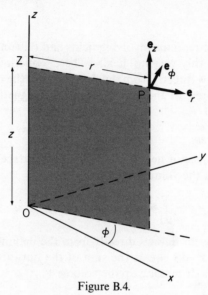

Figure B.4.

At P a vector field **F** is resolved into components F_r, F_ϕ and F_z along the directions of the orthogonal unit vectors $\mathbf{e}_r$, $\mathbf{e}_\phi$ and $\mathbf{e}_z$. The vector $\mathbf{e}_z$ is parallel to the z-axis, $\mathbf{e}_r$ lies in the plane OPZ, perpendicular to $\mathbf{e}_z$, and $\mathbf{e}_\phi$ is perpendicular to the plane OPZ, pointing in the direction of increasing ϕ.

In this coordinate system, the divergence of $\mathbf{F}(r, \phi, z)$ is

$$\operatorname{div}\mathbf{F} = \frac{\partial F_r}{\partial r} + \frac{F_r}{r} + \frac{1}{r}\frac{\partial F_\phi}{\partial\phi} + \frac{\partial F_z}{\partial z}. \tag{B12}$$

The components of curl **F** are

$$(\operatorname{curl}\mathbf{F})_r = \frac{1}{r}\frac{\partial F_z}{\partial\phi} - \frac{\partial F_\phi}{\partial z},$$

$$(\operatorname{curl}\mathbf{F})_\phi = \frac{\partial F_r}{\partial z} - \frac{\partial F_z}{\partial r},$$

$$(\operatorname{curl}\mathbf{F})_z = \frac{\partial F_\phi}{\partial r} + \frac{F_\phi}{r} - \frac{1}{r}\frac{\partial F_r}{\partial\phi}. \tag{B13}$$

The components of the gradient of a scalar field $\psi(r, \phi, z)$ are

$$(\text{grad } \psi)_r = \frac{\partial \psi}{\partial r}, \qquad (\text{grad } \psi)_\phi = \frac{1}{r} \frac{\partial \psi}{\partial \phi}, \qquad (\text{grad } \psi)_z = \frac{\partial \psi}{\partial z}. \qquad (B14)$$

The Laplacian operator is

$$\nabla^2 = \frac{\partial^2}{\partial r^2} + \frac{1}{r} \frac{\partial}{\partial r} + \frac{1}{r^2} \frac{\partial^2}{\partial \phi^2} + \frac{\partial^2}{\partial z^2}. \qquad (B15)$$

B.2   IDENTITIES

There are several equations involving fields and differential operators which hold for any smoothly-varying fields. These equations are *mathematical identities* which follow directly from the definitions of the differential operators. The identities required in the theory of electromagnetism are discussed in this section.

The divergence theorem

The total flux of a vector field **F** out of a closed surface S equals the volume integral of div **F** over the volume enclosed within S:

$$\int_S \mathbf{F} \cdot d\mathbf{S} \equiv \int_V \text{div } \mathbf{F} \, d\tau. \qquad (B16)$$

The divergence theorem follows directly from the definition of div **F** given in section B.1, since $\int_S \mathbf{F} \cdot d\mathbf{S}$ equals the sum of the outward fluxes from all the infinitesimal volumes $d\tau$ making up the volume V.

Stokes' theorem

The line integral of a vector field **F** around a circuit s equals the flux of curl **F** through any surface S which spans s:

$$\oint_s \mathbf{F} \cdot d\mathbf{l} \equiv \int_S \text{curl } \mathbf{F} \cdot d\mathbf{S}. \qquad (B17)$$

The line integral is taken in the sense related by the right-hand rule to the direction of the flux. Stokes' theorem follows directly from the definition of curl given in section B.1, since $\oint_s \mathbf{F} \cdot d\mathbf{l}$ is equal to the sum of the line integrals around the perimeters of the mosaic of elementary surfaces d**S** which make up S.

Identities involving products of the vector operator ∇

The operator ∇ is itself a vector, and obeys the same rules for forming vector products as any other vector. The identities given below follow from these rules: they may also be checked by writing them out in full in terms of their Cartesian components.

(i) For any scalar field ψ,

$$\text{curl grad } \psi = 0,$$

or

$$\nabla \wedge \nabla\psi \equiv 0. \tag{B18}$$

(This identity is true because the operator $\nabla \wedge \nabla$ is zero, since it is the vector product of a vector with itself.)

(ii) For any vector field $\mathbf{F}$, div curl $\mathbf{F} \equiv 0$, or

$$\nabla \cdot (\nabla \wedge \mathbf{F}) \equiv 0. \tag{B19}$$

(Scalar triple products are zero when any two of the vectors in the product are the same.)

(iii) curl curl $\mathbf{F} \equiv$ grad div $\mathbf{F} - \nabla^2\mathbf{F}$, or

$$\nabla \wedge (\nabla \wedge \mathbf{F}) \equiv \nabla(\nabla \cdot \mathbf{F}) - \nabla^2\mathbf{F}. \tag{B20}$$

Identities including products of two fields

Here again the identities follow from the application of the rules of vector algebra. One must be careful about the order in which the fields and operators are written, specifying clearly which fields are operated on by ∇

(i) div $(\psi\mathbf{F}) \equiv \psi$ div $\mathbf{F} + \mathbf{F} \cdot$ grad ψ, or

$$\nabla \cdot (\psi\mathbf{F}) \equiv \psi(\nabla \cdot \mathbf{F}) + \mathbf{F} \cdot \nabla\psi. \tag{B21}$$

(ii) curl $(\psi\mathbf{F}) \equiv \psi$ curl $\mathbf{F} - \mathbf{F} \wedge$ grad ψ, or

$$\nabla \wedge (\psi\mathbf{F}) \equiv \psi\nabla \wedge \mathbf{F} - \mathbf{F} \wedge \nabla\psi. \tag{B22}$$

This identity could also be written as

$$\text{curl}(\psi\mathbf{F}) \equiv \psi \text{ curl } \mathbf{F} + (\text{grad } \psi) \wedge \mathbf{F}.$$

We have preferred to reverse the order of the second term on the right hand side, with a consequent change of sign. In this notation, the operator is then allowed to operate on everything to its right in each term.

(iii) div $(\mathbf{F} \wedge \mathbf{G}) \equiv \mathbf{G} \cdot$ curl $\mathbf{F} - \mathbf{F} \cdot$ curl $\mathbf{G}$, or

$$\nabla \cdot (\mathbf{F} \wedge \mathbf{G}) \equiv \mathbf{G} \cdot (\nabla \wedge \mathbf{F}) - \mathbf{F} \cdot (\nabla \wedge \mathbf{G}). \tag{B23}$$

The scalar triple product $\mathbf{a} \cdot (\mathbf{b} \wedge \mathbf{c})$ represents the volume of a parallelopiped with sides $\mathbf{a}$, $\mathbf{b}$ and $\mathbf{c}$. Its value is independent of the way it is written except that there is a change of sign if the cyclic order of $\mathbf{a}$, $\mathbf{b}$ and $\mathbf{c}$ is altered. The identity written above is just a rearrangement of the components of the triple scalar product; there are two terms on the right hand side because ∇ must be allowed to operate on both the fields $\mathbf{F}$ and $\mathbf{G}$.

C

The derivation of the Biot–Savart law

In this appendix we derive the Biot–Savart law, starting from the differential form of Ampère's Law, Equation (4.40).

From Equations (4.40) and (4.42), we obtain

$$\text{curl curl } \mathbf{A} = \mu_0 \mathbf{j}. \tag{C1}$$

The quantity curl curl $\mathbf{A}$ may be rewritten using the identity (B.20);

$$\text{curl curl } \mathbf{A} \equiv \text{grad div } \mathbf{A} - \nabla^2 \mathbf{A}, \tag{C2}$$

where ∇^2 is the Laplacian operator; $\nabla^2 \mathbf{A}$ is the vector whose three components in Cartesian coordinates are

$$\frac{\partial^2 A_x}{\partial x^2} + \frac{\partial^2 A_x}{\partial y^2} + \frac{\partial^2 A_x}{\partial z^2}$$

$$\frac{\partial^2 A_y}{\partial x^2} + \frac{\partial^2 A_y}{\partial y^2} + \frac{\partial^2 A_y}{\partial z^2}$$

$$\frac{\partial^2 A_z}{\partial x^2} + \frac{\partial^2 A_z}{\partial y^2} + \frac{\partial^2 A_z}{\partial z^2}.$$

With the choice div $\mathbf{A} = 0$, Equations (C1) and (C2) reduce to the equation

$$-\nabla^2 \mathbf{A} = \mu_0 \mathbf{j}. \tag{C3}$$

This equation has three component equations. For example in Cartesian coordinates the x-component is

$$-\frac{\partial^2 A_x}{\partial x^2} - \frac{\partial^2 A_x}{\partial y^2} - \frac{\partial^2 A_x}{\partial z^2} = \mu_0 j_x.$$

Each one of the Cartesian components is an equation similar to Poisson's Equation (3.2), and the solution to the equation for A_x for example is similarly

$$A_x(\mathbf{r}) = \frac{\mu_0}{4\pi} \int_V \frac{j_x(\mathbf{r}')\,\mathrm{d}\tau'}{|\mathbf{r} - \mathbf{r}'|}.$$

The general solution to Equation (C3) is then

$$\mathbf{A}(\mathbf{r}) = \frac{\mu_0}{4\pi} \int_V \frac{\mathbf{j}(\mathbf{r}')\,\mathrm{d}\tau'}{|\mathbf{r} - \mathbf{r}'|}. \tag{C4}$$

When the current is flowing in a thin conductor the current density $\mathbf{j}(\mathbf{r})$ is zero everywhere except along the thin conductors. The term $\mathbf{j}(\mathbf{r}')$ in the integrand above can then be replaced by the term $I\,\mathrm{d}\mathbf{l}'$, where I is the current and the vector $\mathrm{d}\mathbf{l}'$ is in the direction of the current in the wire at the point with position vector $\mathbf{r}'$. The integral over the volume V containing the current distributions now becomes an integral over the closed length s of the conductor, and we have

$$\mathbf{A}(\mathbf{r}) = \frac{\mu_0 I}{4\pi} \oint_s \frac{\mathrm{d}\mathbf{l}'}{|\mathbf{r} - \mathbf{r}'|}.$$

Hence

$$\mathbf{B}(\mathbf{r}) = \frac{\mu_0 I}{4\pi} \operatorname{curl}\left\{ \oint_s \frac{\mathrm{d}\mathbf{l}'}{|\mathbf{r} - \mathbf{r}'|} \right\}.$$

The curl operation is with respect to the coordinates of the point with position vector $\mathbf{r}$, hence the operator can be taken inside the integral. Applying identity (B.22) we find that

$$\operatorname{curl} \frac{\mathrm{d}\mathbf{l}'}{|\mathbf{r} - \mathbf{r}'|} = \frac{1}{|\mathbf{r} - \mathbf{r}'|} \operatorname{curl} \mathrm{d}\mathbf{l}' + \operatorname{grad} \frac{1}{|\mathbf{r} - \mathbf{r}'|} \wedge \mathrm{d}\mathbf{l}'.$$

But $\operatorname{curl} \mathrm{d}\mathbf{l}' = 0$, since the differentiations are with respect to $\mathbf{r}$, and also

$$\operatorname{grad} \frac{1}{|\mathbf{r} - \mathbf{r}'|} = -\frac{(\mathbf{r} - \mathbf{r}')}{|\mathbf{r} - \mathbf{r}'|^3}.$$

Hence finally

$$\mathbf{B}(\mathbf{r}) = \frac{\mu_0 I}{4\pi} \oint_s \frac{\mathrm{d}\mathbf{l}' \wedge (\mathbf{r} - \mathbf{r}')}{|\mathbf{r} - \mathbf{r}'|^3}.$$

This is the Biot–Savart law used in section 4.5.3.

Solutions to problems

PROBLEMS 1

1.1 The force on each charge is 1.6×10^{-2} newtons. The field is zero at the centre of the triangle, and has a magnitude $12\,000$ V/m at the centre of each side.

1.3 10^5 V.

1.4 31.1 MV (MV = megavolt = 10^6 volts). When the terminal is at this voltage the intershield must be at 18.8 MV to prevent breakdown.

1.5 The capacitance is $4\pi\varepsilon_0/(1/a - 1/b)$. When the radius b of the outer sphere becomes large, the capacitance tends to the limiting value $4\pi\varepsilon_0 a$. The same limit is reached when the sphere is surrounded by a distant conductor of any shape: this limit is often referred to as the capacitance to earth of an isolated sphere.

1.6 On the line joining two charges $\pm q$ separated by distance d the magnitude of the field a distance r from one of them is

$$E = \frac{q}{4\pi\varepsilon_0}\left\{\frac{1}{r^2} + \frac{1}{(d-r)^2}\right\}.$$

This field points towards the negative charge. The potential difference V between spheres of radius a can be found from the field on the line joining their centres; if the field is almost the same as for the point charges, then

$$V \simeq \int_a^{d-a} \frac{q}{4\pi\varepsilon_0}\left\{\frac{1}{r^2} + \frac{1}{(d-r)^2}\right\} dr = \frac{q}{4\pi\varepsilon_0}\left\{\frac{1}{a} - \frac{1}{d-a}\right\},$$

and the capacitance

$$C = \frac{q}{V} \simeq \frac{2\pi\varepsilon_0 a(d-a)}{d-2a}.$$

1.7 The electrostatic potential energy of a uniformly charged sphere of radius R carrying a total charge Q is $\frac{3}{5}Q^2/(4\pi\varepsilon_0 R)$. For a uranium nucleus $Q = 92e$ and $R = 10^{-14}$ m, and this energy amounts to 750 MeV. If the nucleus undergoes fission and splits into two equal spherical fragments, then the electrostatic potential energy of the pair of fragments is about 470 MeV. The reduction of 280 MeV is more than the energy actually released in uranium fission (just over 200 MeV) because the increased surface area of the two fragments is energetically unfavourable.

1.8 $\frac{1}{2}\left(V + \dfrac{qd}{2\varepsilon_0 a}\right).$

1.9 The quadrupole potential for $r \gg a$ is $(qa^2/4\pi\varepsilon_0 r^3)(3\cos^2\theta - 1)$. The dipole potential and the quadrupole potential are two of a series of *multipole potentials*, which are successively proportional to higher powers of $(1/r)$. The multipole potentials may be used as the basis of an expansion for any potential, in much the same way as powers of sines and cosines form the basis of a Fourier expansion.

1.10 The couple is 17.7×10^{-5} newton m. This is about the same as the couple exerted by a 1 gm weight on a lever arm 2 cm long, and is easily measurable with a fine suspension. Electrostatic voltmeters are sometimes useful when one wants to measure high voltages without drawing any current.

1.11 The field lines are straight, going radially from the point to the screen. An object on the point thus appears on the screen magnified by the ratio of the radii of curvature, namely 10^7.

  If there is no charge between point and screen, the electric field at a distance r from the centre is proportional to $1/r^2$, i.e. $E = k/r^2$. The potential difference between the point (at r_p) and the screen (at r_s) is

$$V = -\int_{r_p}^{r_s} E\,dr = \frac{k}{r_s} - \frac{k}{r_p} \simeq \frac{k}{r_p},$$

and

$$E = -\frac{V r_p}{r^2}$$

$$= -V/r_p \text{ at the surface of the point.}$$

Since r_p is 10^{-8} m, E reaches the value of 10^8 V/m needed to cause field emission when the screen potential is only one volt. In practice much higher potential differences are required to produce a large enough current of energetic electrons to give a visible image on the screen—typically a potential difference of about 10 kV would be applied.

  Although the simple microscope described here does have a very high magnification, it is of limited application. However, field emission electron sources are used in sophisticated electron microscopes, which include lenses to focus electron beams. The fact that the electron beam from a field emission source is diverging from a very small area is advantageous, and allows the best possible resolution to be attained.

PROBLEMS 2

2.1 The final potential difference is $\dfrac{2V}{\varepsilon + 1}$.

2.2 (i) charge $Q = CV$, energy stored $U = \frac{1}{2}CV^2$, work done by battery
    $W = \frac{1}{2}CV^2$
  (ii) $Q = \varepsilon CV,\ U = \frac{1}{2}\varepsilon CV^2$
    $\Delta Q = (\varepsilon - 1)CV, \qquad \Delta U = \frac{1}{2}(\varepsilon - 1)CV^2, \qquad \Delta W = V\,\Delta Q = (\varepsilon - 1)CV^2$
    The battery does work in pulling the dielectric into the capacitor as well as in storing electrostatic energy.

(iii) $Q = \varepsilon CV$, $U = \frac{1}{2}Q^2/C = \frac{1}{2}\varepsilon^2 CV^2$ i.e. more than the total amount of work $[\frac{1}{2}CV^2 + (\varepsilon - 1)CV^2]$ done by the battery during the sequence of operations. The difference $\frac{1}{2}(\varepsilon - 1)^2 CV^2$ represents work done by gravity.

2.3 If the capacitor is entirely filled with polycarbonate, the thickness of the sheet is about 5×10^{-6} m. When $50\,V$ is applied across the sheet, it experiences a compressional pressure of $800\,\text{N/m}^2$.

2.4 $E_{\text{local}}/E \approx 6.6$. Notice that the Clausius–Mossotti formula (Equation (2.16)) is hopelessly inaccurate for a polar liquid like water. In water the dipole moments of neighbouring molecules interact strongly with one another, and the problem of making a theoretical estimate of the average local field becomes very difficult.

2.5 The force between two dipoles varies as $1/r^4$. The potential energy of the two dipoles is $(-0.14)\,\text{eV}$.

2.7 At the surfaces $\pm a$ there is a polarization charge density

$$\sigma_p = \pm \rho_f a(1 - 1/\varepsilon).$$

2.8 The density of polarization charge on the surface of the slab is

$$\sigma_p = \pm \varepsilon_0 E \cos\theta(1 - 1/\varepsilon).$$

PROBLEMS 3

3.1 Let us choose the z-axis to be perpendicular to the plane of the silicon disc: put the uncovered face of the silicon at the origin and the gold layer at $z = a$, say. There is no variation of the potential in the x- and y-directions, and within the depletion layer Poisson's equation becomes

$$\frac{\mathrm{d}^2\phi}{\mathrm{d}z^2} = -\frac{N_D e}{\varepsilon\varepsilon_0},$$

which has the solution $\phi = A + Bz - N_D e Z^2/2\varepsilon\varepsilon_0$. The electric field in the depletion layer is everywhere positive, since mobile electrons must be swept back towards the origin. Hence $E = -\partial\phi/\partial z \geqslant 0$ for positive z, and if the depletion layer extends throughout the silicon disc, the minimum potential difference occurs when $B = 0$, and is

$$V = -N_D e d^2/\varepsilon\varepsilon_0 = -190 \text{ volts}.$$

(The potential barrier at the gold–silicon junction is 0.3 volts, small compared with the externally applied bias).

3.2 (i) If the cathode is at $\phi = 0$, then at the point where the potential is ϕ the electron speed v satisfies the equation $\frac{1}{2}m_e v^2 = e\phi$.

(ii) The magnitude of the current density is $j = Nev$, and

$$\rho = -Ne = -j\sqrt{m_e/2e\phi}.$$

(iii) Poisson's equation is

$$\frac{\mathrm{d}^2\phi}{\mathrm{d}z^2} = \frac{j}{\varepsilon_0}\sqrt{\frac{m_e}{2e\phi}}.$$

The solution which satisfies the boundary conditions is

$$\phi^{3/2} = \frac{9j}{4\varepsilon_0}\sqrt{\frac{m_e}{2e}}z^2.$$

Thus when a potential difference V is maintained across plates a distance d apart, the current density is

$$j = \frac{4\varepsilon_0 V^{3/2}}{9d^2} \sqrt{\frac{2e}{m_e}}.$$

3.3 The electrostatic force attracting the proton towards the image charge $(-e)\,2\,\text{cm}$ away is 36 times the gravitational force acting on it.

3.4 5.3 pF/m.

3.5 If the sphere carries no net charge it can be represented by q' and an additional charge of magnitude $+qR/a$ situated at the centre of the sphere. The force on q is

$$\frac{q^2 R}{4\pi\varepsilon_0 a} \left\{ \frac{1}{(a-b)^2} - \frac{1}{a^2} \right\}.$$

3.6 In the previous problem the charge q and its image q' have different magnitudes. It follows that two equal and opposite image charges cannot exactly represent the two charged spheres. However, since their separation is a good deal larger than their radii, it is not too bad an approximation to represent them by point charges $\pm Q$ located at their centres. In this approximation the potential difference between the spheres along the line of their centres is 85 kV when the field at the surface is on the point of breakdown. A better approximation is to represent each sphere by a charge $\pm 0.2Q$ at a distance 0.4 cm from the centre (this is the image of $\mp Q$ located at the centre of the other sphere) and an additional $\pm 0.8Q$ at the centre to preserve the correct magnitude of the charge. This leads to an estimate of 81 kV for the breakdown voltage. A further approximation can be made by taking the images of both these charges, and continuing the process the spheres can be exactly represented by an infinite series of image charges. Only a few terms in the series are needed to give a good answer, since we have seen that the first and second approximations differ by only about 3%.

3.7 The induced charge density on each plate is $-\sigma/2$.

3.9 The charge on a conductor is proportional to the number of field lines ending on it, and the potential difference between one conductor and another is proportional to the number of equipotentials between them. The pattern of equipotentials and field lines is preserved by a conformal transformation, and hence the capacitance is unaltered.

3.10 Assuming that the field lines are straight, the total energy stored between the plates per unit length of the capacitor is

$$U = \frac{\varepsilon_0 a V^2}{2d} \ln 2,$$

and hence

$$C = \frac{2U}{V^2} = \frac{\varepsilon_0 a}{d} \ln 2,$$

a good approximation to the value $\varepsilon_0 \ln 2/\theta$ found in the last problem. An approximate field pattern always leads to a result for the stored energy a little greater than the exact value; the pattern of an electrostatic field is the one which minimizes the stored energy.

3.11 The Fourier coefficients are:

$$A = \tfrac{1}{4}V_0; \qquad A_m = \frac{2V_0}{m\pi a^m} \sin \frac{m\pi}{4}.$$

The charge density is

$$\sigma = \varepsilon_0 \frac{\partial V}{\partial r}\bigg|_{r=a},$$

and the total charge on the left hand quadrant is

$$\int_{3\pi/4}^{5\pi/4} \sigma a \, d\phi = \frac{2\varepsilon_0 V_0}{\pi} \sum_{m=1}^{\infty} \frac{1}{m} \sin\frac{m\pi}{4}\left(\sin\frac{5m\pi}{4} - \sin\frac{3m\pi}{4}\right)$$

$$= \frac{\varepsilon_0 V_0}{\pi} \sum_{m=1}^{\infty} \frac{1}{m}\left(2\cos m\pi - \cos\frac{m\pi}{2} - \cos\frac{3m\pi}{2}\right)$$

$$= -\frac{\varepsilon_0 V_0 \ln 2}{\pi}.$$

PROBLEMS 4

4.1 After a collision with an atom in the gas the momentum of an electron is, on average, zero. Hence the mean speed is equal to one half of the speed gained between collisions, which can be worked out by calculating the acceleration of the electrons due to the applied electric field. Remember that the thermal speed are much greater than the speeds acquired by acceleration. The mean drift speed is 0.5 m sec^{-1}. The current is given by Equation (4.3) to be 2.1×10^{-8} A.

4.2 See Problem (1.6). The resistance between each sphere and Earth is $\rho/2\pi a$. The resistance between the spheres equals $\rho/\pi a$.

4.3 The magnetic dipole moment equals $LQ/2M$. The nucleus of an atom which has atomic number Z is quite well represented by a uniformly charged sphere carrying total charge $+Ze$. If it contains A nucleons (protons or neutrons) its mass is $\sim m_N A$, where m_N is the nucleon mass (nearly the same for protons and neutrons). The angular momentum L of a rotating nucleus $= R\hbar$ where R is an integer and $\hbar$ is Planck's constant divided by 2π (see section 4.4.3). Hence the magnetic dipole moment $\mathbf{m}$ of the rotating nucleus is equal to $R\hbar Ze/2m_N A$. This may be rewritten as

$$\mathbf{m} = \mathbf{R}\mu_N\frac{Z}{A},$$

where μ_N, the nuclear magneton, is given by

$$\mu_N = \frac{e\hbar}{2m_N},$$

and $\mathbf{R}$ is a vector in the direction of the angular momentum. The factor Z/A is called the rotational gyromagnetic factor, and the magnetic moment may be written as

$$\mathbf{m} = g\mu_N\mathbf{R}.$$

4.4 The maximum field gradient in the x–y plane is 2 tesla m^{-1}. The force on a moving atom which is neutral, but which possesses a magnetic dipole moment, will be a maximum if the atom is moving in the x-direction at that value of the coordinate y where the field gradient is a maximum. Use is made of this in the Stern–Gerlach experiment.

4.5 The field can most easily be obtained by solving the equation curl $\mathbf{B} = \mu_0\mathbf{j}$ inside and outside the sheet, and noting that the field must be independent of the coordinates x and z. The field then has an x-component only, which varies with coordinate y as shown in Figure D1. The force on an electron moving near the speed of light perpendicular to a field $\pi \times 10^{-2}$ tesla is 1.51×10^{-12} newtons. In some circular machines that accelerate particles to high energies a large current is pulsed through a conducting sheet at the time a burst of beam comes near. The magnetic field produced deflects the charged particle beam out of its path in the accelerator and into experimental apparatus.

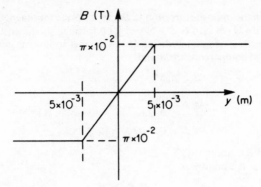

Figure D.1.

4.7 The principle of superposition is true for magnetic fields. The field of a cable with a hole drilled in it and carrying a uniform current density **j** is the same as the field of a solid cable carrying that current density *plus* the field of a current density $-\mathbf{j}$ in the opposite direction occupying the space where the hole was.
    Answer: (i) $\pi \times 10^{-6}$ teslas   (ii) $\pi \times 10^{-4}$ teslas.
4.9 Assume that the field from the large coil is constant over the small coil and calculate the mutual inductance using Equation (4.33). The force is given by Equation (4.57).
    Answer: 1.4×10^{-4} N.
4.12 The motion of the electron parallel to the magnetic field is unaffected by the magnetic force. The motion in the plane perpendicular to the field is circular with radius given by Equation (4.63). The electrons thus follow the magnetic field lines, spiralling about them in a helix. The number of revolutions made is equal to 8.8×10^{10}.

PROBLEMS 5

5.1 About 3×10^{-23} A m^2. The diamagnetic correction is about 15%.
5.3 The field is given by the intersection of the line

$$B = 0.4\pi - 8\pi^2 \times 10^{-5}H$$

with the $B–H$ curve plotted from the data given in the question. The intersection gives $B = 1.0$ T.

The current I required to produce a given field B is obtained from the equation (derived from Ampère's integral law)

$$0.2\pi H + \frac{10^{-3}B}{\mu_0} = 200I.$$

This gives I_1 and I_2. The power required is proportional to I^2, whence power ratio is 0.21.

5.4 When one coil is disconnected, the yoke forms a complete magnetic circuit of low reluctance. Almost all the flux is trapped in the yoke, and the field in the gap falls to a low value.
5.6 The field B inside the magnet is related to the magnetic intensity H and the magnetization M by

$$B = \mu_0(H + M).$$

The field H is in the opposite direction to B and M inside the magnet (in order to make the line integral $\oint \mathbf{H} \cdot \mathbf{dl}$ zero over a closed loop through the magnet). The field H will be of the same order of magnitude inside the magnet to just outside; consequently it can be neglected compared with M and

$$B \simeq \mu_0 M.$$

$$M = \frac{\text{moment}}{\text{volume}} = \frac{75 \times 4}{\pi \times (0.05)^2 \times 0.1}$$

$$= 1.2 \times 10^6 / \pi \, \text{A m}^{-1}.$$

Hence the field B is about 0.48 T.

5.8 The total field at the point P is

$$B = B_0 + B_M,$$

using the same symbols as in the text.

$$B_0 = \tfrac{1}{2}\mu_0 \frac{NI}{L} (\cos \alpha - \cos \beta),$$

from the Biot–Savart law. The uniform magnetization of the rod is equivalent to a surface current $I_M = M$ per unit length around the surface of the rod. Hence

$$B_M = \tfrac{1}{2}\mu_0 M (\cos \alpha - \cos \beta)$$

in the same way. The total field is therefore

$$B = \tfrac{1}{2}\mu_0 \left(\frac{NI}{L} + M \right) (\cos \alpha - \cos \beta).$$

The field inside the solenoid to a good approximation is given by

$$B_{\text{in}} = \mu_0 \left(\frac{NI}{L} + M \right),$$

hence the uniform magnetization of the rod is

$$M = \frac{NI}{L} \frac{\chi_B}{1 - \chi_B},$$

since $M = \chi_B B_{\text{in}}/\mu_0$. The field B at the point P is now given by substituting for M in the above equation for B,

$$B = \tfrac{1}{2}\mu\mu_0 \frac{NI}{L} (\cos \alpha - \cos \beta),$$

where

$$\mu = \frac{1}{1 - \chi_B}.$$

PROBLEMS 6

6.1 The force on a charge q a distance r from the axis of the rod is equal to $q\omega r B$ in the direction of r increasing. This is equivalent to an electric field in this direction of magnitude $\omega r B$. The polarization of the rod at distance r is thus $\mathbf{P} = \chi_E \omega B \mathbf{r}$, and this is equivalent to a volume distribution of charge $-\operatorname{div} \mathbf{P} = -\partial P_r/\partial r - P_r/r$

$= -2\chi_E\omega B$ throughout the rod, plus a surface distribution of charge $P_n = \chi_E\omega aB$ over the surface of the rod. See Equations (2.3) and (2.17).

6.2 The field on axis due to the dipole **M** is given by Equation (4.26)

$$B = \frac{H_0}{4\pi}\frac{2M}{x^3}$$

where x is the distance of the particle from the loop. The flux Φ through the loop is thus

$$\Phi = \frac{2\pi a^2\mu_0 M}{4\pi x^3}$$

and the e.m.f. induced in the loop is

$$-\frac{\partial\Phi}{\partial t} = -\frac{3a^2\mu_0 Mv}{2x^4}.$$

The current I in the loop is then

$$I = \frac{3a^2\mu_0 Mv}{2x^4 R},$$

and the magnetic dipole moment associated with the current in the loop has the value

$$\pi a^2 I = \frac{3\pi a^4\mu_0 Mv}{2x^4 R}.$$

This produces a field at the position of the particle

$$B' = \frac{3\mu_0^2 a^4 Mv}{4x^7 R},$$

and the force on the particle is $M\,\partial B'/\partial x$, which is equal to $21\,\mu_0^2 a^4 M^2 v/4x^7 R$.

6.3 The force attracting the armature is equal to $+\mathrm{grad}\,U$ from Equation (6.33) where $U = \frac{1}{2}\int_V \mathbf{B}\cdot\mathbf{H}\,\mathrm{d}\tau$ is the total magnetic energy. This integral should be calculated for a separation x of the iron and current I through the coils. Then

$$F = \frac{\partial U}{\partial x}\Big|_{x=0.01} = -10,$$

and this gives the current I necessary to close the relay. The answer is 2.1 A. For the armature to open again, we put

$$F = \frac{\partial U}{\partial x}\Big|_{x=0.0} = -10,$$

giving a current of 0.08 A.

6.5 (i) Magnetic flux increased, hence self-inductance increases. There are some induced (eddy) currents flowing circularly around the axis of the bar. These tend to reduce the increase of flux and also dissipate energy, causing a slight increase in apparent resistance of the solenoid.

(ii) Large eddy currents flow in the high-conductivity copper, in such a direction that the flux change is reduced, leading to a smaller self-inductance. There is also a slight increase in apparent resistance.

(iii) Similar to (ii).

(iv) No change in self-inductance or apparent resistance because induced currents cannot now flow.

6.6 Initially the currents in both coils have to be zero, and the equations obtained by adding up voltages around the two coils at $t = 0$ are

$$L_1 \frac{dI_1}{dt} + M \frac{dI_2}{dt} = E$$

and

$$L_2 \frac{dI_2}{dt} + M \frac{dI_1}{dt} = 0.$$

These give the equation

$$\left(L_1 - \frac{M^2}{L_2}\right) \frac{dI_1}{dt} = E.$$

6.7 The voltage difference between the rim and axis of the rotating flywheel shown in Figure D2 is

$$V = \frac{1}{q} \int \mathbf{F} \cdot d\mathbf{l} = \int_0^a \omega r B \, dr = \tfrac{1}{2}\omega a^2 B.$$

When a load of $10^{-3} \, \Omega$ is connected the initial current is thus $\omega a^2 B/2R$ which in this example is $2.25\pi \times 10^5 \, \text{A}$.

The rate at which the flywheel slows down can be calculated from the principle of conservation of energy which states that the rate of dissipation of heat in the load resistor is equal to the rate of loss of kinetic energy of the flywheel. Hence

$$\frac{d}{dt}(\tfrac{1}{2}\mathscr{I}\omega^2) + I^2 R = 0,$$

where $\mathscr{I}$ is the moment of inertia of the flywheel, I the current and ω the angular frequency at time t.

$$I = \frac{\omega a^2 B}{2R},$$

and so

$$\frac{d\omega}{\omega} = -\frac{a^4 B^2}{4R\mathscr{I}} \, dt,$$

from which

$$\omega = \omega_0 \exp(-t/\tau),$$

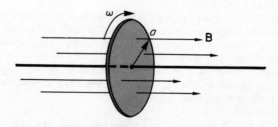

Figure D.2.

where

$$\tau = \frac{4R\mathscr{I}}{a^4 B^2}.$$

The energy of the flywheel at time t is thus given by

$$E = E_0 \exp(-2t/\tau),$$

where E_0 is the initial energy. The angular frequency thus falls to one half of its initial value in a time $t_{1/2}$ where

$$t_{1/2} = \tau \ln 2.$$

In this example $t_{1/2}$ is 6.16 seconds.

6.9 The flux Φ through the coil (Figure D3) is

$$\Phi = AB\cos\theta.$$

The e.m.f. induced in the coil as it rotates is

$$-\frac{\partial\Phi}{\partial t} = AB\omega\sin\omega t,$$

where ω is the angular speed of rotation. Adding the voltages around the coil we obtain

$$IR + L\frac{dI}{dt} = AB\omega\sin\omega t,$$

from which the current I is found to be

$$I = \frac{AB\omega}{(R^2 + \omega^2 L^2)^{1/2}}\sin(\omega t + \phi)$$

with

$$\tan\phi = -\frac{\omega L}{R}.$$

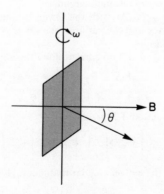

Figure D.3.

The energy dissipated in the resistance of the coil per cycle is $(AB\omega)^2 R/2(R^2 + \omega^2 L^2)$, and this is equal to $2\pi T$, where T is the mean torque required to rotate the coil.
6.12 χ_B is 1.2×10^{-5}.

PROBLEMS 7

7.1 (a) The amplitude

$$V_{\text{out}} = \frac{\omega C R V_0}{\sqrt{1 + \omega^2 C^2 R^2}}.$$

V_{out} leads V_0 by the phase angle $\phi = \tan^{-1}(1/\omega CR)$.
(b) The amplitude

$$V_{\text{out}} = V_0 \sqrt{1 + \omega^2 C^2 R^2}.$$

V_{out} lags behind V_0 by the phase angle $\phi = \tan^{-1}(\omega CR)$.
7.3 The power factor is 3.2×10^{-3}.
7.6 A $4\,\mu$F capacitor gives critical damping: $Q = \omega_0 L/R = \frac{1}{2}$ for any critically damped circuit.
7.7 If the coil has 65 turns its inductance is $350\,\mu$H, allowing the circuit to be tuned over the required range. At 1151 kHz the value of $\omega_0 L/R$ is greater than 10^4: in practice Q will be less than the ideal value because of other heat losses besides those due to the resistance of the windings. However, the tuning will be sharp enough not to mix 1151 and 1214 Hz signals.

PROBLEMS 8

8.1 $\frac{5}{6}\Omega$.
8.2 If the load resistor R_L is reduced, I_2 increases and V_{out} drops. Eliminating R_L, find V_{out} in terms of I_2. Then

$$Z_{\text{out}} = -\frac{\partial V_{\text{out}}}{\partial I_2} = \frac{2r_p}{\mu + 2 + 2r_p/R} \simeq \frac{2r_p}{\mu}.$$

For a pentode, $2r_p/\mu$ is small, and the anode follower is a suitable output stage for a valve circuit.
8.3 $Q = \omega_0 L/R$ as for the series resonant circuit.
8.5 $L = CR_1 R_2$, $R = R_1 R_2/R_3$. As with the Owen bridge, the balance conditions are independent of frequency and the variable quantities C and R_3 each appear in only one of the balance equations.
8.7 The conditions to be satisfied if no current flows through the detector are

$$R(S + R_2 + R_3) = \omega^2 M_{12} M_{34},$$

$$(L_1 + L_2)R + M_{12}S = 0.$$

8.10 The voltage transfer function $\mathbf{T}$ is given by

$$\frac{1}{\mathbf{T}} = \left[1 - 2\left(\frac{\omega}{\omega_c}\right)^2\right] + 2j\frac{\omega}{\omega_c}\left[1 - \left(\frac{\omega}{\omega_c}\right)^2\right]$$

where $\omega_c = 2/\sqrt{LC}$ is the cut-off angular frequency. Hence

$$|\mathbf{T}|^2 = \left[1 - 4\left(\frac{\omega}{\omega_c}\right)^4 + 4\left(\frac{\omega}{\omega_c}\right)^6\right]^{-1}$$

For frequencies up to $(\omega/\omega_c) = \frac{1}{3}$, **T** differs from unity by less than 3%, and low-frequency signals are less distorted by the T-section than by the L-section described in section 8.4. The reason for this is that at frequencies below cut-off the matching impedance for the T-section is a pure resistance

$$R = \sqrt{\frac{L}{C}}\left[1 - \left(\frac{\omega}{\omega_c}\right)^2\right]^{1/2}$$

which differs from the terminating resistance of the network only to the second order in (ω/ω_c).

8.12 The coupling coefficient $k = 0.95$.

PROBLEMS 9

9.1 Since the amplitudes of the incident and reflected waves are equal, the terminating impedance is purely reactive. For a node 75 cm from the end of the line, the reflection coefficient **K** $= \exp(-j\pi/2)$, requiring **Z** $= -jZ_0$. The impedance $-jZ_0$ corresponds to a capacitance of 8 pF.

9.2 $\frac{50}{3}\,\Omega$.

9.3 The input impedance of the line is inductive, and the source of e.m.f. therefore sees a parallel resonance circuit with $L = CZ_0^2$. The resonant condition is $L = 1/\omega^2 C$ i.e. $C = (\omega Z_0)^{-1} = 300$ pF.

9.4 The real part of the input admittance is almost equal to $(200\,\Omega)^{-1}$ one-sixth of a wavelength from the end of the line. The imaginary part can be cancelled by the input admittance of a short-circuited stub of length l, where $\tan kl = \sqrt{3}/2$.

9.5 3/5.

9.6 Signals leaving arm 1 in different directions reach arm 3 with a phase difference of π, and there is no output from this arm. The outputs from 2 and 4 are of equal magnitude but opposite phase. The hybrid ring thus allows high frequency signals to be divided without incurring power losses. With other choices for the impedances on the arms, a hybrid ring can also be used as a signal mixer, as a phase-changing or as a phase-measuring device.

PROBLEMS 10

10.2 The conduction electrons have inertia and will not follow very rapidly changing fields. The relation $\mathbf{j} = \sigma\mathbf{E}$ breaks down at frequencies greater than about 10^{11} Hz. (Classical theory is not often useful at such high frequencies). In this problem it is best to begin with the equation of motion of an electron and include a frictional damping force to account for the retarding action of the surroundings. Hence

$$m\frac{d\mathbf{v}_i}{dt} = -e\mathbf{E} - k\mathbf{v}_i,$$

where the label i refers to a specific electron, and we use the symbol m for the electron mass. If we apply this equation to the N free electrons in unit volume of the metal and average, after multiplying by e we have,

$$eNm\frac{d\mathbf{v}}{dt} = -Ne^2\mathbf{E} - Nek\mathbf{v},$$

where $\mathbf{v}$ is now the mean drift velocity of the electrons in the direction of the field $\mathbf{E}$.

Now

$$\mathbf{j} = -Ne\mathbf{v}$$

from equation (4.1), and so

$$-m\frac{d\mathbf{j}}{dt} = -Ne^2\mathbf{E} + \mathbf{j}k.$$

For steady currents $\mathbf{j} = \sigma\mathbf{E}$ and $\mathbf{v}$ is constant, giving $k = Ne^2/\sigma$. The above equation then becomes

$$-m\frac{d\mathbf{j}}{dt} = -Ne^2\mathbf{E} + \frac{Ne^2}{\sigma}\mathbf{j}.$$

The field $\mathbf{E}$ can be replaced using Equation (1.16), div $\mathbf{E} = \rho/\varepsilon_0$, and we obtain

$$-m\frac{d}{dt}(\text{div }\mathbf{j}) = -\frac{Ne^2\rho}{\varepsilon_0} + \frac{Ne^2}{\sigma}\text{div }\mathbf{j}.$$

We have assumed that the relative permittivity of the metal is unity. Using the equation of continuity (10.1) this equation can be rewritten

$$m\frac{\partial^2\rho}{\partial t^2} = -\frac{Ne^2\rho}{\varepsilon_0} - \frac{Ne^2}{\sigma}\frac{\partial\rho}{\partial t},$$

or

$$\frac{\partial^2\rho}{\partial t^2} + \frac{Ne^2}{\sigma m}\frac{\partial\rho}{\partial t} + \frac{Ne^2}{\varepsilon_0 m}\rho = 0.$$

This equation tells us how the charge density ρ varies with time. It represents damped simple harmonic motion if $\varepsilon_0 Ne^2/2m\sigma^2 < 1$, which is true for metals. The solution is

$$\rho = \rho_0 \exp(-\beta t)\cos\omega t$$

where

$$\beta = \frac{Ne^2}{2m\sigma},$$

and ω, the plasma frequency, is approximately given by

$$\omega^2 = \frac{Ne^2}{\varepsilon_0 m}.$$

Substitution of the values for copper in these equations gives

$$\beta \sim 3.85 \times 10^{13}\,\text{s}^{-1}; \qquad \omega \sim 5.5 \times 10^{15}\,\text{s}^{-1},$$

assuming that there is one free electron for each copper atom, and using the value of σ given in Table 4.1. The oscillations thus die away to one half of their amplitude in a time $\sim 1.8 \times 10^{-14}$ seconds.

10.3 The electric dipole moment is

$$\mathbf{d}_{\text{mol}} = \int_{V_m} \rho_d \mathbf{r}\,d\tau,$$

where V_m is the volume occupied by the molecule.

The magnetic dipole moment has two contributions, one from the intrinsic dipoles given by

$$\mathbf{m}_{int} = \int_{V_m} \mathbf{M}_a \, d\tau,$$

the other $\mathbf{m}_{current}$ from the distribution of currents within the molecule. The magnetic dipole moment of a current I in a small closed loop of area ΔS is given by

$$\Delta \mathbf{m} = I \, \Delta \mathbf{S},$$

where $\Delta \mathbf{S}$ is a vector normal to the plane or the loop, in a direction given by the right hand screw rule. It can be shown that

$$\Delta \mathbf{S} = \tfrac{1}{2} \oint \mathbf{r} \wedge d\mathbf{l},$$

hence

$$\Delta \mathbf{m} = \tfrac{1}{2} \oint \mathbf{r} \wedge I \, d\mathbf{l}.$$

For the current distributed over a volume V_m the vector $I \, d\mathbf{l}$ can be replaced by the vector $\mathbf{j}_a \, d\tau$ and

$$\mathbf{m}_{current} = \tfrac{1}{2} \int_{V_m} \mathbf{r} \wedge \mathbf{j}_a \, d\tau.$$

The total magnetic dipole moment of the molecule is thus

$$\mathbf{m}_{mol} = \int_{V_m} (\mathbf{M}_a + \tfrac{1}{2}\mathbf{r} \wedge \mathbf{j}_a) \, d\tau$$

$$= \int_{V_m} (\mathbf{M}_a + \tfrac{1}{2}\mathbf{r} \wedge \mathbf{v}_a \rho_a) \, d\tau.$$

If the molecule is considered to consist of point particles the electric dipole moment becomes

$$\mathbf{d}_{mol} = \sum_i e_i \mathbf{r}_i,$$

where the sum is over all the particles in the molecule. The magnetic dipole moment becomes

$$\mathbf{m}_{mol} = \sum_i (\mathbf{m}_i + \tfrac{1}{2}\mathbf{r}_i \wedge \mathbf{v}_i e_i)$$

$$= \sum_i \left(\mathbf{m}_i + \frac{e_i \mathbf{l}_i}{2m_i} \right),$$

where $\mathbf{l}_i$ is the orbital angular momentum of particle i. The intrinsic magnetic dipole moments $\mathbf{m}_i$ are related to the intrinsic spins by

$$\mathbf{m}_i = g_{s_i} \mathbf{s}_i$$

as discussed in Chapter 5. g_{s_i} is a constant for particles of the same type. Hence

$$\mathbf{m}_{mol} = \sum_i (g_{s_i} \mathbf{s}_i + g_{l_i} \mathbf{l}_i)$$

where g_{l_i} is a constant for particles of the same type.

The electric dipole moment of an isolated atom or molecule usually vanishes, whereas the magnetic dipole moment is usually non-zero.

PROBLEMS 11

11.1 The displacement current density $\mathbf{j}_D$ is equal to

$$\partial \mathbf{D}/\partial t = \varepsilon_0 \, \partial \mathbf{E}/\partial t = j\varepsilon_0 \omega \mathbf{E}.$$

Hence

$$\langle j_D^2 \rangle^{1/2} = \varepsilon_0 \omega E_0 / \sqrt{2}.$$

This gives E_0 equal to $2.54 \times 10^{-3} \, \text{V m}^{-1}$, and B_0 equal to $8.47 \times 10^{-12} \, \text{T}$.

11.4 If the resistance of the wire is R and the capacitance is C the capacitor decays according to the expression $q = q_0 \exp(-t/RC)$. The field between the plates, assumed uniform, is thus given by $E = E_0 \exp(-t/RC)$, where $E_0 = V_0/l$, V_0 being the initial voltage across the plates and l their separation. By symmetry the magnetic field at the surface of the wire is tangential to the surface and has constant magnitude at a fixed time. By Ampère's law

$$2\pi a H = I,$$

where a is the radius of the wire and I the current. We have neglected the displacement current through the wire; it is negligible if the conductivity $\sigma \gg \varepsilon_0 \omega$, where ω may be taken to be of the order of $(RC)^{-1}$. The above equation may be written

$$2\pi a H = \pi a^2 \sigma E$$

and hence $H = \sigma a E/2$. The Poynting vector $\mathbf{E} \wedge \mathbf{H}$ is constant over the surface of the wire and points inwards. It has magnitude

$$EH = \frac{\sigma a E^2}{2}.$$

The total energy flowing into the wire is thus

$$U = \int_0^\infty 2\pi a l \cdot \frac{\sigma a E^2}{2} \cdot dt.$$

This can be worked out to be equal to $CV_0^2/2$, the initial energy stored in the capacitor.

The use of the static field solution in time dependent problems can sometimes lead to difficulties. For example if we have a large parallel plate capacitor discharging through an external resistor the static field solution would suggest that curl $\mathbf{E} = 0$ near the middle of the capacitor. From Faraday's law this implies that $\partial \mathbf{B}/\partial t = 0$, and since the field $\mathbf{B}$ was initially zero it remains zero. Hence curl $\mathbf{H}$ would remain zero if our assumptions were correct. However this would imply that $\partial \mathbf{D}/\partial t$ were zero, since curl $\mathbf{H} = \mathbf{j} + \partial \mathbf{D}/\partial t$, and $\mathbf{j}$ is zero inside the capacitor. This conclusion contradicts our initial supposition. The way out of the puzzle of course is to realize that the static field solution is not always a sufficiently good approximation to the actual fields existing in time dependent situations.

11.5 $2.8 \times 10^{-2} \, (\Omega \, \text{m})^{-1}$.

11.6 The resonant angular frequency ω is $3.16 \times 10^6 \, \text{s}^{-1}$. Calculate the skin depth in copper at this frequency: the answer is about $10^{-4} \, \text{m}$. Assuming that the current flows uniformly in an outer layer of the wire of thickness equal to the skin depth, the resistance R is estimated to be about $1 \, \Omega$. If L is the self-inductance of the coil, Equation (7.25) gives the Q of the circuit to be $\omega L/R$, which is about 3000.

11.8 $0.796 \, \text{m}$.

11.9 If the electric field is $\mathbf{E} = \mathbf{e}_x E_0 \cos \omega t$ the equation of motion of an electron of mass

m is

$$m\ddot{x} + eE_0 \cos \omega t = 0,$$

whence

$$x = \frac{eE_0}{m\omega^2} \cos \omega t$$

plus other terms which depend on the initial conditions. Since we are going to average over the n_0 electrons in unit volume to obtain the polarization $\mathbf{P}$, these other terms average to zero and may be ignored.

$$\mathbf{P} = -\frac{n_0 e^2}{m\omega^2} \mathbf{E}.$$

Hence

$$\mathbf{D} = \varepsilon_0 \mathbf{E} + \mathbf{P} = \mathbf{E}\left[\varepsilon_0 - \frac{n_0 e^2}{m\omega^2}\right]$$

and

$$\varepsilon = 1 - \frac{n_0 e^2}{m\omega^2 \varepsilon_0}.$$

This gives the relative permittivity less than unity, and so the phase velocity of a wave in the ionosphere is greater than the speed of light in free space.

11.11 The water molecule is polar, and the permanent electric dipole moment contributes to the electric field as long as the molecule can rotate and keep aligned with the applied field. As soon as the frequency passes into the IR region the molecules can no longer rotate quickly enough to follow the field, because of their inertia. The relative permittivity thus begins to decrease. The nitrogen molecule is non-polar and its contribution to an electric field arises solely from the induced dipole moment. The frequencies at which such contributions begin to fall off depend on the electrons in the molecule and are in the visible or UV region of the electromagnetic spectrum.

11.13 Let the boundary be at coordinate $z = 0$ and the electric wave in the metal be

$$\mathbf{E_T} = \mathbf{e}_x E_{OT} \exp\left[j(\omega t - \alpha z)\right] \exp(-\alpha z),$$

as in Equation (11.61). This electric wave produces a current density $\mathbf{j} = \sigma \mathbf{E}$ in the metal; $\mathbf{j} = -N e \mathbf{v}$, where N is the number of free electrons per unit volume and $\mathbf{v}$ is their mean drift velocity in the direction of the field $\mathbf{E}$. The average force on an electron is thus $-e\mathbf{v} \wedge \mathbf{B}$ arising from the magnetic field in the wave. In a thickness dz of the metal at coordinate z the number of electrons per unit area is $N\, dz$ and the force on them is thus $-N e \mathbf{v} \wedge \mathbf{B}\, dz$, which is equal to $\sigma \mathbf{E} \wedge \mathbf{B}\, dz$. The total force on unit area of the metal is

$$\mathbf{F} = \sigma \int_0^\infty \mathbf{E} \wedge \mathbf{B}\, dz.$$

The fields $\mathbf{E}$ and $\mathbf{B}$ can be substituted from the expressions given in section 11.5.3, and the average force calculated. The answer obtained is

$$\overline{\mathbf{F}} = \mathbf{e}_z \frac{\sigma E_{OT}^2}{4\omega}.$$

We now have to return to the considerations of section 11.5.3 to determine the relationship between E_{OT} and E_{OI}, the amplitude of the electric field in the wave incident on the metal. As the conductivity σ tends to infinity we find

$$E_{OT}^2 = E_{OI}^2 \frac{4\omega\varepsilon_0}{\sigma}$$

and hence $\bar{F} = \varepsilon_0 E_{OI}^2$, the same result as Equation (11.72).

PROBLEMS 12

12.2 If a is the length of a side of the waveguide, from equation (12.18) we see that no wave at all will be propagated without attenuation of $\pi/a > k$. Hence the minimum value of a is given by $\pi/k = \lambda/2$. The TE_{11}, TE_{02} or TE_{20} waves all obey the same equation

$$\frac{2\pi^2}{a^2} = k^2 - k_g^2.$$

Hence if $2\pi^2/a^2 > k^2$ none of these waves will propagate, i.e. if $a < \pi\sqrt{2}/k$ no wave other than the $TE_{01}(TE_{10})$ will propagate. Hence the maximum value of a is $\lambda/\sqrt{2}$.

12.3 A similar calculation was done on page 413. The answer is 0.79 MW.

12.6 We can make an estimate of the losses by assuming that the currents in the walls flow only in a thin layer of thickness equal to the skin depth δ, and that the current density is uniform within this layer. Applying Ampère's integral law (5.29) over a loop of length dz and height δ lying along the z-direction in the top or bottom walls, we have

$$H_z\, dz = -j_y\, dz \quad \text{or} \quad j_y = -H_z/\delta.$$

Applying the law to a loop of length dy, height δ lying along the y-direction in the top or bottom wall,

$$H_y\, dy = j_z\, dz \quad \text{or} \quad j_z = H_z/\delta.$$

Applying the law to a loop of length dz height δ lying along the z-direction in either of the side walls at $y = 0$ or $y = a$,

$$H_z\, dz = -j_x\, dz \quad \text{or} \quad j_x = -H_z/\delta$$

The average power loss $d\bar{W}$ in the strip of thickness δ length dz around the guide is

$$d\bar{W} = -\frac{1}{\sigma} \int \bar{j}^2\, d\tau$$

$$= -\frac{2\, dz}{\sigma\delta} \int_0^b (\bar{H}_z^2 + \bar{H}_y^2)\, dy - \frac{2\, dz}{\sigma\delta} \int_0^a \bar{H}_z^2\, dx,$$

where the first term comes from the top and bottom walls, and the second term comes from the side walls. Integrating, and using equations (12.27) and (12.28) to obtain H_y and H_z, gives

$$\frac{d\bar{W}}{dz} = -\frac{C^2\pi^2}{\omega^2 b^2 \sigma\delta\mu_0^2}\left[\frac{b}{2}\left(k_g^2 + \frac{\pi^2}{b^2}\right) + \frac{\pi^2 a}{b^2}\right].$$

From equation (12.18)

$$\frac{d\bar{W}}{dz} = -\frac{C^2\pi^2}{\omega^2 b^2 \sigma\delta\mu_0^2}\left(\frac{bk^2}{2} + \frac{\pi^2 a}{b^2}\right)$$

The average power $\overline{W}$ flowing down the guide is

$$\overline{W} = \int_0^a \int_0^b N_z \,\mathrm{d}x \,\mathrm{d}y$$

$$= \frac{1}{4}\frac{C^2\pi^2 k_g}{\omega b^2 \mu_0}ab.$$

Hence

$$\frac{\mathrm{d}\overline{W}}{\overline{W}} = -\delta\left(\frac{k^2}{k_g a} + \frac{2\pi^2}{k_g b^3}\right)\mathrm{d}z,$$

where we have used Equation (11.41) for the skin depth δ. The above equation leads to an attenuation constant β in the equation

$$\overline{W} = \overline{W}_0 \exp\left(-\beta z\right)$$

given by

$$\beta = \delta\left(\frac{k^2}{k_g a} + \frac{2\pi^2}{k_g b^3}\right).$$

Substitution of the appropriate values for 10^{10} Hz in a brass guide with $a = 1.0$ cm, $b = 2.0$ cm gives β equal to 0.06 m^{-1}, i.e. the power falls to one half over a distance of about $0.7/\beta \sim 11$ m.

12.7 The answer is three. In order of increasing frequency they are the TE_{011}, TE_{012} and TE_{111} resonances.

PROBLEMS 13

13.1 Answer is about 400 W.

The frequency chosen has to be above the international long-range communications band—the 10 m band. There are only narrow frequency ranges allotted for taxis, police etc. A frequency of 10^8 Hz is a suitable frequency from the point of view of cost and simplicity of electronics, and size of antennas. The wavelength of 3 m also ensures that there is sufficient diffraction of the waves around buildings to give signals everywhere.

13.2 See sections 9.2 and 9.4.

13.3 This problem is the four slit problem in optics where four narrow slits radiate in phase. The pattern is thus like that of a four line diffraction grating.

13.4 An antenna of length l shorter than half a wavelength has only a small oscillating current. This is because the current has to be zero at the ends, and the amplitude increases towards the middle according to the formula

$$I = I_0 \cos\frac{2\pi}{\lambda}\left(\frac{\lambda}{4} - l + z'\right).$$

One way of making the current larger is to capacitatively load the ends of the aerial, when the current at the ends is no longer zero. This is often done by stringing wires from the top of a vertical antenna. The reflection in the earth is used as the second half of the antenna, and typically a vertical height of ~ 40 m would suffice for a local station with a range of ~ 50 km.

13.5 If the dipoles are separated by a small distance Δs, the electric field at large distances is

$$E_\theta = \frac{j\omega^2 p_0}{4\pi\varepsilon_0 c^2} \cdot \frac{\sin\theta}{r} \cdot \exp[j(\omega t - kr)]\{\exp(jk\,\Delta s\cos\alpha) - \exp(-jl\,\Delta s\cos\alpha)\}$$

$$= \frac{j2\omega^3 p_0\,\Delta s\cos\alpha}{4\pi\varepsilon_0 c^3} \cdot \frac{\sin\theta}{r} \cdot \exp[j(\omega t - kr)],$$

where α is the angle between the direction of observation and the line joining the centre of the dipoles. We have taken the origin to be at the mid point of this line.

13.6 If V_1 is the e.m.f. from the short wire and V_2 from the loop

$$\frac{V_1}{V_2} = \frac{lc}{S\omega},$$

where ω is the angular frequency of the radiation, and c the speed of light. If $S \sim l^2$,

$$\frac{V_1}{V_2} = \frac{c}{l\omega}.$$

PROBLEMS 14

14.1 In the moving frame the wires of length a appear to be charged, and are shortened by the FitzGerald contraction. The magnitude of the coil's magnetic moment is $m = Iab$, and the electric dipole moment in the moving frame is

$$\mathbf{p} = \frac{1}{c^2}\mathbf{m} \wedge \mathbf{v}.$$

14.2 3.2×10^9 V/m.

14.3 The charge at the origin generates no magnetic field along the x-axis. The second charge is acted upon only by the electric field directed along the x-axis. However, the charge at the origin is itself moving perpendicularly to the magnetic field generated by the other charge, and experiences a force in the y-direction. Action and reaction are not equal and opposite, implying that the momentum of the particles is not conserved. This is not as disturbing as it first appears. The fields carry energy and momentum, and they are changing in such a way that momentum is conserved for the whole system of particles and fields.

14.4 The force is readily derived in the same form as Equation (4.59). The contribution from two elements dl_1 and dl_2 to the force is also in a form which does not lead to action and reaction being equal and opposite. However, a *steady* current can only flow in a complete circuit, and the total forces acting on the two coils are indeed equal and opposite.

Further reading

B. I. Bleaney and B. Bleaney, *Electricity and Magnetism*, Oxford University Press, 1965. More material on microscopic aspects, and on practical electricity and magnetism.

E. M. Purcell, *Electricity and Magnetism*, McGraw-Hill, 1967. Less advanced text which covers fewer topics but has more discussion of them.

D. R. Corson and P. Lorrain, *Introduction to Electromagnetic Fields and Waves*, W. H. Freeman, 1962. Has much more on electromagnetic waves.

W. K. H. Panofsky and M. Phillips, *Classical Electricity and Magnetism*, Addison-Wesley, 1962. More advanced formal text.

J. A. Stratton, *Electromagnetic Theory*, McGraw-Hill, 1941. More advanced formal text.

S. Ramo and J. R. Whinnery, *Fields and Waves in Modern Radio*, J. Wiley and Sons, 1953. More details on transmission lines, waveguides and antennas.

R. Becker and F. Sauter, *Electromagnetic Fields and Interactions*, Vol. I. Blackie, 1964. More on electromagnetism and relativity.

R. P. Feynman, *Lectures on Physics*, Vol. II, Addison-Wesley, 1964.

Index

Stray impedance 287, 322
Strip line 331
Superconducting magnets 206
Superposition
  of electric fields 4
  of magnetic fields 493
  of potentials 31
Surface
  charge density 14
  currents 387
  integral 15
Susceptibility
  electric 57
  magnetic 187, 246

T-section filter 320
TEM waves 402
TE waves 405
TM waves 407
Tesla (unit) 125
Time dilatation 444
Thermo-electric e.m.f.s 286
Transformation
  linear 449
  Lorentz 443, 450
  of fields 455
  of forces 458
Transformer 306
  equivalent circuits 310, 313, 315
  ideal 307
  losses 315
  ratio-arm bridge 291

real 312
Transients 274
Transmission line 323
  coaxial cable 329
  parallel wire 328
  strip line 331
Transmission coefficient 391
Two-dimensional potentials 105

Uniqueness theorem 86
Units 468

Vector
  field 8, 479
  potential 148
Velocity of light
  see Speed
Volt 6, 30
Voltmeter, electrostatic 46
Volume integral 10

Wave equation 326, 362
Wave front 365
Waveguide 408
  equation 410
  slotted 416
Wavelength 365
Wavenumber 327, 365
  in waveguide 405, 410
Wave velocity 375, 406
Wheatstone bridge 284, 289